Perovskites

Perovskites

Structure–Property Relationships

RICHARD J. D. TILLEY

Professor Emeritus, Cardiff University, UK

WILEY

This edition first published 2016

First Edition published in 2016

Registered Office
John Wiley & Sons, Ltd, The Atrium, Southern Gate, Chichester, West Sussex, PO19 8SQ, United Kingdom

For details of our global editorial offices, for customer services and for information about how to apply for permission to reuse the copyright material in this book please see our website at www.wiley.com.

Library of Congress Cataloging-in-Publication data applied for

ISBN: 9781118935668

A catalogue record for this book is available from the British Library.

Set in 10.5/13.5pt Times by SPi Global, Pondicherry, India

Printed in the UK

Contents

Preface

Perovskites are a class of compounds with structures related to that of the mineral Perovskite, $CaTiO_3$, and can be considered to be derived from a parent phase of general formula ABX_3. They have been intensively studied since the middle of the twentieth century because of their innate properties: initially dielectric, piezoelectric and ferroelectric. This range of behaviour has been expanded into areas which include magnetic ordering, multiferroic properties, electronic conductivity, superconductivity and thermal and optical properties. Apart from these purely physical aspects, the phases show a wide range of chemical attributes. They are used as electrode materials for solid oxide fuel cells where materials with high oxide ion conductivity, electronic conductivity and mixed ionic/electronic conductivity are required. Many perovskite phases show useful catalytic and redox behaviour, often dependent upon the presence of chemical defects in the phase.

This complexity is a result of two prime factors. Firstly the crystal structures encompassed by the term 'perovskite' embrace a huge range, from the simple cubic 'aristotype' $SrTiO_3$ to cation- and anion-deficient phases, modular phases including the cuprate superconductors and hexagonal perovskites related to $BaNiO_3$. In addition, both the chemical and physical properties of any member of these structural forms can be tuned over wide ranges by relatively simple substitution into all or some of the A-, B- and X-sites. This wide-ranging flexibility includes the formation of perovskites in which the A cation is replaced by an organic molecule, typified by the perovskite methylammonium lead iodide, now intensely studied as the core of 'perovskite' photovoltaic cells. Additionally, the properties of thin films, superlattices and nanoparticles show new and totally unexpected responses, when compared to the behaviour of the separate bulk phases.

The aim of this book is to provide a compact overview of this large body of knowledge. An outline of the structures of these phases is of primary importance as a prerequisite to an understanding of many physical properties. This material is contained in the earlier chapters. For the purposes of providing an overall vista, crystal structures are mostly represented as idealised forms. This has the advantage of bringing out the structural relationships between the phases described, but the obverse side is that it does obscure small details that may be of significance. However, the detailed crystallographic parameters of all the phases mentioned in this book are available via the CrystalWorks database (http://www.cds.rsc.org). This source will also provide literature references to the originally published crystallographic data of all the phases listed. Crystal structures are followed by a survey of physical and chemical properties. The properties emphasised are, where possible, those unique to perovskites or at least manifested in a singular way by these compounds. They are grouped into broad categories – chemical, dielectric, magnetic, electronic, thermal and optical properties – although these classifications cannot truly be cleanly separated from each other.

In a book of modest size, it is necessary to be somewhat selective in material content. For this reason, two areas have been omitted. The first of these concerns preparation techniques. In the main these are the normal techniques of solid-state chemistry, physics and ceramic science and are not unique to perovskites. Secondly catalysis has also been omitted. Again, the bulk of the catalytic reactions studied are not unique to perovskites and are better described and discussed within the broader perspective of catalysis rather than via the narrower standpoint of perovskites.

Although large numbers of papers are published each month detailing some aspect or other of perovskite physics and chemistry, it is felt that an extensive literature reference section would overwhelm any reader seeking a broad overview of the field. Because of this, the Further Reading sections contain mainly reviews or selected recent references that expand material in the text. These are sufficient to provide an entry point to the literature base for those needing additional information. In addition a few sources that explain the basic concepts of crystallography and structure–property relationships are added, and two appendices are included that explain two rather more specialist aspects of nomenclature.

As ever, I am grateful to my wife Anne for support and tolerance during the assembly of this work, without which the project could not have been undertaken. In addition I thank my family for continual encouragement. Finally I am indebted to the staff of the Trevithick Library, University of Cardiff, who helped with literature resources and related matters.

Richard Tilley
January 2016

1

The ABX_3 Perovskite Structure

1.1 Perovskites

Perovskite is a mineral of formula $CaTiO_3$. It was discovered in 1839 by the Prussian mineralogist Gustav Rose in mineral deposits in the Ural Mountains and named after the Russian mineralogist Count Lev Aleksevich von Petrovski. Natural crystals have a hardness of 5.5–6 and a density of 4000–4300 kg m^{-3}. They are usually dark brown to black, due to impurities, but when pure are clear with a refractive index of approximately 2.38. The crystal structure of this compound, initially thought to be cubic, was later shown to be orthorhombic (Table 1.1).

As with many minerals, Perovskite has given its name to a family of compounds called *perovskites*, which have a general formula close to or derived from the composition ABX_3. At present many hundreds of compounds are known that adopt the perovskite structure. In fact a perovskite structure mineral, Bridgmanite (Fe,Mg)SiO_3, is the most abundant solid phase in the Earth's interior, making up 38% of the total. The phase occurs between depths of approximately 660–2900 km but is only stable at high temperatures and pressures so that it is not found at the surface of the Earth.

To some extent the multiplicity of phases that belong to the perovskite family can be rationalised by assuming that perovskites are simple ionic compounds, where A is usually a large cation, B is usually a medium-sized cation and X is an anion.

Perovskites: Structure–Property Relationships, First Edition. Richard J. D. Tilley.

Table 1.1 *Representative ABX_3 perovskite phases*[a]

Phase	Space group[b]	Unit cell		
		a (nm)	*b* (nm)	*c* (nm)
1, 2				
$AgMgF_3$	C, *Pm*$\bar{3}$*m* (221)	0.41162		
$CsPbI_3$	C, *Pm*$\bar{3}$*m* (221)	0.62894		
$KCuF_3$	T, *I4/mcm* (140)	0.56086		0.76281
$KMgF_3$	C, *Pm*$\bar{3}$*m* (221)	0.39897		
$KZnF_3$	C, *Pm*$\bar{3}$*m*	0.40560		
$NaMgF_3$	O, *Pbnm* (62)	0.48904	0.52022	0.71403
$NaFeF_3$	O, *Pnma* (62)	0.56612	0.78801	0.54836
NH_4ZnF_3	C, *Pm*$\bar{3}$*m* (221)	0.41162		
1, 5				
$KTaO_3$	C, *Pm*$\bar{3}$*m* (221)	0.40316		
$KNbO_3$	O, *Amm2* (38)	0.3971	0.5697	0.5723
2, 4				
$SrTiO_3$	C, *Pm*$\bar{3}$*m* (221)	0.3905		
$BaTiO_3$	T, *P4mm* (99)	0.39906		0.40278
$CaTiO_3$	O, *Pbmn* (62)	0.54035	0.54878	0.76626
$BaSnO_3$	C, *Pm*$\bar{3}$*m* (221)	0.4117		
$CdSnO_3$	O, *Pnma* (62)	0.52856	0.74501	0.51927
$CaIrO_3$	O, *Pbnm* (62)	0.52505	0.55929	0.76769
$PbTiO_3$	T, *P4mm* (99)	0.3902		0.4143
$PbZrO_3$	O, *Pbam* (55)	0.58822	1.17813	0.82293
$SrCoO_3$	C, *Pm*$\bar{3}$*m* (221)	0.3855		
$SrMoO_3$	C, *Pm*$\bar{3}$*m* (221)	0.39761		
$SrRuO_3$	O, *Pnma* (62)	0.55328	0.78471	0.55693
$(Fe,Mg)SiO_3$	O, *Pnma* (62)	0.5020	0.6900	0.4810
3, 3				
$BiFeO_3$	Tr, *R3c* (161)	0.55798		1.3867
$BiInO_3$	O, *Pnma* (62)	0.59546	0.83864	0.50619
$ErCoO_3$	O, *Pbnm* (62)	0.51212	0.54191	0.73519
$GdFeO_3$	O, *Pbnm* (62)	0.53490	0.56089	0.76687
$HoCrO_3$	O, *Pnma* (62)	0.5518	0.7539	0.5245
$LaAlO_3$	Tr, *R3c* (161)	0.53644		1.31195
$LaCoO_3$	Tr, *R*$\bar{3}$*c* (167)	0.54437		1.30957
$LaMnO_3$	O, *Pbnm* (62)	0.55367	0.57473	0.76929
$LaTiO_3$	O, *Pbnm* (62)	0.5576	0.5542	0.7587
$NdAlO_3$	Tr, *R*$\bar{3}$*c* (167)	0.53796		1.31386
$PrRuO_3$	O, *Pnma* (62)	0.58344	0.77477	0.53794
$YbMnO_3$	O, *Pbnm* (62)	0.52208	0.58033	0.73053
4, 5				
$ThTaN_3$	C, *Pm*$\bar{3}$*m*	0.4020		

[a] Many of these phases are polymorphic, and lattice parameters vary with temperature and pressure.
[b] The crystal system, here and throughout the other tables in this book, is abbreviated thus: C, cubic; H, hexagonal; M, monoclinic; O, orthorhombic; T, tetragonal; Tr, trigonal (often specified in terms of a hexagonal unit cell); Tri, triclinic.

Naturally the overall ionic structure must be electrically neutral. If the charges on the ions are written as q_A, q_B and q_X, then

$$q_A + q_B = -3q_X$$

Frequently encountered (but not exclusive) combinations are

$$\begin{aligned}
q_X &= -1, \quad (q_A, q_B) = (1,2); \text{for example}, KNiF_3; \\
q_X &= -2, \quad (q_A, q_B) = (1,5); \text{for example}, NaNbO_3; \\
&\qquad\quad (q_A, q_B) = (2,4); \text{for example}, CaTiO_3; \\
&\qquad\quad (q_A, q_B) = (3,3); \text{for example}, LaAlO_3; \\
q_X &= -3, \quad (q_A, q_B) = (4,5); \text{for example}, ThTaN_3.
\end{aligned}$$

The importance of perovskites became apparent with the discovery of the valuable dielectric and ferroelectric properties of barium titanate, $BaTiO_3$, in the 1940s. This material was rapidly employed in electronics in the form of capacitors and transducers. In the decades that followed, attempts to improve the material properties of $BaTiO_3$ lead to intensive research on the structure – property relations of a large number of nominally ionic ceramic perovskite-related phases with overall compositions ABO_3, with a result that vast numbers of new phases were synthesised.

It was soon realised that, as a group, these materials possessed very useful physical and chemical properties far broader than those shown by $BaTiO_3$, and research widened to include a range of structures and phases that could all be related structurally to the perovskite family, including nominally ionic nitrides and oxynitrides. In addition, a number of materials which are better described as alloys, of formula A_3BX, where A and B are metals and X is an anion or semimetal, typically C, N, O and B are known. These are often said to adopt the so-called *antiperovskite* or *inverse perovskite* structure, because the metal A atoms occupy the positions corresponding to the anions in the ionic perovskites and the B and X atoms occupy sites corresponding to those occupied by the cations. The flexibility of the perovskite framework also allows it to include cations such as NH_4^+, which can often be considered to be spherical at normal temperatures. More complex phases, such as the inorganic–organic hybrid compounds $(CH_3NH_3)PbX_3$, where X is typically Cl, Br, I or a combination of these anions, have also been synthesised.

As well as phases with an ABX_3 composition, large numbers of modular structures have been prepared, all of which are built up, at least in part, from fragments, usually slabs, of perovskite-like structure. The formulae of these are not easily reconciled with a composition of ABX_3 until the structural building principles have been found and the nature of the interfaces between the various slabs is clarified.

For example, $Bi_2Ca_2Sr_2Cu_3O_{10+\delta}$, a superconducting oxide, is built from slabs of perovskite type separated by slabs of composition Bi_2O_2.

As would be expected, there is a close correlation between chemical and physical properties in these complex materials. It is this flexibility that makes the perovskites as a group, important, as the facile replacement of any of the atoms in this range of structures can be used to modify important physical properties in a controlled way. The flexibility comes at a structural cost. The ABX_3 perovskite structure is beset by structural variations that depend upon exact composition as well as temperature and pressure, all of which have a profound significance for physical properties. Moreover, many multi-cation or anion materials show an intricate microdomain structure when examined by transmission electron microscopy. These microdomains are small volumes of differing structural complexity that exist within a coherent anion matrix. Often they show ordering of atoms over several unit cell volumes with the pattern of order changing from one microdomain to its neighbours. When these microdomains are arranged throughout the crystal in a more or less random fashion, dependent upon the symmetry of the phase, the microscopic ordering is hidden from normal X-ray and neutron diffraction structure solving methods and may not feature in the refined structure of the macroscopic crystal studied. This level of order is generally revealed by high-resolution transmission electron microscopy. Because of this divergence, exact structural details of many perovskite phases of complex composition are open to question, although the overall broad-brush structure is known.

Fortunately much of this diversity can be understood or rationalised in terms of an ideal cubic perovskite structure. In this chapter the ideal ABX_3 perovskite structure is described together with some of the structural variations that occur which have significance for chemical and physical properties and which make precise structure determination a difficult task.

1.2 The Cubic Perovskite Structure: $SrTiO_3$

The idealised or *aristotype* perovskite structure is cubic and is adopted by $SrTiO_3$ at room temperature (but not at all temperatures). There are two general ways of listing the atoms in the cubic unit cell. The standard crystallographic description places the choice of origin at the Sr atom:

$SrTiO_3$: cubic; $a = 0.3905$ nm, $Z = 1$; space group, $Pm\bar{3}m$ (No. 221);

Atom positions: Sr: 1 (*a*) 0,0,0;
Ti: 1 (*b*) ½,½,½;
O: 3 (*c*) ½,½,0; ½,0,½; 0,½,½;

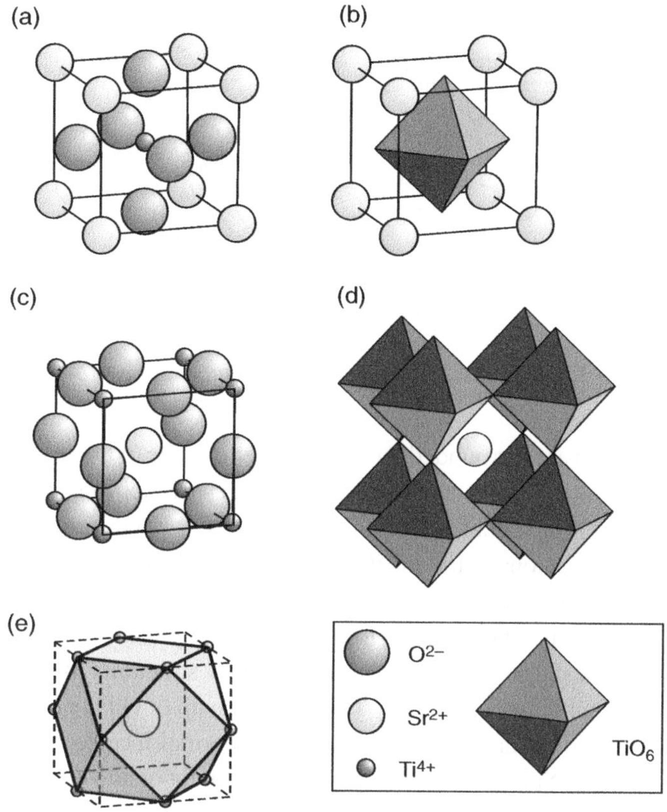

Figure 1.1 The idealised perovskite structure of $SrTiO_3$: (a) atom positions with Sr^{2+} at cell origin; (b) TiO_6 octahedral coordination polyhedron; (c) atom positions with Ti^{4+} at cell origin; (d) TiO_6 octahedral polyhedron framework with Sr^{2+} at the cell centre; (e) cuboctahedral cage site

The Sr^{2+} ions lie at the corners of the unit cell. The Ti^{4+} ions lie at the cell centre and are surrounded by a regular octahedron of O^{2-} ions (Figure 1.1a and b). For some purposes it is useful to translate the cell origin to the Ti^{4+} ions:

$$\begin{aligned}\text{Atom positions: } & \text{Ti}:1(a)\,0,0,0;\\ & \text{Sr}:1(b)\,½,½,½;\\ & \text{O}:3(d)\,½,0,0;\ 0,½,0;\ 0,0,½;\end{aligned}$$

The large Sr^{2+} ions are coordinated to 12 O^{2-} ions and are now situated at the unit cell centre (Figure 1.1c). For a discussion of the chemical and physical properties of this (and other) perovskites, it is convenient to think of the structure as built-up from an array of corner sharing TiO_6 octahedra (Figure 1.1d). The large Sr^{2+} ions are located at

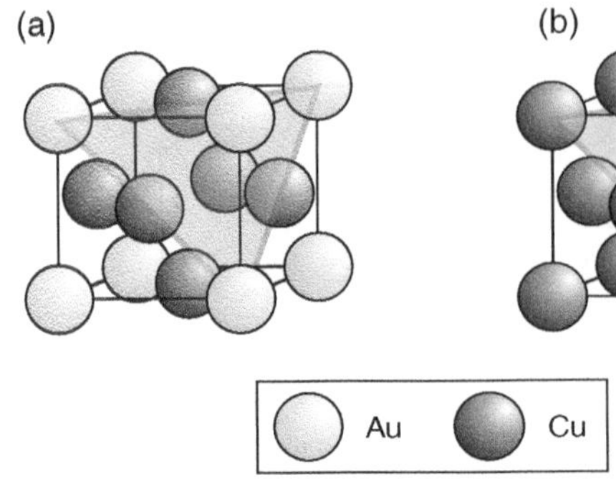

Figure 1.2 *(a) The $AuCu_3$ and (b) the Cu (A1, fcc) structure*

the unit cell centre and are surrounded by a cuboctahedral cage of O^{2-} ions (Figure 1.1e). The TiO_6 framework is regular and the octahedra are parallel to each other. All the Ti^{4+}—O^{2-} bond lengths are equal and the six O^{2-}—Ti^{4+}—O^{2-} bonds are linear.

The Sr^{2+} and O^{2-} positions in the $SrTiO_3$ structure are identical to that of the Au and Cu positions in the alloy Cu_3Au, and if the difference between the Sr^{2+} and O^{2-} ions (or Cu and Au atoms) is ignored, they form a cubic array identical to that of the Cu structure (Figure 1.2a and b). This latter is the simple A1 structure type, often described as the face-centred cubic (fcc) structure, which is made up of (111) planes that lie normal to the cell body diagonal [111], stacked in the normal face-centred sequence … ABCABC…. Thus the $SrTiO_3$ structure can also be thought of as a built-up from close-packed layers of (111) planes containing ordered Sr^{2+} and O^{2-} ions that lie normal to the cubic unit cell body diagonal [111]. The charge balance needed to maintain charge neutrality in this skeleton structure is provided by an ordered distribution of the Ti^{4+} ions in the available octahedral interstices that are bounded by O^{2-} ions only (Figure 1.3).

It is often convenient when describing the structures of more complex perovskite-related phases (Chapters 2 and 3) to display the structure as linked ideal TiO_6 octahedra. The conventional view of the ideal perovskite structure (Figure 1.4a) is often shown tilted to make the (111) layers almost or exactly horizontal (Figure 1.4b and c). More often the alkaline earth atoms are omitted and just the octahedral framework is shown (Figure 1.4d and e). Other projections, such as down [111], show the octahedra projected as hexagonal outlines or down [110] as diamond outlines (Figure 1.4f and g).

1.3 The Goldschmidt Tolerance Factor

From a crystallographic perspective, the ideal perovskite structure is inflexible, as the unit cell has no adjustable atomic position parameters, so that any compositional change must be accommodated by a change in lattice parameter. This is a

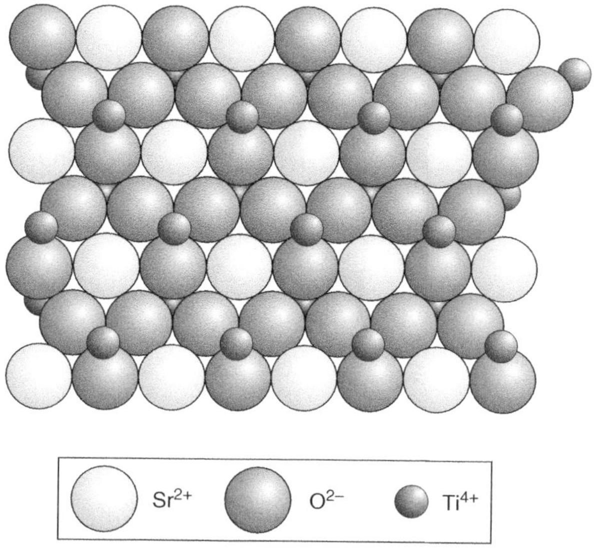

Figure 1.3 *A single SrO_3 (111) plane in $SrTiO_3$. The Ti^{4+} ions, above and below the SrO_3 layer, occupy octahedral interstices that are bounded by six O^{2-} ions*

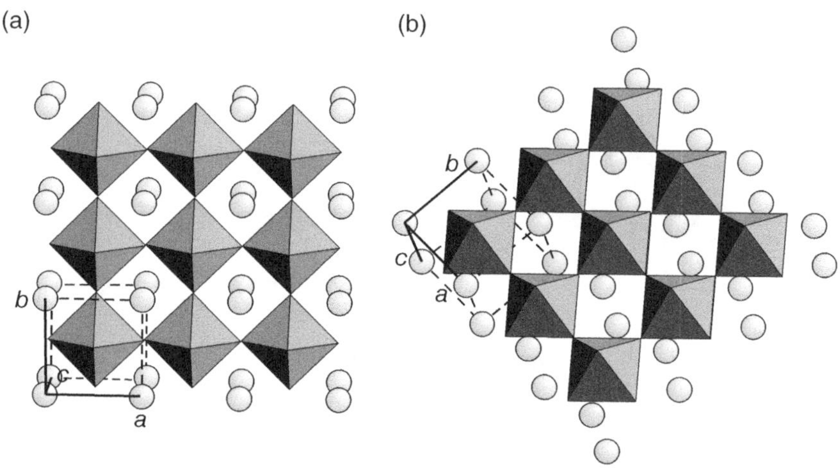

Figure 1.4 *The cubic $SrTiO_3$ perovskite structure (a) conventional view with $3 \times 3 \times 1$ unit cells displayed; (b) the same rotated by approximately 45°; (c) the same rotated further so that one set of SrO_3 planes lies normal to the plane of the page; (d and e) as (b and c) showing only the TiO_6 octahedral framework; (f) octahedral framework projected down [111]; (g) octahedral framework projected down [110]*

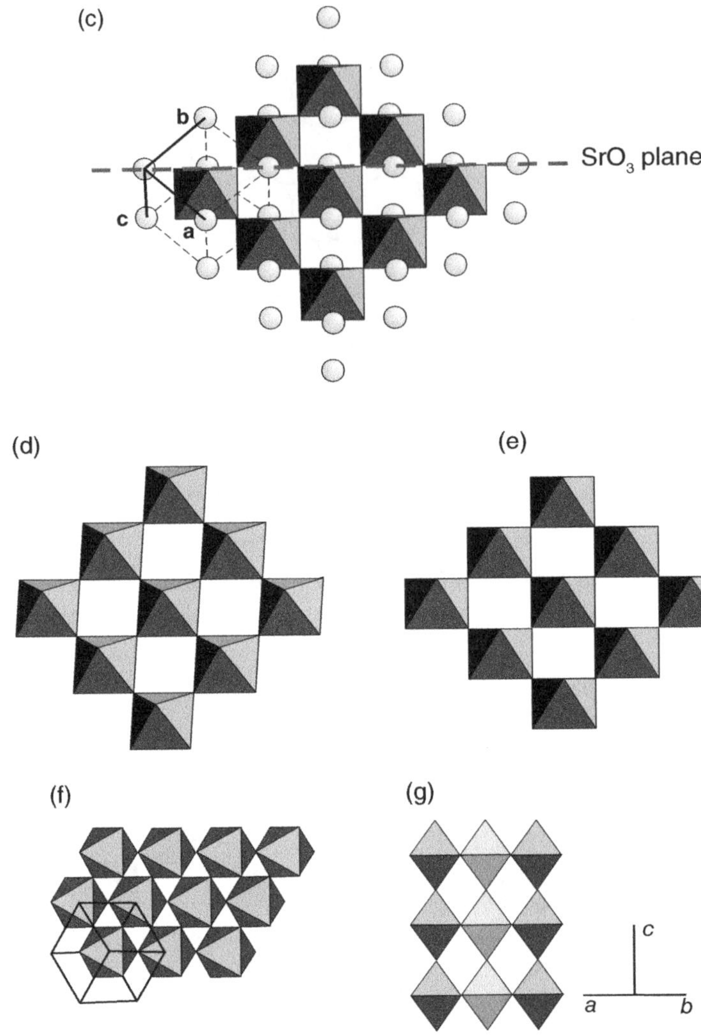

Figure 1.4 (Continued)

simple sum of anion – cation bond lengths. The cubic unit cell edge, a, is equal to twice the B—X bond length:

$$2(B - X) = a$$

The width of the cuboctahedral cage site, $\sqrt{2}a$, is equal to twice the A—X bond length:

$$2(A - X) = \sqrt{2}a$$

This means that the ideal structure forms when the ratio of the bond lengths is given by:

$$\frac{(\mathrm{A{-}X})}{(\mathrm{B{-}X})} = \sqrt{2}$$

or

$$\frac{(\mathrm{A{-}X})}{\sqrt{2}(\mathrm{B{-}X})} = 1$$

This relationship was first exploited by Goldschmidt, in 1926, who suggested that it could be used to predict the likelihood that a pair of ions would form a perovskite structure phase. When this was initially proposed, very few crystal structures had been determined and so ionic radii were used as a substitute for measured bond lengths. For this purpose, it is assumed that for a stable structure to form the cations, just touch the surrounding anions (Goldschmidt's rule), then:

$$\frac{(r_\mathrm{A} + r_\mathrm{X})}{(r_\mathrm{B} + r_\mathrm{X})} = \sqrt{2}$$

or

$$t = \frac{(r_\mathrm{A} + r_\mathrm{X})}{\sqrt{2}(r_\mathrm{B} + r_\mathrm{X})} = 1\,(\text{ideal structure})$$

where t is called the *tolerance factor*, r_A is the radius of the cage site cation, r_B is the radius of the octahedrally coordinated cation and r_X is the radius of the anion. Goldschmidt's proposal was that a perovskite structure phase would form if the value of the tolerance factor, t, was close to 1.0.

Note that it is necessary to use ionic radii appropriate to the coordination geometry of the ions. Thus r_A should be appropriate to 12 coordination, r_B to octahedral coordination and r_X to linear coordination. Furthermore, it is best to use radii scales that mirror the X anion present, as radii appropriate to oxides, although a reasonable approximation for fluorides, are poor when applied to chlorides and sulphides.

Because many perovskite structures have been described, it is now usual to use the measured bond lengths in the crystal rather than ionic radii to give an *observed tolerance factor* t_obs:

$$t_\mathrm{obs} = \frac{(\mathrm{A{-}X})}{\sqrt{2}(\mathrm{B{-}X})}$$

where (A—X) is the average of the measured bond lengths between the A cation and the surrounding 12 anions and (B—X) is the average of the measured bond lengths between the B cation and the surrounding six anions. It is found that for specific groups of perovskites (e.g. $ATiO_3$ titanates, $AAlO_3$ aluminates), there is a linear relationship between t_{obs} and t which varies slightly from one family to another.

Despite its simplicity, the tolerance factor has reasonable predictive power, especially for oxides, where ionic radii are known with greatest precision. Ideally t should be equal to 1.0 and it has been found empirically that if t lies in the approximate range 0.9–1.0 a cubic perovskite structure is a reasonable possibility. If $t > 1$, that is, large A and small B, a hexagonal packing of the AX_3 layers is preferred and hexagonal phases of the $BaNiO_3$ type form (Chapter 3). In cases where t of the order of 0.71–0.9, the structure, particularly the octahedral framework, distorts to close down the cuboctahedral coordination polyhedron, which results in a crystal structure of lower symmetry than cubic. For even lower values of t, the A and B cations are of similar size and are associated with the ilmenite, $FeTiO_3$, structure or the C-type rare earth Ln_2O_3 structure.

In the case of chlorides and sulphides, the tolerance factor tends to move downwards compared to that for oxides and fluorides so that cubic and distorted cubic phases form for t in the range 0.8–0.9, and hexagonal perovskites form if t is greater than 0.9.

The concept of the tolerance factor can be extended to perovskites with more complex compositions by using an average value for ionic radii or bond length. For example, for an A-site substituted phase $A_{1-x}A'_xBX_3$, one can write

$$t = \frac{\left[(1-x)r_A + xr_{A'} + r_X\right]}{\sqrt{2}(r_B + r_X)}$$

and in the case of B-site substitution $AB_{1-x}B'_xX_3$:

$$t = \frac{(r_A + r_X)}{\sqrt{2}\left[(1-x)r_B + xr_{B'} + r_X\right]}$$

Similarly, in terms of bond lengths:

$$t_{obs} = \frac{\langle A\text{–}X \rangle}{\sqrt{2}\,\langle B\text{–}X \rangle}$$

where <A–X> signifies the average bond length for the links A–X, A′–X, etc. and <B–X> signifies the average bond length for the links B–X, B′–X, etc. Both equations can be generalised to more complex structures (A, A′, A″...) (B, B′, B″...) $(X, X', X''...)_3$.

1.4 ABX_3 Perovskite Structure Variants

The BX_6 octahedra are the root of many of the important physical properties of perovskites, such as magnetic and ferroelectric responses to external fields. This is because these are often mediated by the electron configurations of the B cations, which themselves are modified by the surrounding geometry of the six anions. The A cations, although they cannot be ignored, tend to be closed-shell ions with a fixed valence and so less responsive to chemical manipulation with a view to modification of chemical and physical properties. Thus it is useful to place the distortions that occur in perovskites due to the shape and relative orientation of the BX_6 polyhedra into a crystallographic framework.

The simplest change that can be envisaged is that in which the BX_6 octahedra remain perfect (or nearly so) and the cations are simply displaced away from the centre of the octahedron. Cation displacement is usually associated with cations that are 'too small' for the octahedral site, leading to a tolerance factor significantly less than 1 (although in fact B-cation size is only one of several factors of importance in this distortion). The structure now becomes tetragonal, trigonal or orthorhombic, depending upon the direction of cation displacement and the magnitude of the displacements which occur. In addition, the displacements produce permanent electric dipoles in the unit cell and can give rise to pyroelectric, ferroelectric and antiferroelectric effects (Chapter 6).

A second structural response which preserves the perfect (or nearly perfect) BX_6 octahedral geometry is octahedral tilt or rotation. This response is mostly associated with A cations that are too small for the cuboctahedral cage site, and so the BX_6 octahedra twist so as to effectively reduce the cavity dimensions, again allowing the structure to accommodate values of t less than 1. As with cation displacement, rotation also lowers the symmetry of the crystal (Section 1.7) and has a profound influence on the physical properties of these phases.

Finally, the BX_6 octahedron itself can distort to give elongated or flattened octahedra, which in extreme cases can lead to square planar or square pyramidal coordination. These distortions are a result of interactions between the cation electron orbitals and the surrounding anions, typified by the Jahn–Teller effect (Section 1.6). Octahedral distortion can also be caused by cation valence changes such as disproportionation:

$$2B^{n+} \rightarrow B^{n+1} + B^{n-1}$$

The two different-sized cations then may adjust to the surroundings by a distortion of one or both of the cation-centred BX_6 octahedra to give rise to two different-sized octahedra.

These three modifications, namely, B-cation displacement, BX_6 tilt/rotation and BX_6 distortion, are not mutually exclusive and they can occur independently or, often, in combination with one another. Moreover, the resulting changes may be cooperative in that they affect all octahedra in a similar way, or non-cooperative, in which case the distortions may cancel out at a macroscopic level although they may still influence microscopic properties.

The amounts of distortion are generally small and are readily influenced by the ambient conditions. Thus changes in temperature, pressure, crystallite size or form may alter the degree of distortion or the type of distortion present. The majority of perovskite phases manifest a series of symmetry changes as the temperature or pressure is changed, usually resulting in a cubic form at higher temperatures and pressures. For example, the perovskite $SrSnO_3$, which is orthorhombic, space group *Pmna*, at room temperature, changes as the temperature increases to orthorhombic, space group *Imma*, at 905 K, to tetragonal, space group *I4/mcm* at 1062 K and finally to cubic space group $Pm\bar{3}m$ at 1295 K.

For many of these lower symmetry forms, the shift from the ideal cell is small or can be neglected for some purposes, and in such cases it is often convenient to refer these structures to an idealised pseudocubic structure, of cell length a_p (equivalent to a for ideal $SrTiO_3$, 0.39 nm), that can be used as a first (and often sufficient) approximation in describing the properties of the phases.

1.5 Cation Displacement: $BaTiO_3$ as an Example

Cation displacement in $BaTiO_3$ is attributed to the second order (or pseudo) Jahn–Teller effect, acting upon the d^0 Ti^{4+} ion. This mechanism applies to nonlinear molecules with a nondegenerate electronic ground state accompanied by a very low-lying excited state. Under these conditions, distortions of symmetry occur that facilitate the mixing of the electronic ground and excited states in a way to lower the ground-state energy. In perovskites, the appropriate symmetry change is provided by cation displacement. (The first-order Jahn–Teller effect, which affects d^n cations, is described in Section 1.6.)

To preserve a maximum degree of symmetry, cation displacement needs to be along one of the symmetry axes of the octahedron. There are three tetrad (fourfold) axes of rotational symmetry, each passing through opposite vertices of the octahedron; four triad (threefold) axes of rotational symmetry, each passing through the centres of opposing triangular faces of the octahedron and six diad (twofold) axes

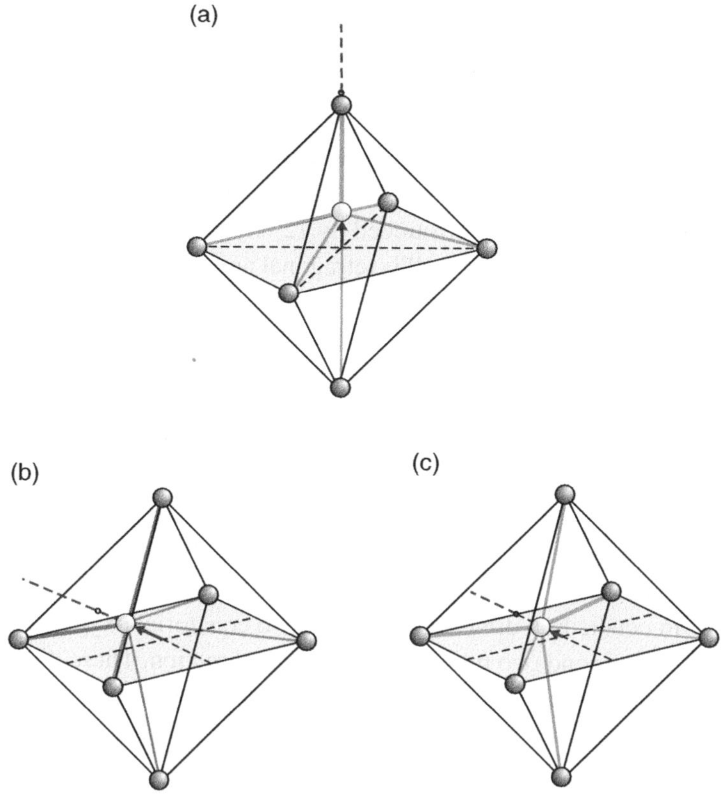

Figure 1.5 Cation displacement in a BX_6 octahedron: cation displacement along (a) a tetrad axis; (b) a triad axis; (c) a diad axis

of rotational symmetry, each passing though the centres of opposite edges of the octahedron (see Figure 1.10). The symmetry axis parallel to the direction of the displacement is maintained after the displacement. Cation displacement parallel to one of the tetrad axes ideally gives rise to one longer bond, one shorter bond and four bonds of intermediate length (Figure 1.5a). Cation displacement along a triad axis ideally gives rise to three shorter bonds and three longer bonds (Figure 1.5b), while cation displacement along a diad ideally gives rise to two long and two short bonds in the plane containing the diad and two bonds of intermediate length roughly perpendicular to that plane (Figure 1.5c). The simplest structures are found when all of the cation displacements are identical and in such cases, B-cation displacement is usually accompanied by smaller A-cation displacement.

The effects of B-cation displacement upon the structures and dielectric properties of perovskite phases have been extensively studied in the important dielectric/ferroelectric perovskite, $BaTiO_3$. The phase shows transformations from trigonal to orthorhombic (at 183 K), orthorhombic to tetragonal (at 263 K) and tetragonal to cubic (at 393 K), all due to cation displacement.

The high-temperature phase, stable above 393 K, adopts the ideal perovskite structure: space group $Pm\bar{3}m$ (221), $a = 0.39732$ nm (Figure 1.6a). The medium-sized Ti^{4+} ions are situated at the centre of an octahedron of oxygen ions with equal bond lengths of approximately 0.2 nm, while the large Ba^{2+} cations are in the centres of the cuboctahedral cage sites, surrounded by 12 oxygen ions.

Between 393 and 268 K the unit cell is tetragonal and ferroelectric. As the crystal cools through the cubic – tetragonal transition temperature, the cubic cell expands slightly along one edge to produce the tetragonal **c**-axis and is slightly compressed along the other two edges to form the tetragonal **a**- and **b**-axes: space group *P*4*mm* (99). $a = b = 0.39910$ nm, $c = 0.40352$ nm (Figure 1.6b). The change from cubic to tetragonal is accompanied by an off-centre movement of the octahedrally coordinated Ti^{4+} ions along the +**c**-axis, which is one of the tetrad axes of the octahedron (Figure 1.6c). The O^{2-}—Ti^{4+} bond lengths parallel to the **c**-axis are now ≈0.22 and ≈0.18 nm, while the equatorial bond lengths remain at ≈0.2 nm. The change in the Ba^{2+} positions is almost negligible. The Ti^{4+} displacement is accompanied by a slight change in octahedron dimensions so that two equatorial oxygen atoms move parallel to the +**c**-axis and two move in the opposite direction. These displacements give rise to electric dipoles, p, one in each octahedron (Figure 1.6d) that are the source of the ferroelectric properties of tetragonal $BaTiO_3$. There is no preference as to which of the original cubic axes becomes the polar direction, and so this can take one of six equivalent directions, so that a single crystal invariably becomes heavily twinned on cooling.

Between the temperatures 263 and 183 K, the tetragonal structure undergoes a further transition with the Ti^{4+} cation displaced along a diad axis (Figure 1.6e) giving four Ti—O bonds approximately 0.198 nm and two bonds approximately 0.201 nm in length, with once again a slight distortion of the octahedra, producing an ordered array of electronic dipoles. This results in an elongation along a face diagonal of the pseudocubic unit cell to give an orthorhombic phase: space group *Amm*2 (38), $a = 0.39594$ nm, $b = 0.56266$ nm, $c = 0.56435$ nm (Figure 1.6f).

Below 183 K the Ti^{4+} cation displacement is along the triad axis, giving rise to three shorter Ti—O bonds of approximately 0.21 nm and three bonds of ≈0.19 nm in length (Figure 1.6g). The pseudocubic cell is now elongated along the cell body diagonal to produce a rhombohedral cell: space group *R*3*m* (160), $a = 0.40043$ nm, $\alpha \approx 89.86°$ (Figure 1.6h). This cell is very close in dimensions to the cubic high-temperature structure but again displays a ferroelectric response to applied electric fields, which the cubic phase does not.

All of the phases described and the cations displacements and octahedral distortions are temperature and pressure sensitive. The phase diagram for $BaTiO_3$ at

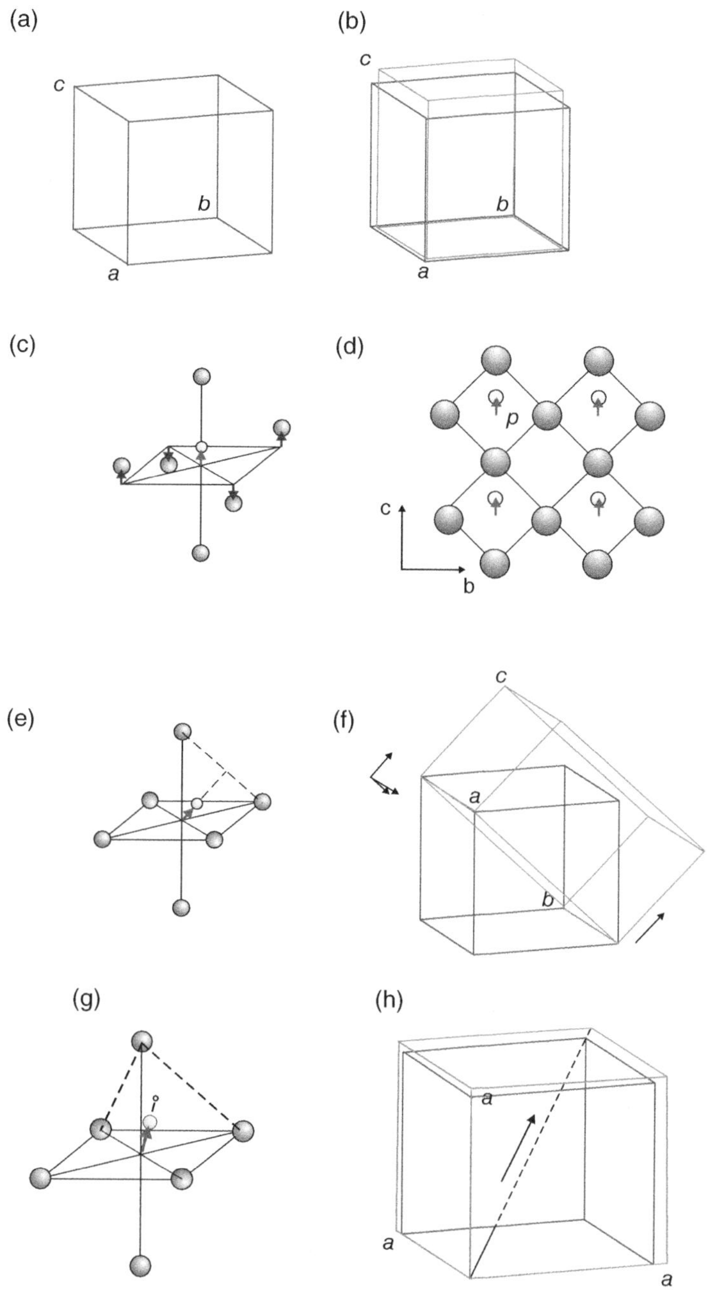

Figure 1.6 *Cation displacements (exaggerated for clarity) in* $BaTiO_3$*: (a) T > 398 K, cubic; (b) 398 > T > 278 K, tetragonal; (c) ion displacements in each* TiO_6 *octahedron; (d) electric dipoles generated by* Ti^{4+} *ion displacements; (e)* Ti^{4+} *ion displacement in the orthorhombic phase; (f) orthorhombic unit cell compared cubic unit cell; (g)* Ti^{4+} *ion displacement in the trigonal phase; (h) rhombohedral unit cell compared to cubic unit cell*

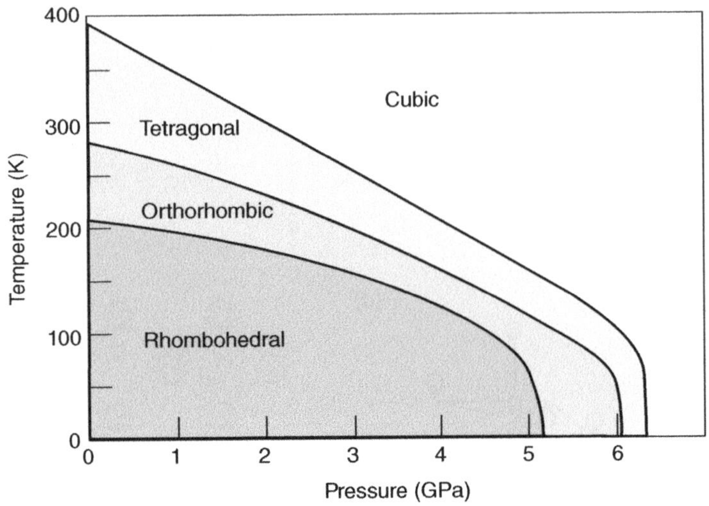

Figure 1.7 *The temperature–pressure phase diagram of* $BaTiO_3$ *(Original data in Hayward and Salje (2002) and Ishidate* et al. *(1997))*

moderate temperatures and pressures illustrates the symmetry changes encountered (Figure 1.7). It is seen that, at 300 K and ambient pressure (1 atm = 0.000101 GPa), $BaTiO_3$ will exist as the tetragonal form. When the applied pressure reaches approximately 2 GPa, both the cubic and tetragonal forms will coexist. A slight increase in either pressure or temperature will tip the balance in favour of the cubic form, which will then exist alone.

Note that these changes are also influenced by particle size and morphology. Nanoparticles often present a different pattern of crystallographic behaviour than that illustrated.

1.6 Jahn–Teller Octahedral Distortion: $KCuF_3$ as an Example

The best known examples of octahedral distortion in perovskites are Jahn–Teller distortions. These arise from the (first order) Jahn–Teller effect, which, in the broadest sense, applies to all nonlinear molecules. The Jahn–Teller theorem says that a molecule with a symmetrical atomic configuration and a degenerate electronic ground state is unstable and will distort so as to remove the electronic degeneracy. With respect to perovskite structures, the Jahn–Teller theorem applies to octahedrally coordinated transition metal B-site cations. It implies that perovskites containing ions with certain d-electron configurations (so-called Jahn–Teller ions) will have an enhanced stability when located in

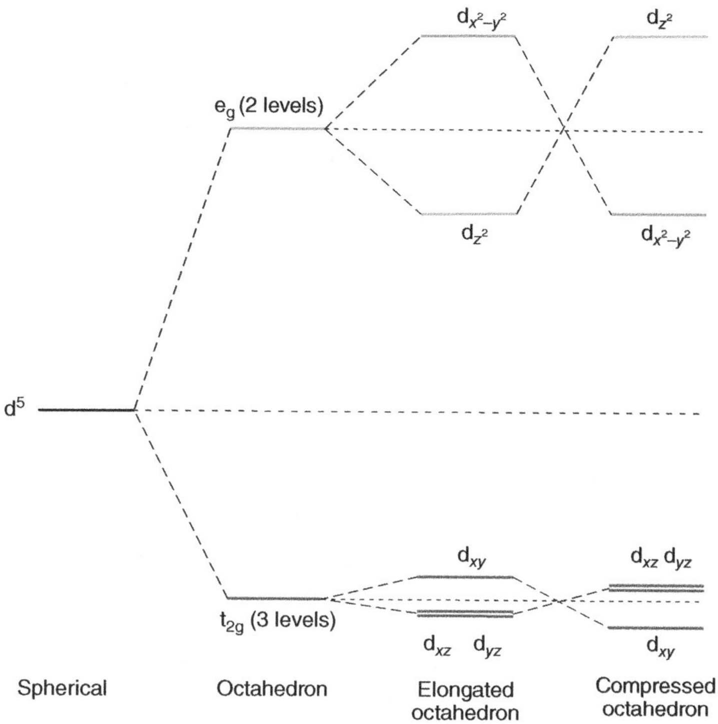

Figure 1.8 *The effect of a Jahn–Teller distortion on the d-electron energy levels of a B-site cation in an octahedron*

a distorted octahedral environment compared to that in an undistorted one. The distortion may take the form of an elongation of each octahedron along one of its fourfold tetrad symmetry axes, giving a coordination polyhedron with four shorter B—X bonds and two longer B—X bonds or the converse, with four longer B—X bonds and two shorter B—X bonds, leading to a compressed octahedron.

The degenerate ground state comes about in the following way. In an ideal octahedron the d-electron energy levels split into two groups, a lower energy triply degenerate pair, labelled t_{2g}, consisting of the d_{xy}, d_{yz} and d_{xz} orbitals, and an upper doubly degenerate pair labelled e_g, consisting of the orbital pair $d_{x^2-y^2}$ and d_{z^2} (Figure 1.8). A transition metal ion with an odd number of electrons in the upper (e_g) pair of d orbitals, such as high-spin Cr^{2+}, Mn^{3+} (d^4, $t_{2g}^3e_g^1$); low-spin Co^{2+}, Ni^{3+} (d^7, $t_{2g}^6e_g^1$) and Cu^{2+} (d^9, $t_{2g}^6e_g^3$) will have a degenerate ground state because there are a two equivalent electron distributions possible. For example, with Mn^{3+} one electron can be placed in either the $d_{x^2-y^2}$ or d_{z^2} orbital to give the degenerate energy configurations t_{2g}^3, $d_{x^2-y^2}^1d_{z^2}^0$ or t_{2g}^3 $d_{z^2}^1d_{x^2-y^2}^0$. The same degeneracy will occur for the other ions listed. An elongation or compression of the octahedron will remove the degeneracy by a further splitting of the both the t_{2g} and e_g groups (Figure 1.8). The change in the t_{2g} group is rather small and is not of greatest importance for perovskites. The splitting of the e_g group is, however, significant and does lead to octahedral distortion. It is not possible to determine from these qualitative descriptions whether the favoured distortion will be octahedral elongation or compression, but by far the commonest in perovskites is an elongation of the octahedron, which in extreme cases can give rise to square pyramidal or square planar coordination.

In general the A and B ions remain in the centres of the coordination polyhedra in spite of the distortion, but the symmetry of the unit cell drops to tetragonal or orthorhombic. The distortion is both temperature and pressure sensitive, and perovskites showing Jahn–Teller distortion generally return to cubic symmetry at higher temperatures and pressures.

An example that has been investigated in detail is $KCuF_3$, containing the Jahn–Teller ion Cu^{2+}. Here the electronic energy level degeneracy in a perfect octahedron corresponds to the alternative electron distributions $d^2_{x^2-y^2}d^1_{z^2}$ or $d^1_{x^2-y^2}d^2_{z^2}$. The Jahn–Teller distortion is apparent in the elongated form of the CuF_6 octahedra.

$KCuF_3$: tetragonal; $a = b = 0.58550$ nm, $c = 0.78456$ nm, Z = 4; space group, *I4/mcm* (No. 140), at 295 K and normal pressure;

Atom positions: Kr: 4 $(a)\,0,0,\tfrac{1}{4};\,0,0,\tfrac{3}{4}$;
Cu: 4 $(d)\,0,\tfrac{1}{2},0;\,\tfrac{1}{2},0,0$;
F1: 4 $(b)\,0,\tfrac{1}{2},\tfrac{1}{4};\,\tfrac{1}{2},0,\tfrac{1}{4}$;
F2: 8 $(h)\,x,\,x+\tfrac{1}{2},0;\,-x,-x+\tfrac{1}{2},0;\,-x+\tfrac{1}{2},x,0;\,x+\tfrac{1}{2},-x,0$;
$x = 0.2281$

The CuF_6 octahedra have one axis elongated to give a Cu—F bond length of 0.225 nm, while the other two are shortened unequally to 0.196 and 0.189 nm (Figure 1.9a and b). The sheets of octahedra lie on planes at $z = 0$ and $z = \tfrac{1}{2}$. All are positioned so that one of the short diameters lies parallel to [001], but within any one sheet of structure parallel to (001), the long dimensions are arranged so that they lie perpendicular to those of the neighbouring octahedra (Figure 1.9c). The perovskite pseudocubic cell has dimensions $a_p = a/\sqrt{2} = c/2 \approx 0.40$ nm.

The distortion varies with temperature and pressure in a uniform way. When the pressure increases to 8 GPa, the four shorter bonds become almost equal in length and the long bonds shorten considerably to give an approximately regular octahedral coordination polyhedron.

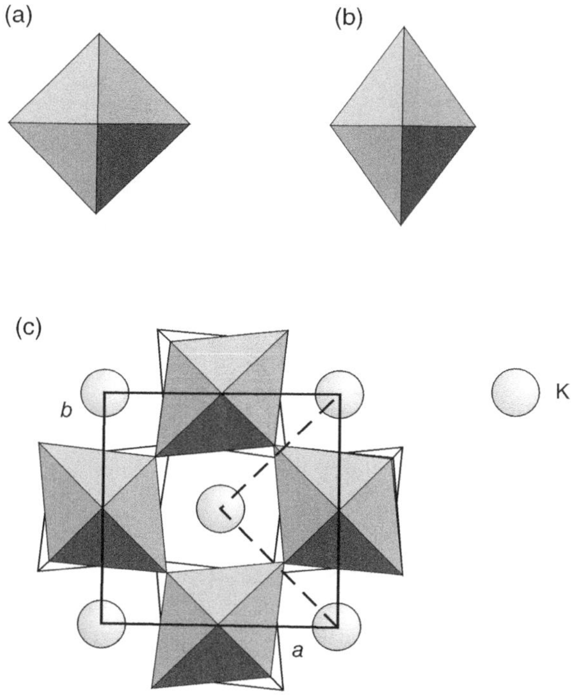

Figure 1.9 *The $KCuF_3$ structure: (a) a regular CuO_6 octahedron; (b) CuO_6 octahedron elongated along one tetrad axis; (c) the $KCuF_3$ structure projected down [001]. The distorted octahedra in the lower layer are shown in outline, and those in the upper layer are shown shaded. The unit cell is outlined and the pseudocubic cell is dotted*

1.7 Octahedral Tilting

1.7.1 Tilt Descriptions

The (ideally rigid) apex-linked BX_6 octahedra can be imagined to be joined by flexible hinges that allow for a certain amount of rotation or tilting. In many cases, slight tilting of the octahedra, so as to enhance stability, is associated with A cations that are slightly too small to fit the ideal structure, as in the example of perovskite, $CaTiO_3$, itself, which displays an orthorhombic symmetry compared to cubic $SrTiO_3$. The tilting contrives to reduce the size of the A-site cavity slightly and so is a mechanism of compensation for a mismatch in the A–X/B–X bond lengths. Tilting can occur about the (a) tetrad, (b) diad and (c) triad axes of the octahedron (Figure 1.10).

The corner connectivity of the BX_6 octahedra means that the tilt of one octahedron completely determines the tilting in the plane perpendicular to the tilt axis

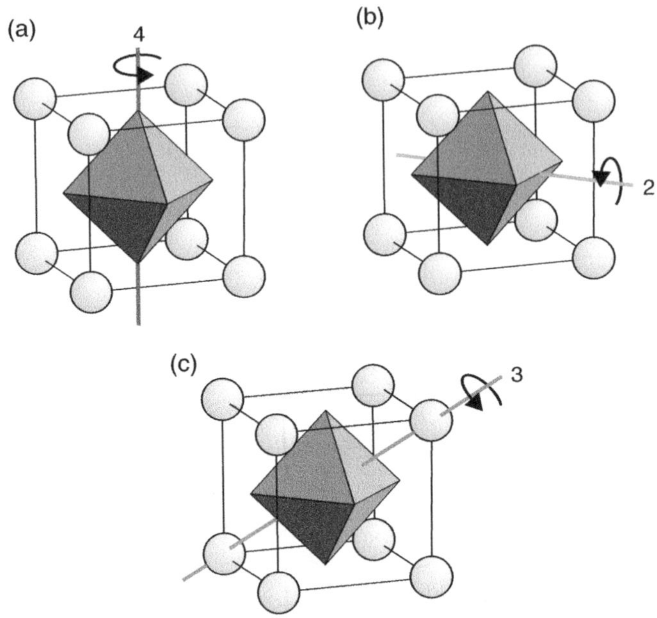

Figure 1.10 *Tilt axes of an octahedron: (a) one of three tetrad axes; (b) one of six diad axes; (c) one of four triad axes*

(Figure 1.11). In addition, the tilting in an adjacent plane must imitate that in the first, but can be in phase with it (i.e. exactly the same when viewed down the tilt axis) or out of phase with it (i.e. in the opposite direction).

The tilting of the octahedra can also be envisaged in terms of rotations compounded of three component tilts along the x-, y- and z-axes of the parent ideal perovskite structure. The system of nomenclature most frequently encountered is that of Glazer (1972). A rotation is described by a symbol a, b or c for tilt axes parallel to x, y and z. If there is no tilt, the superscript 0 is added, and if the tilt is exactly the same about two axes, then the letter symbol is repeated. In this way, the ideal perovskite structure is given the symbol $a^0a^0a^0$, signifying no tilt about x, y or z (Figure 1.12a). A rotation around z can give two layers of octahedra with exactly the same relative orientation, $a^0a^0c^+$ (Figure 1.12b), or with opposite relative orientations $a^0a^0c^-$ (Figure 1.12c). Similarly, three equal tilts, $a^-a^-a^-$, give a structure with each layer of octahedra with the same relative orientation (Figure 1.12d) while the opposite relative orientation of the two layers of octahedra is produced by $a^-a^-c^+$ (Figure 1.12e). Furthermore, three equal tilts correspond to a rotation $\varphi°$ about triad [111]-type axis while two equal tilts corresponds to rotation $\theta°$ about a diad [110]-type axis, provided that tilt angles are fairly small, say less than about 15°, and one tilt corresponds to a rotation $\varphi°$ about a tetrad [100]-type axis.

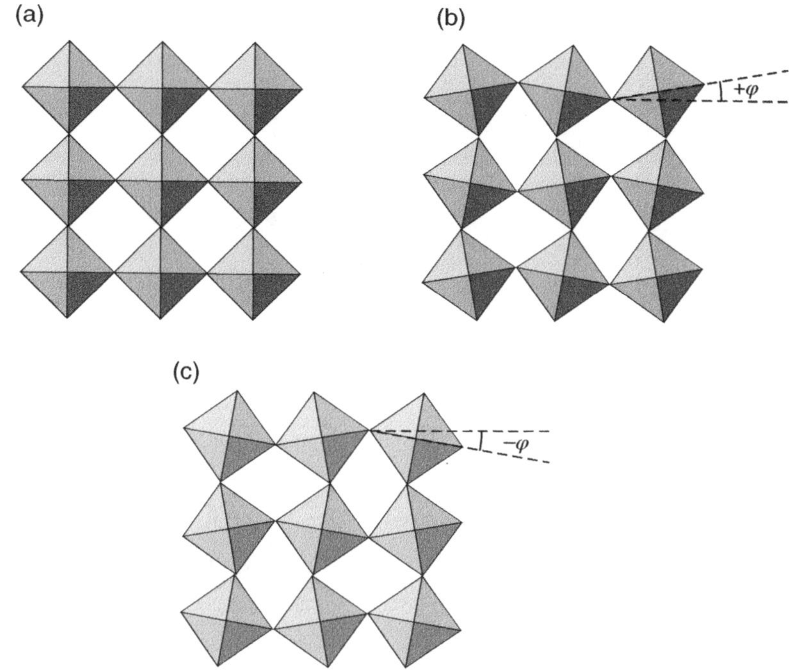

Figure 1.11 *Tilting of a plane of corner-linked octahedra about a tetrad axis perpendicular to the plane of the figure through one octahedron: (a) no tilt; (b) a tilt of $+\varphi$; (c) a tilt of $-\varphi$*

In the simplest case, and considering only a two-octahedron repeat along each of the cubic or pseudocubic axes, 10 possible tilt combinations are possible:

three tilts; $a^+b^+c^+$, $a^+b^+c^-$, $a^+b^-c^-$, $a^-b^-c^-$;
two tilts; $a^0b^+c^+$, $a^0b^+c^-$, $a^0b^-c^-$;
one tilt; $a^0a^0c^+$, $a^0a^0c^-$;
no tilts; $a^0a^0a^0$;

In order to relate these to crystallographic structures, it is necessary to take into account the relative magnitudes of the tilts. In this case, three unequal tilts produce three unequal pseudocubic axes, two equal tilts and one different tilt produce two equal pseudocubic axes and one different and three equal tilts produce three equal pseudocubic axes. Moreover, whenever a superscript is 0 or +, a mirror plane must exist perpendicular to the tilt axis. By examination of the 10 tilt combinations described in terms of these and related constraints, 23 different tilt systems can be derived together with the pseudocubic unit cell dimensions and the ideal space group of the four octahedra unit cell (Table 1.2). It is, though, important to

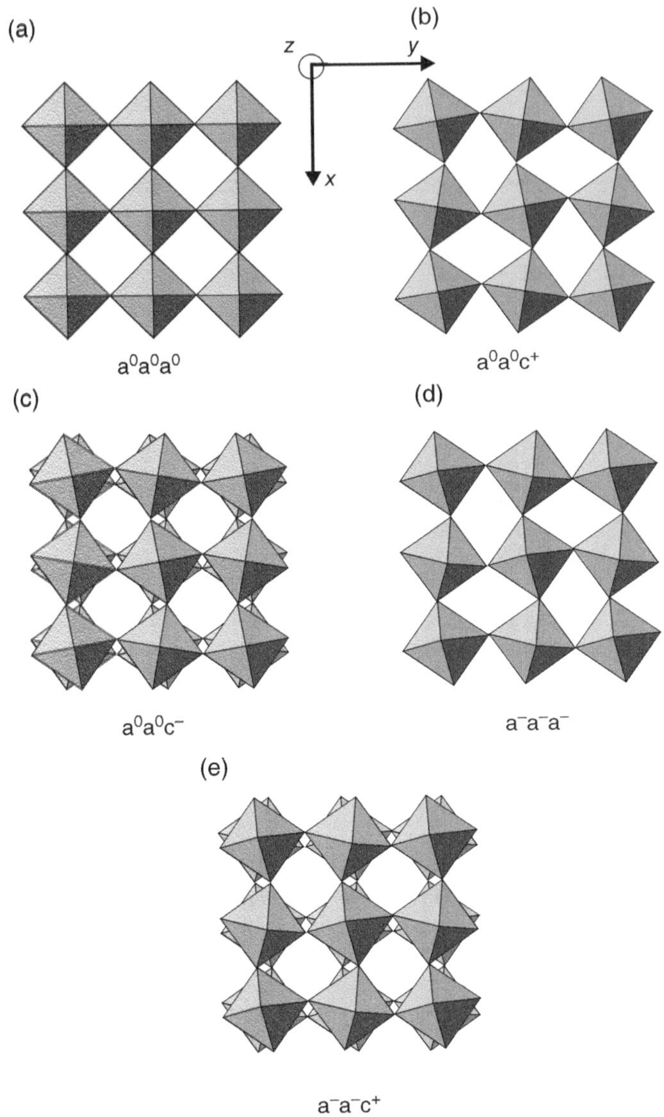

Figure 1.12 *Glazer tilt notation: (a) $a^0a^0a^0$; (b) $a^0a^0c^+$; (c) $a^0a^0c^-$; (d) $a^-a^-a^-$; (e) $a^-a^-c^+$*

remember should a perovskite phase have a repeat of more than two octahedra in each axial direction, a more complex notation will be needed.

The lattice parameters of many of the resulting unit cells can be related to the B–O bond distance, d, and the angles of rotation, φ about a tetrad axis [001], θ about a diad axis [110] and φ about a triad axis [111]. The simplest of these are given in Table 1.3.

***Table 1.2** Tilt systems for BX_6 octahedra in ABX_3 perovskites*

Number	Symbol	Pseudocubic unit cell	Space group
Three tilts			
1	$a^+b^+c^+$	$a_p \neq b_p \neq c_p$	*Immm* (71)
2	$a^+b^+b^+$	$a_p \neq b_p = c_p$	*Immm* (71)
3	$a^+a^+a^+$	$a_p = b_p = c_p$	$Im\bar{3}$ (204)
4	$a^+b^+c^-$	$a_p \neq b_p \neq c_p$	*Pmmn* (59)
5	$a^+a^+c^-$	$a_p = b_p \neq c_p$	$P4_2/nmc$ (137)
6	$a^+b^+b^-$	$a_p \neq b_p = c_p$	*Pmmn* (59)
7	$a^+a^+a^-$	$a_p = b_p = c_p$	*Pmmn* (59)
8	$a^+b^-c^-$	$a_p \neq b_p \neq c_p$ $\alpha \neq 90°$	$P2_1/m$ (11)
9	$a^+a^-c^-$	$a_p = b_p \neq c_p$ $\alpha \neq 90°$	$P2_1/m$ (11)
10	$a^+b^-b^-$	$a_p \neq b_p = c_p$ $\alpha \neq 90°$	*Pnma* (62)
11	$a^+a^-a^-$	$a_p = b_p = c_p$ $\alpha \neq 90°$	*Pnma* (62)
12	$a^-b^-c^-$	$a_p \neq b_p \neq c_p$ $\alpha \neq \beta \neq \gamma \neq 90°$	$P\bar{1}$ (2)
13	$a^-b^-b^-$	$a_p \neq b_p = c_p$ $\alpha \neq \beta \neq \gamma \neq 90°$	*C2/c* (15)
14	$a^-a^-a^-$	$a_p = b_p = c_p$ $\alpha = \beta = \gamma \neq 90°$	$R\bar{3}c$ (167)
Two tilts			
15	$a^0b^+c^+$	$a_p < b_p \neq c_p$	*Immm* (71)
16	$a^0b^+b^+$	$a_p < b_p = c_p$	*I4/mmm* (139)
17	$a^0b^+c^-$	$a_p < b_p \neq c_p$	*Cmcm* (63)
18	$a^0b^+b^-$	$a_p < b_p = c_p$	*Bmmb* (63)
19	$a^0b^-c^-$	$a_p < b_p \neq c_p$ $\alpha \neq 90°$	*C2/m* (12)
20	$a^0b^-b^-$	$a_p < b_p = c_p$ $\alpha \neq 90°$	*Imma* (74)
One tilt			
21	$a^0a^0c^+$	$a_p = b_p < c_p$	*P4/mbm* (127)
22	$a^0a^0c^-$	$a_p = b_p < c_p$	*I4/mcm* (140)
Zero tilt			
23	$a^0a^0a^0$	$a_p = b_p = c_p$	$Pm\bar{3}m$ (221)

These tilts and rotations are extremely sensitive to external conditions, including temperature, pressure and strain. This latter aspect is particularly important in the case of thin films which are epitaxially grown onto a suitable substrate. Considerable tensile or compressive strain in the film often arises because of mismatch between the lattice parameter of the substrate and the lattice parameter of the film. Such effects can cause considerable (and sometimes surprising) changes in the physical properties of the film itself or of the interface region. In addition, chemical substitution at A- and B-sites in a perovskite may significantly alter the tilt and rotation of

Table 1.3 Lattice parameter equations for some tilt systems

Tilt system	Space group	Lattice parameters
$a^0a^0a^0$ (23)	$Pm\bar{3}m$ (321)	$a=2d$
$a^0a^0c^-$ (22)	$I4/mcm$ (140)	$a=(\sqrt{8})d\cos\varphi$, $c=4d$
$a^0a^0c^+$ (21)	$P4/mbm$ (127)	$a=(\sqrt{8})d\cos\varphi$, $c=2d$
$a^0b^-b^-$ (20)	$Imma$ (74)	$a=(\sqrt{8})d$, $b=4d\cos\theta$, $c=(\sqrt{8})d\cos\theta$
$a^0b^-c^+$ (17)	$Cmcm$ (63)	$a=4d\cos\theta$, $b=2d(\cos\theta+1)$, $c=2d(\cos\theta+1)$
$a^0b^+b^+$ (16)	$I4/mmm$ (139)	$a=2d(1+\cos\theta)$, $c=4d\cos\theta$
$a^-a^-a^-$ (14)	$R\bar{3}c$ (167)	$a=(\sqrt{8})d\cos\phi$, $c=(\sqrt{48})d$
$a^+a^+a^+$ (3)	$Im\bar{3}$(204)	$a=(8\cos\phi+4)/3$

the linked octahedra. For example, the octahedral tilt in such series as the lanthanoid nicklates, $LnNiO_3$, varies directly in proportion to the size of the lanthanoid cation and can be tuned by suitable A-site doping.

1.7.2 Trigonal Symmetry: $LaAlO_3$ as an Example

Trigonal perovskites, typified by the (3, 3) phase $LaAlO_3$ at room temperature, are produced by a rotation of the BO_6 octahedra by approximately 6° about a trigonal axis normal to a triangular octahedron face compared to the cubic parent structure (Figure 1.13a and b). (The crystal class of these phases is trigonal, while the axes are termed rhombohedral. The axes are often more conveniently described in terms of a hexagonal unit cell.) In terms of the tilts in Table 1.2, these compounds belong to the tilt system (14), $a^-a^-a^-$, with the octahedra in each successive layer tilted in opposite directions (Figure 1.13c).

$LaAlO_3$: trigonal, hexagonal cell; $a_H=b_H=0.5365$ nm, $c_H=1.3111$ nm, $\alpha=\beta=90°$, $\gamma=120°$, $Z=6$; space group, $R\bar{3}c$ (167); using a hexagonal cell:

$$\begin{aligned}\text{Atom positions: } & \text{La}: 6(a)\, 0,0,\tfrac{1}{4};\ 0,0,\tfrac{3}{4};\\ & \text{Al}: 6(b)\, 0,0,0;\ 0,0,\tfrac{1}{2};\\ & \text{O}: 18(c)\, x,0,\tfrac{1}{4};\ 0,x,\tfrac{1}{4};\ -x,-x,\tfrac{1}{4};\ -x,0,\tfrac{3}{4};\ 0,-x,\tfrac{3}{4};\\ & \quad x,x,\tfrac{3}{4};\ x=0.5254 \text{ at } 320\text{ K}.\end{aligned}$$

The rotation angle diminishes as the temperature increases and becomes zero, giving the cubic structure (Figure 1.13a), at approximately 813 K with a lattice parameter of $a=0.38113$ nm at 850 K. The octahedra are found to be slightly flattened down the triad (hexagonal c, rhombohedral [111]) axis, but the Al—O bond lengths are all equal at 0.190 nm.

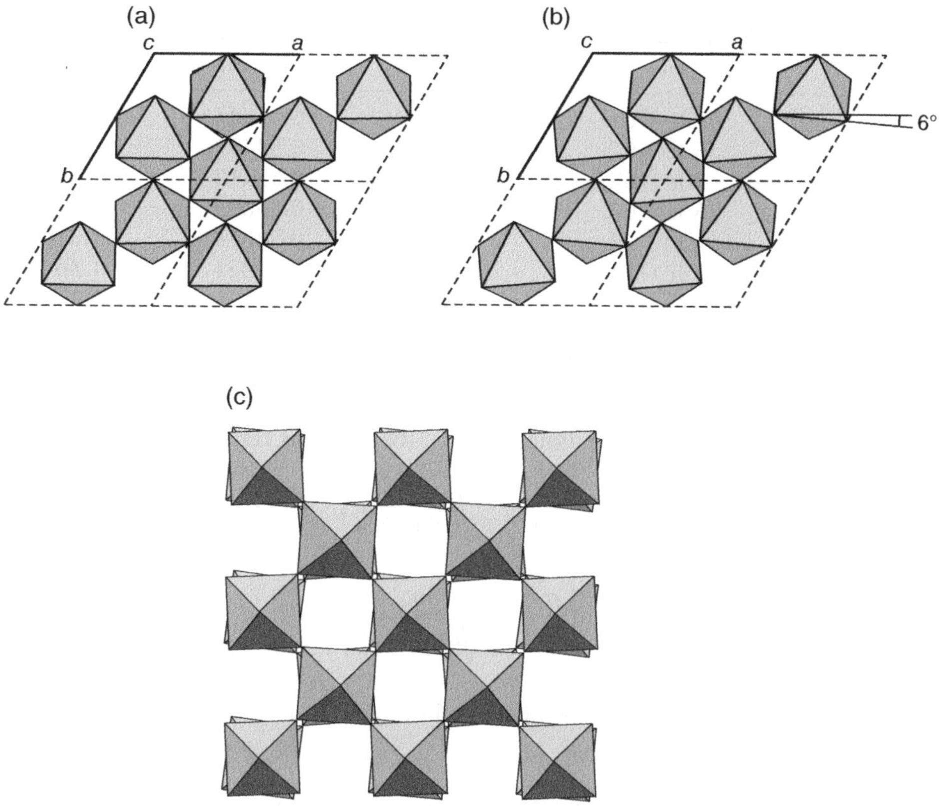

Figure 1.13 The $LaAlO_3$ structure: (a) projection of ideal cubic structure down [001]; (b) projection of the $LaAlO_3$ structure down [001], with a rotation angle of 6°; (c) projection showing octahedral layer tilts

It is possible to estimate the unit cell parameters using the formulae in Table 1.3. Taking d as half the $SrTiO_3$ unit cell edge, 0.195 nm:

$$a = \left(\sqrt{8}\right) d \cos\phi = \left(\sqrt{8}\right) \times 0.195 \times \cos 6 = 0.548 \text{ nm}$$

$$\mathrm{c} = \left(\sqrt{48}\right) d = 1.316 \text{ nm}$$

which are in good agreement with the experimental values.

The structure can also be described in terms of a primitive rhombohedral cell:

$$a_{\mathrm{RP}} = b_{\mathrm{RP}} = c_{\mathrm{RP}} = 0.5357 \text{ nm}, \alpha = \beta = \gamma = 60.1°, Z = 2$$

Table 1.4 *Phases with the trigonal $LaAlO_3$ structure*

Phase	Space group	*a* (nm)	*b* (nm)	*c* (nm)
$LaAlO_3$	*R*3*c* (161)	0.53644	0.53644	1.31195
$LaCoO_3$	$R\bar{3}c$ (167)	0.54437	0.54437	1.30957
$LaMnO_3$ [a]	$R\bar{3}c$ (167)	0.55312	0.55312	1.3488
$La_{0.6}Sr_{0.4}MnO_3$	$R\bar{3}c$ (167)	0.55005	0.55005	1.33575
$LaNiO_3$	$R\bar{3}c$ (167)	0.54524	0.54524	1.31572
$LiNbO_3$	*R*3*c* (161)	0.51381	0.51381	1.38481
$LiTaO_3$	*R*3*c* (161)	0.51528	0.51528	1.37795

[a] Also reported as adopting the $GdFeO_3$ structure (Section 1.7.3).

The relationship between the dimensions of the two unit cells is

$$a_{\rm RP} = \frac{1}{3}\sqrt{\left(3a_{\rm H}^2 + c_{\rm H}^2\right)}$$

$$\sin\left(\frac{\alpha}{2}\right) = \frac{3a_{\rm H}}{2\sqrt{\left(3a_{\rm H}^2 + c_{\rm H}^2\right)}}$$

Finally the structure can also be described in terms of a face-centred rhombohedral cell:

$$a_{\rm RF} = b_{\rm RF} = c_{\rm RF} = 0.7581\,\text{nm},\ \alpha = \beta = \gamma = 90.1°,\ Z = 8$$

The relationship between the dimensions of the hexagonal and face-centred rhombohedral unit cells is

$$a_{\rm RF} = \frac{1}{3}\sqrt{\left(12a_{\rm H}^2 + c_{\rm H}^2\right)}$$

$$\cos\alpha = \frac{\left(c_{\rm H}^2 - 6a_{\rm H}^2\right)}{\left(c_{\rm H}^2 + 12a_{\rm H}^2\right)}$$

Despite the tilting, the unit cell is almost cubic; the pseudocubic cell having dimensions $a_{\rm p}$ = half the $a_{\rm RF}$ axis, 0.379 nm (Table 1.4).

Many of these perovskites revert to the ideal cubic structure at higher temperatures.

1.7.3 Orthorhombic Symmetry: $GdFeO_3$ and $CaTiO_3$ as Examples

The classical orthorhombic distorted perovskites are often referred to as being of the $GdFeO_3$ type, choosing the (3, 3) phase as the structure type. The octahedral rotations in these phases are about the diad axes, [110], belonging to tilt system 10, $a^+b^-b^-$. $GdFeO_3$ has the crystallographic parameters:

$GdFeO_3$: orthorhombic; $a = 0.5349$ nm, $b = 0.5609$ nm, $c = 0.7669$ nm, $Z = 4$; space group, *Pbmn* (No. 62, nonstandard setting);

$$\begin{aligned}
\text{Atom positions: } & \text{Gd}: 4(a)\, 0.9846, 0.0629, \tfrac{1}{4} \\
& \text{Fe}: 4(b)\, \tfrac{1}{2}, 0, 0 \\
& \text{O1}: 4(c)\, 0.1116, 0.4569, \tfrac{1}{4} \\
& \text{O2}: 8(d)\, x, y, z;\ x + \tfrac{1}{2}, -y + \tfrac{1}{2}, -z;\ -x, -y, z + \tfrac{1}{2}; \\
& \quad -x + \tfrac{1}{2}, y + \tfrac{1}{2}, -z + \tfrac{1}{2};\ -x, -y, -z;\ -x + \tfrac{1}{2}, y + \tfrac{1}{2}, z; \\
& \quad x, y, -z + \tfrac{1}{2};\ x + \tfrac{1}{2}, -y + \tfrac{1}{2},\ z + \tfrac{1}{2}; \\
& \quad x = 0.6960,\ y = 0.3021,\ z = 0.0521.
\end{aligned}$$

Another example is provided by Perovskite itself, which is typical of the (2, 4) phases:

$CaTiO_3$: orthorhombic; $a = 0.5379$ nm, $b = 0.5436$ nm, $c = 0.7639$ nm, $Z = 4$; space group, *Pbmn* (No. 62, nonstandard setting);

$$\begin{aligned}
\text{Atom positions: } & \text{Ca}: 4(a)\, 0.9922, 0.0357, \tfrac{1}{4} \\
& \text{Ti}: 4(b)\, \tfrac{1}{2}, 0, 0 \\
& \text{O1}: 4(c)\, 0.0736, 0.4828, \tfrac{1}{4} \\
& \text{O2}: 8(d)\, x, y, z;\ x + \tfrac{1}{2}, -y + \tfrac{1}{2}, -z;\ -x, -y, z + \tfrac{1}{2};\ -x + \tfrac{1}{2}, y \\
& \quad + \tfrac{1}{2}, -z + \tfrac{1}{2};\ -x, -y, -z;\ -x + \tfrac{1}{2}, y + \tfrac{1}{2}, z;\ x, y, -z \\
& \quad + \tfrac{1}{2};\ x + \tfrac{1}{2}, -y + \tfrac{1}{2}, z + \tfrac{1}{2}; \\
& \quad x = 0.7113,\ y = 0.2893,\ z = 0.0375.
\end{aligned}$$

The nonstandard setting of the space group is often (but not universally) used for these structures (Table 1.5). In this setting, the orthorhombic unit cell is given the space group *Pbnm*, with $a < b < c$; $Z = 4$; A atoms are in the positions 4*c*; B atoms in 4*b*; O1 atoms in 4*c*; O2 atoms in 8*d*. In the standard setting, the orthorhombic unit cell is given the space group *Pnma*, with $a < b > c$; $Z = 4$; A atoms in positions 4*c*; B atoms in 4*b*; O1 atoms in 4*c*; O2 atoms in 8*d*.

At room temperature the structure of orthorhombic $CaTiO_3$ is due to the rotation of the TiO_6 octahedra about a diad axis, corresponding to the tilt group $a^+b^-b^-$ (Figure 1.14). As the temperature increases these distortions gradually diminish and eventually the orthorhombic phase transforms to tetragonal at 1498 K: space group *I4/mcm* (140), $a = b = 0.54984$ nm, $c = 0.77828$ nm. The octahedra are now much more regular, but still tilted, with a tilt system $a^0a^0c^-$. As the temperature continues to increase, these

Table 1.5 *Phases with the orthorhombic $GdFeO_3$ structure*

Phase	Space group	*a* (nm)	*b* (nm)	*c* (nm)
$NaFeF_3$	*Pnma* (62)	0.566119	0.788006	0.548359
$YAlO_3$	*Pbnm* (62)	0.518	0.531	0.735
$CaTiO_3$	*Pbnm* (62)	0.5379	0.5436	0.7639
$NdTiO_3$	*Pbnm* (62)	0.552532	0.565945	0.779066
$SmTiO_3$	*Pbnm* (62)	0.54647	0.56712	0.77291
$GdTiO_3$	*Pbnm* (62)	0.54031	0.57009	0.77133
$YTiO_3$	*Pbnm* (62)	0.53210	0.56727	0.75949
$TbCoO_3$	*Pbnm* (62)	0.51995	0.53945	0.74102
$DyCoO_3$	*Pbnm* (62)	0.51655	0.54143	0.77866
$CaRuO_3$	*Pnma* (62)	0.5544	0.7649	0.5354
$SrRuO_3$	*Pbnm* (62)	0.55754	0.55405	0.78546
$CaMnO_3$	*Pnma* (62)	0.52781	0.74542	0.52758
$Nd_{0.99}Ca_{0.1}CoO_3$	*Pnma* (62)	0.533379	0.755034	0.534640
$EuNiO_3$	*Pbnm* (62)	0.52945	0.54681	0.75404
$GdFeO_3$	*Pbnm* (62)	0.5349	0.5609	0.7669
$LaTiO_3$	*Pbnm* (62)	0.563676	0.561871	0.791615
$LaCrO_3$	*Pbnm* (62)	0.55163	0.54790	0.77616
$LaFeO_3$	*Pbnm* (62)	0.55563	0.55630	0.78535
$SrSnO_3$	*Pbnm* (62)	0.57042	0.57113	0.80674
$SrZrO_3$	*Pbnm* (62)	0.57093	0.57053	0.80676
$La(Fe_{0.5}V_{0.5})O_3$	*Pbnm* (62)	0.5552	0.5558	0.7836
$LaMnO_3$[a]	*Pbnm* (62)	0.55397	0.54891	0.77928
$CeMnO_3$	*Pbnm* (62)	0.5537	0.5557	0.7821
$PrMnO_3$	*Pnma* (62)	0.5608	0.7634	0.5442
$NdMnO_3$	*Pnma* (62)	0.57233	0.7587	0.54209
$SmMnO_3$	*Pbnm* (62)	0.53584	0.57959	0.74608
$EuMnO_3$	*Pbnm* (62)	0.53437	0.58361	0.746186
$GdMnO_3$	*Pbnm* (62)	0.53160	0.58683	0.74252
$DyMnO_3$	*Pbnm* (62)	0.531606	0.582304	0.738443
$HoMnO_3$	*Pbnm* (62)	0.52565	0.58329	0.73602
$ErMnO_3$	*Pbnm* (62)	0.52402	0.58228	0.73436
$TmMnO_3$	*Pbnm* (62)	0.52310	0.58157	0.73218
$YbMnO_3$	*Pbnm* (62)	0.52190	0.58038	0.73027
$LuMnO_3$	*Pbnm* (62)	0.518956	0.578498	0.72815
$La_{0.5}Ca_{0.5}MnO_3$	*Pnma* (62)	0.54182	0.76389	0.54269
$LaFe_{0.5}V_{0.5}O_3$	*Pbnm* (62)	0.55597	0.55571	0.78523
$NdFe_{0.5}V_{0.5}O_3$	*Pbnm* (62)	0.54497	0.55682	0.77516
$EuFe_{0.5}V_{0.5}O_3$	*Pbnm* (62)	0.53699	0.55975	0.76745
$YFe_{0.5}V_{0.5}O_3$	*Pbnm* (62)	0.52839	0.55969	0.75966

[a] Also reported as adopting the $LaAlO_3$ structure (Section 1.7.2).

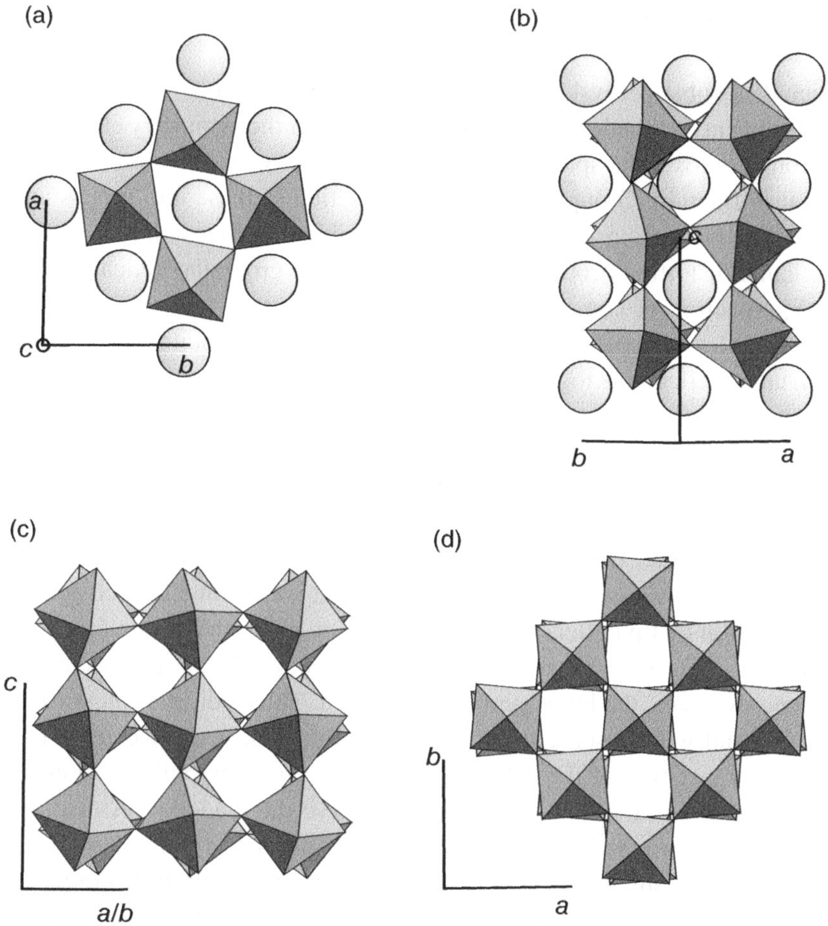

Figure 1.14 *The $GdFeO_3$-type structure of $CaTiO_3$: (a) projection down [001]; (b) projection down [110]; (c) projection down [110] showing octahedra only; (d) projection down [001] showing octahedra only*

remaining rotations diminish further, and the unit cell becomes cubic, space group $Pm\bar{3}m$ (221), $a=0.38967$ nm, with a tilt system $a^0a^0a^0$, via a transition at 1634 K.

For many purposes it is adequate to describe $GdFeO_3$-type structures in terms of a pseudocubic cell, which in the $Pbnm$ setting is $a \approx \sqrt{2}a_p$, $b \approx \sqrt{2}a_p$, $c \approx 2a_p$ with $a_p \approx 0.39$ nm. The B cations are nearly central in the octahedra, which are only slightly distorted (Figure 1.14a). The octahedral tilting, $a^+b^-b^-$, is noticeable in projections along [110] and [001] directions of the unit cell (Figure 1.14b–d). The octahedral tilting causes the ideal cuboctahedral sites to be reduced nearer to cubic antiprism geometry, with the A cation located in a space with 8 nearest neighbour oxygen ions and 4 anions at a greater distance rather than 12 equidistant anions.

1.8 Symmetry Relationships

When an ideal cubic perovskite, space group $Pm\bar{3}m$, drops in symmetry due to cation displacement, octahedral tilting and so on, some of the symmetry elements of the space group are lost. The space groups listed previously tend to obscure this aspect of the transformation. Group theoretical organisation of these into groups and subgroups makes the relationships clear and the possible subgroups that can apply when any particular distortion occurs can be enumerated by group theoretical methods.

This approach is of considerable value when new structures have to be determined. The diffraction patterns from slightly distorted perovskites show splitting of the main cubic reflections together with the appearance of new superstructure lines. However, when these distortions are very small, diffraction pattern changes are minute and the new superstructure reflections are weak. A listing of the possible subgroups simplifies the crystallographic problem by limiting the number of space groups that are compatible with any assumed distortions.

This procedure can be illustrated with respect to B-site cation displacements, described previously, assuming that the cation shift is the same in each octahedron. Using a notation that parallels the rotation notation given in Section 1.7.1 allows a displacement along x, y and z to be written as (a, b, c). If the displacement is the same along two axes, the letter symbol is repeated. Thus a displacement (0, 0, a) is along [001] and will preserve the fourfold rotation axis. A displacement (a, a, 0), obtained by translation of the cation by an amount a along both [100] and [010], will give a resultant along [110] and will preserve the diad axis. A displacement (a, a, a), obtained by translation of the cation by an amount a along [100], [010] and [001], will have a resultant along [111] and will preserve the triad axis (Section 1.5).

For example, displacement along [001] in original space group $Pm\bar{3}m$ will retain the fourfold rotation axis and so lead to a tetragonal space group, of which there are 68. Group theoretical analysis of the effects of the distortion limits this to just one, *P4mm* (99). Similar analysis shows that all permissible cation displacements lead to a hierarchy of just six subgroups (Figure 1.15).

A similar hierarchy can be constructed for the octahedral tilts described earlier. Group theoretical analysis of the symmetry of these tilt systems has led to the conclusion that 8 of the total of 23 tilts listed previously cannot be observed in real crystals and so are redundant in experimental terms. The remaining 15 are $a^0a^0a^0$ (23), $a^0a^0c^+$ (21), $a^0b^+b^+$ (16), $a^+a^+a^+$ (3), $a^+b^+c^+$ (1), $a^0a^0c^-$ (22), $a^0b^-b^-$ (20), $a^-a^-a^-$ (14), $a^0b^-c^-$ (19), $a^-b^-b^-$ (13), $a^-b^-c^-$ (12), $a^0b^+c^-$ (17), $a^+b^-b^-$ (10), $a^+b^-c^-$ (8), and $a^+a^+c^-$ (5). The hierarchy of these subgroups shows a decent in symmetry from the cubic parent $Pm\bar{3}m$ space group via tetragonal and orthorhombic to triclinic (Figure 1.16).

The 15 allowed transformations are imagined to operate alone. In reality many perovskite phases show additional octahedral cation displacements or distortions which will change things. For example, the tilt system 14, $a^-a^-a^-$ has an ideal trigonal space

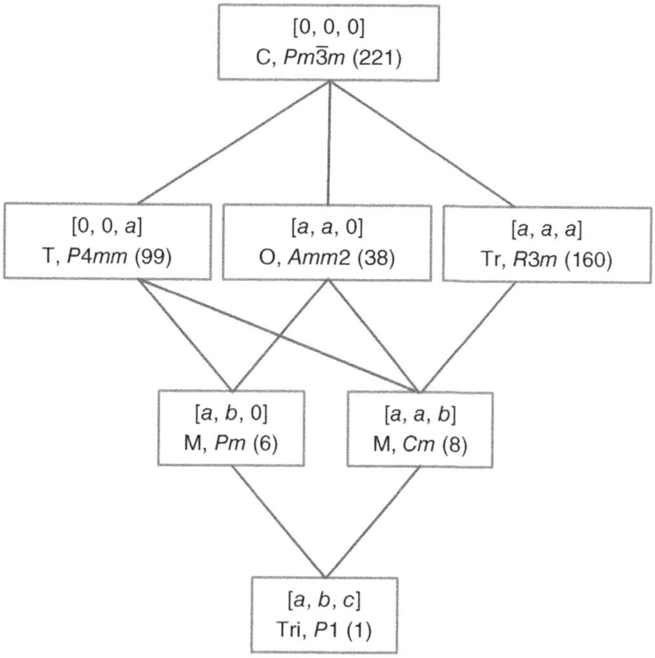

Figure 1.15 *Subgroups derived by displacement of the B-site cation from the parent space group* Pm3̄m. *The cation displacement along* x, y *or* z *is denoted by* a, b *or* c

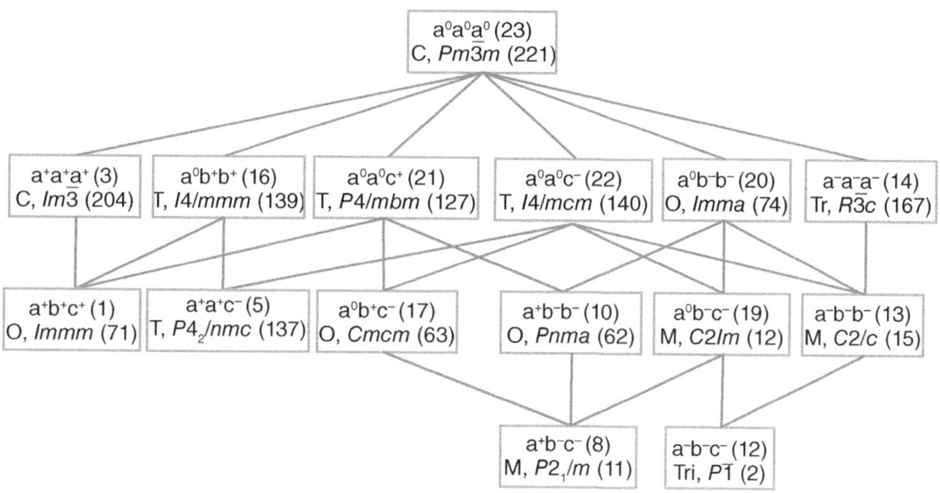

Figure 1.16 *The space groups allowed by the possible tilt systems of octahedra in pseudocubic perovskites*

group $R\bar{3}c$ (167) with a centre of symmetry. If the B cations that centre the octahedra are displaced, then the centre of symmetry is lost and the space group remains trigonal but is changed to $R3c$ (161). Jahn–Teller distortions have a similar effect. $KCuF_3$ has no octahedral tilting and belongs to the tilt system 23, $a^0a^0a^0$ but does not possess the cubic space group $Pm\bar{3}m$ (221) of $SrTiO_3$ because of the elongation of the CuO_6 octahedra, as described earlier, to give a tetragonal cell, space group $I4/mcm$ (140).

Table 1.6 *Organic–inorganic hybrid and related perovskites*

Phase	Space group	*a* (nm)	*b* (nm)	*c* (nm)	Angles (°)	Temp (K)
$CsSnI_3$	O, *Pnma* (62)	0.86885	1.23775	0.86384		300
$CsSnI_3$	T, *P4/mbm* (127)	0.87182		0.61908		350
$CsSnI_3$	C, $Pm\bar{3}m$ (221)	0.62057				478
$MAGeCl_3$	M, $P2_1/n$ (14)	1.09973	0.72043	0.82911	α, 90.47	2
$MAGeCl_3$	O, *Pmna* (62)	1.11567	0.73601	0.82936		250
$MAGeCl_3$	Tr, *R3m* (160)	0.56784			α, 90.95	370
$MAGeCl_3$	C, $Pm\bar{3}m$ (221)	0.56917				475
$CD_3ND_3GeCl_3$	M, $P2_1/n$ (14)	1.09973	0.72043	0.82911	β, 90.47	2
$CD_3ND_3GeCl_3$	O, *Pnma* (62)	1.11567	0.73601	0.82936		250
$CD_3ND_3GeCl_3$	Tr, *R3m* (160)	0.56584			α, 90.95	370
$CD_3ND_3GeCl_3$	C, $Pm\bar{3}m$ (221)	0.56917				475
$MASnCl_3$	Tri, *P1* (1)	0.5726	0.8227	0.7910	α, 90.40 β, 93.08 γ, 90.15	297
$MASnCl_3$	M, Pc (7)	0.5718	0.8326	0.7938	β, 93.03	318
$MASnCl_3$	Tr, *R3m* (160)	0.5734			α, 91.90	350
$MASnCl_3$	C, $Pm\bar{3}m$ (221)	0.5760				478
$MASnBr_3$	O, $Pmc2_1$ (26)	0.58941	0.83862	0.82406		215
$MASnI_3$	T, *P4mm* (99)	0.62302		0.6231		293
$MASnI_3$	T, *I4cm* (108)	0.87577		1.2429		200
$MAPbI_3$	T, *P4mm* (99)	0.63115		0.63161		400
$MAPbI_3$	T, *I4cm* (108)	0.8849		1.2642		293
$MAPbI_3$	O, *Pmna* (62)	0.88362	1.25804	0.85551		100
$FASnI_3$	O, *Amm2* (38)	0.63286	0.89554	0.89463		340
$FASnI_3$	O, *Imm2* (44)	1.25121	1.25171	1.25099		180
$CH_3ND_3PbCl_3$	O, *Pnma* (62)	1.11747	1.13552	1.12820		80
$CH_3ND_3PbCl_3$	C, $Pm\bar{3}m$ (221)	0.5669				280
$CH_3ND_3PbBr_3$	O, *Pnma* (62)	0.79434	1.18499	0.85918		11
$MAPbI_3$	O, *Pnma* (62)	0.88362	1.25804	0.85551		100
$MAPbI_3$	T, *I4/mcm* (140)	0.8851		1.2444		298
$MAPbI_3$	C, $Pm\bar{3}m$ (221)	0.6274				333
$FAPbI_3$	Tr, *P3m1* (156)	0.89817		1.1006	γ, 120	293

1.9 Hybrid Organic–Inorganic Perovskites

The A-sites in the perovskite structure are large enough to accommodate a number of complex ions including ammonium $(NH_4)^+$; methyl ammonium $(CH_3NH_3)^+$ frequently written as MA; tetramethylammonium, $[(CH_3)_4N]^+$, frequently written as TMA and formamidinium $(NH_2{=}CHNH_2)^+$, written as FA. The most important of these are the compounds formed with the Group 14 elements Ge, Sn and Pb, together with a halogen Cl, Br and I (Table 1.6). The valence state of the organic ion is +1, and this necessitates the B cation to adopt the +2 state. It is unusual to find these large B^{2+} ions in the octahedral B-sites, and they are more frequently associated with A-sites, as in lead titanate, $PbTiO_3$. A large electropositive cation is needed in the A-site to achieve this, which can include the inorganic cation Cs^+ as well as the ions above, and it is convenient to group the Cs halide perovskites with these inorganic–organic phases. The structure of these materials is typically slightly distorted cubic (Figure 1.17).

The organic groups can order or disorder and may have preferred orientations that could lead to orientational order–disorder transformations, with the concurrent formation of superlattices and smaller microdomains of order coherently

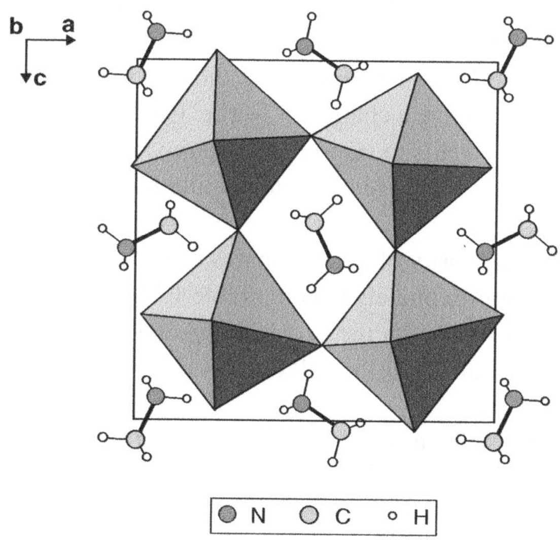

Figure 1.17 The structure of $(CH_3NH_3)PbCl_3$ $(MAPbCl_3)$. One layer of the structure is shown. In preceding and succeeding layers, the organic molecules are rotated by 180° compared to that drawn

intergrowth within the crystal matrix (also see Section 2.6). This A-site ordering is sensitive to temperature. In addition, octahedral tilting and distortion, as described earlier, may also be present. As is usual, many of these structures show variation with temperature. The high-temperature polymorph is cubic, and as the temperature falls the symmetry drops as ordering and octahedral tilting becomes pronounced.

1.10 Antiperovskites

1.10.1 Cubic and Related Structures

The antiperovskite or inverse perovskite structure is adopted by many compounds with an overall composition A_3BX, where A and B are metals and X is typically C, N, O and B. There are two broad categories. The first, not described here, are essentially alloys containing interstitial nonmetal atoms in octahedral interstices; typical examples being Mn_4N, derived from the face-centred cubic structure of Mn, and Fe_3PtN, derived from the alloy Fe_3Pt, with the Cu_3Au structure.

The materials described here have structures quite different from a notional parent metallic phase and are more akin to the oxide and oxynitride perovskites already described. They can be described by a formula AMX_3, where A is from a wide range of metals including Al, Ga, In, Zn, Ge, Sn, Cu and others; M is either N or C and X is a metal, typically Mn, Cr, Fe, Ni, Ca, Ln and so on. The ionic model of use for the oxide perovskites is not, though, particularly appropriate for these phases; taking X to have a charge of 2+ (i.e. Ca^{2+}) or 3+ (i.e. La^{3+}, Mn^{3+}) and N to have a charge of 3− leads to improbable charge states for many A atoms (i.e. $CuNMn_3$), besides which, the large N^{3-} anion would not be expected to occupy the octahedral sites.

The structure of the majority of known phases is cubic, space group $Pm\bar{3}m$, with a lattice parameter of approximately 0.4 nm (Table 1.7). The structure is completely analogous to that of $SrTiO_3$, with the C or N nonmetal atom occupying an octahedron of A atoms, with X atoms at the cell corners (Figure 1.18).

$CuNMn_3$: cubic; $a = 0.3907$ nm, Z = 1; Space group, $Pm\bar{3}m$ (No. 221);

$$\begin{aligned} \text{Atom positions: } & \text{Cu}: 1(a)\, 0,0,0; \\ & \text{N}: 1(b)\, ½,½,½; \\ & \text{Mn}: 3(c)\, ½,½,0;\ ½,0,½;\ 0,½,½; \end{aligned}$$

The Mn atoms lie at the corners of the unit cell. The N atoms lie at the cell centre and are surrounded by a regular octahedron of Cu atoms.

Table 1.7 Antiperovskite phases

Phase	Space group	*a* (nm)	*b* (nm)	*c* (nm)
$CuNMn_3$	C, $Pm\bar{3}m$ (221)	0.3907		
$AlNTi_3$	C, $Pm\bar{3}m$ (221)	0.4110		
$GaNCr_3$	C, $Pm\bar{3}m$ (221)	0.3879		
$GaNMn_3$	C, $Pm\bar{3}m$ (221)	0.3903		
$BiNCa_3$	C, $Pm\bar{3}m$ (221)	0.48884		
$SbNCa_3$	C, $Pm\bar{3}m$ (221)	0.48541		
$PbNCa_3$	C, $Pm\bar{3}m$ (221)	0.49550		
$SnNCa_3$	C, $Pm\bar{3}m$ (221)	0.49460		
$GeNCa_3$	C, $Pm\bar{3}m$ (221)	0.47573		
$SnNLa_3$	C, $Pm\bar{3}m$ (221)	0.50948		
$SnNCe_3$	C, $Pm\bar{3}m$ (221)	0.50159		
$SnNPr_3$	C, $Pm\bar{3}m$ (221)	0.49753		
$SnNNd_3$	C, $Pm\bar{3}m$ (221)	0.49470		
$SnNSm_3$	C, $Pm\bar{3}m$ (221)	0.48835		
$InNCe_3$	C, $Pm\bar{3}m$ (221)	0.50416		
$PNCa_3$	O, *Pnma* (62)	0.67091	0.94518	0.66581
$AsNCa_3$	O, *Pbnm* (62)	0.67249	0.67196	0.95336
$AsNCr_3$	T, *I4/mcm* (140)	0.536		0.8066
$BiNSr_3$	C, $Pm\bar{3}m$ (221)	0.520691		
$MgCNi_3$	C, $Pm\bar{3}m$ (221)	0.38106		
$SbNSr_3$	C, $Pm\bar{3}m$ (221)	0.51725		
$BiNBa_3$	T, $P6_3/mmc$ (194)	0.76111		0.667919
$SbNBa_3$	T, $P6_3/mmc$ (194)	0.75336		0.66431
$NaNBa_3$	T, $P6_3/mmc$ (194)	0.84414		0.69817

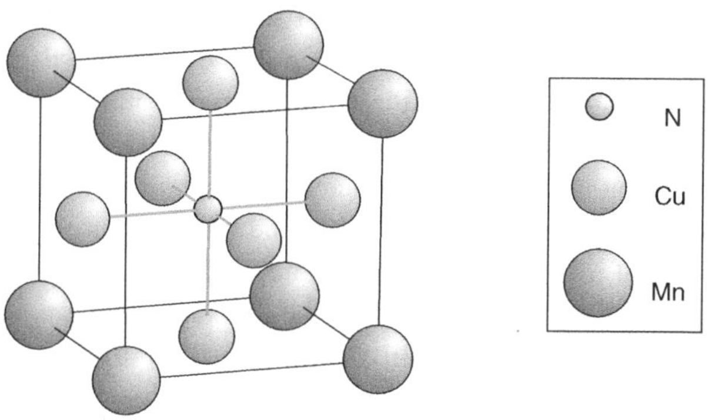

Figure 1.18 *The cubic antiperovskite structure of* $CuNMn_3$

The nitride antiperovskites show other structures, analogous to those displayed by oxide perovskites, but the number of known examples of these phases is minimal to date. The phases $PNCa_3$ and $AsNCa_3$ have tilted octahedra and belong to the $GdFeO_3$ structure type, tilt system $a^+b^-b^-$. $AsNCr_3$ is tetragonal with tilt system $a^0a^0c^-$.

Table 1.8 *Phases in the $BiNSr_{3-x}Ba_x$ and $SbNSr_{3-x}Ba_x$ systems*

Phase	Temperature (°C)	Cubic	4H	9R	2H
$BiNSr_{3-x}Ba_x$	875	$0<x<0.9$	$1.55<x<2.10$	$2.50<x<2.55$	$2.75<x<3.0$
$SbNSr_{3-x}Ba_x$	710	$0<x<1.3$	$1.83<x<2.45$	$2.56<x<2.60$	$2.80<x<3.0$

1.10.2 Other Structures

Two phases, $BiNBa_3$ and $SbNBa_3$, have been found to adopt the 2H—$BaNiO_3$ structure (Section 3.1). This structure, the hexagonal equivalent of the ideal perovskite structure, can form a bewildering array of phases consisting of intergrowths between hexagonal and cubic stacking in the oxide perovskites (Chapter 3, especially Section 3.5.1). In these antiperovskite materials, only two of these structure types are known, both formed in the cubic–hexagonal systems $BiNSr_3$ (cubic)–$BiNBa_3$ (hexagonal) and $SbNSr_3$ (cubic)–$SbNBa_3$ (hexagonal). These are the 4H structure and the 9R structure (for the structures see Section 3.5.1). The phase ranges of these materials at 875°C are given in Table 1.8.

1.11 Structure-Field Maps

Graphical displays of the way in which structures evolve as various physical parameters change have long been used to systematise structural relationships. It is convenient to use this idea to separate perovskite phases from other ABX_3 structure types and, if possible, to separate the different ABX_3 perovskite variants from each other. For this purpose the simplest parameters are the A- and B-cation radii. The relationship between these radii and the structure type found can be conveniently displayed on structure-field maps (see Muller and Roy 1974). These are equivalent to a graphical representation of the tolerance factor and usually refer to the structure stable under normal conditions (Figure 1.19a and b). There are a number of points to remember. First of all, the radii appropriate to the y-axis, associated with the A cations, should refer to 12-fold coordination. As such radii are not widely available, the figures here use eightfold radii on this axis. Similarly, the radii appropriate to the x-axis, associated with the B cations, should refer to octahedral coordination. Thus, for the $A^{3+}B^{3+}O_3$ diagram (Figure 1.19a), a lanthanoid ion on the y-axis will have a different radius than one on the x-axis. Surprisingly there is no extensive cubic structure phase field. The major phase field is that of compounds isostructural (or nearly so) with the $GdFeO_3$ structure type, but for the smallest B cations coupled with the largest A cations, the non-cubic $LaAlO_3$ structure predominates. Note that

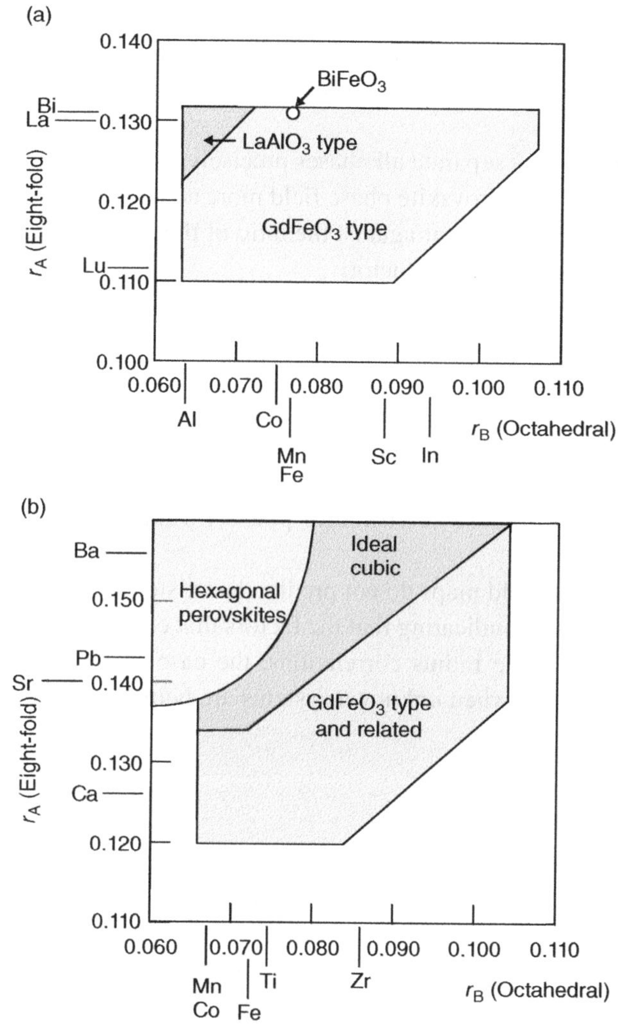

Figure 1.19 *Structure-field maps, schematic; r_A versus r_B: (a) $A^{3+}B^{3+}O_3$; (b) $A^{2+}B^{4+}O_3$ (Original data in Muller and Roy (1974))*

$BiFeO_3$ is somewhat anomalous. (Most Bi^{3+} compounds adopt the pyrochlore structure under normal conditions of temperature and pressure, not a perovskite-related structure. However, high pressures tend to convert these pyrochlores to distorted perovskite phases.)

The equivalent diagram for $A^{2+}B^{4+}O_3$ phases (Figure 1.19b) shows an extensive cubic phase field. There is a certain amount of overlap between the cubic phase field and the $GdFeO_3$-type phase field that is not shown in the figure drawn. A notable difference is the appearance of a hexagonal perovskite phase field for small B cations coupled with large A cations (Chapter 3). Finally note that Pb^{2+} phases are similar to those of Bi^{3+} in that pyrochlore structure compounds form rather than perovskite types. The phases $PbZrO_3$ and $PbHfO_3$ are important exceptions to this generalisation.

These diagrams do not separate all phases precisely, and a number of efforts have been made to clarify the perovskite phase field more accurately. Recent efforts have used plots of the tolerance factor against the ratio of the radii of the B-site cation to the X-site anion as an octahedral factor:

$$\mu = \frac{r_B}{r_X}$$

for A_mO_n—B_mO_n oxide systems (Figure 1.20a) and AX—BX_2 halide systems (Figure 1.20b). These diagrams successfully demarcate most known AB combinations that lead to perovskite structure phase formation from those systems that do not.

Although structure-field maps do not predict the existence of the perovskite form with complete certainty, indicating that the factors that endow stability to this structure lie outside of simple radius correlations, the ease of the method makes it a simple and useful guide when unknown systems are being explored.

1.12 Theoretical Calculations

Information on formation energies or the electronic structure of single-phase perovskites can be assessed by quantum mechanical calculations. In the main, these take two forms, atomistic simulations and density functional theory calculations. Atomistic simulations are useful for the investigation of defect structures. Density functional methods are well-known tools for the estimation of stability and electronic structures. For example, experiment shows that $CaMnO_3$ has an orthorhombic $GdFeO_3$-type structure, in which all of the MnO_6 octahedra are apex linked; $BaMnO_3$ has the 2H-$BaNiO_3$ structure in which all of the MnO_6 octahedra are face sharing (Section 3.1). $SrMnO_3$ lies between these two extremes, with the 4H-structure, in which pairs of face-sharing octahedra are linked by apex sharing (Section 3.5). Density functional calculations reproduce these structures.

$CaMnO_3$: calculated, orthorhombic: $a = 0.53298$ nm, $b = 0.74837$ nm, $c = 0.52828$ nm
experimental, orthorhombic: $a = 0.5281$ nm, $b = 0.74542$ nm, $c = 0.52758$ nm

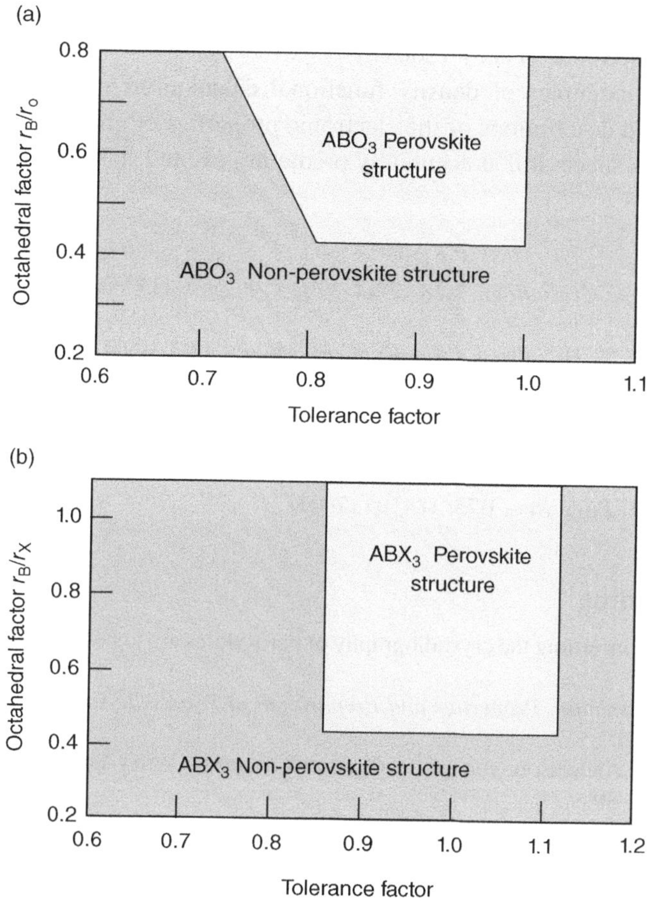

Figure 1.20 *Structure-field maps, schematic; tolerance factor* t *versus octahedral factor* μ*: (a) ABO_3 oxides; (b) ABX_3 halides (Original data in Li* et al. *(2008))*

The calculated stability order for $CaMnO_3$ is orthorhombic $GeFeO_3$, 4H-, cubic, 2H, in agreement with the experimental findings.

$SrMnO_3$:	calculated, 4H:	$a = 0.54893$ nm,	$c = 0.91143$ nm
	experimental, 4H:	$a = 0.54432$ nm,	$c = 0.90704$ nm

The calculated stability order is 4H, 2H, cubic so that the 4H structure is most stable, as found.

$BaMnO_3$:	calculated, 2H:	$a = 0.57781$ nm,	$c = 0.48217$ nm
	experimental, 2H:	$a = 0.56691$ nm,	$c = 0.48148$ nm

The calculated stability order is 2H, 4H, cubic in agreement with experimental data. (Data from Søndenå *et al.* (2007).)

Many other examples of density functional calculations mirror this example. These give good descriptions of the electronic properties of ground-state solids but still remain less successful at accurately predicting excited state properties.

References

A. M. Glazer, *Acta Crystallogr.*, **B28**, 3384–3392 (1972); Acta Crystallog., **A31**, 756–762 (1972).

S. A. Hayward, E. K. H. Salje, *J. Phys. Condens. Matter*, **14**, L599–L604 (2002).

T. Ishidate *et al.*, *Phys. Rev. Lett.*, **78**, 2397 (1997).

C. Li *et al.*, *Acta Crystallogr.*, **B64**, 702–707 (2008).

O. Muller, R. Roy, *The Major Ternary Structural Families*, Springer-Verlag, Berlin (1974); p 175–196.

R. Søndenå *et al.*, *Phys. Rev.*, **B75**, 184105 (2007).

Further Reading

Earlier studies concerning the crystallography of perovskites are to be found in the following four volumes:

F. S. Galasso, *Structure, Properties and Preparation of Perovskite-Compounds*, Pergamon, London (1969).

B. G. Hyde, S. Andersson, *Inorganic Crystal Structures*, Wiley-Interscience, New York (1989); p 295–302.

H. Megaw, *Crystal Structures: A Working Approach*, Saunders, Philadelphia (1973); p 217–221, p 285–304.

O. Muller, R. Roy, *The Major Ternary Structural Families*, Springer-Verlag, Berlin (1974); p175–196.

More recent compilations are:

F. S. Galasso, *Perovskites and High T_c Superconductors*, Gordon & Breach, New York (1990).

R. H. Mitchell, *Perovskites: Modern and Ancient*, Almaz Press, Thunder Bay (2002).

For information on the early development of ferroelectric perovskite phases see:

L. E. Cross, R. E. Newnham, History of Ferroelectrics, in W. D. Kingery (Ed.), *Ceramics and Civilization, Vol III, High-Technology Ceramics–Past, Present and Future*, American Ceramic Society, Westerville (1987); p 289–305.

For information on the tolerance factor and structure-field maps see:

L. M. Feng *et al.*, *J. Phys. Chem. Solids*, **69**, 967–974 (2008).

V. S. Goldschmidt, *Naturwissenschaften*, **21**, 447–485 (1926).

C. Li *et al.*, *Acta Crystallogr.*, **B64**, 702–707 (2008).

S. Sasaki, C. T. Prewitt, R. C. Liebermann, *Am. Mineral.*, **68**, 1189–1198 (1983).

Y. M. Zhang *et al.*, *Mater. Focus*, **1**, 1–8 (2012).

Ionic radii are discussed and tabulated by:
R. D. Shannon, *Acta Crystallogr.*, **A32**, 751–756 (1976).
R. D. Shannon, C. T. Prewitt, *Acta Crystallogr.*, **B25**, 925–946 (1969); Acta Crystallogr., **B26**, 1046 (1970).

The systematics of octahedral tilt in perovskites is given by H. Megaw above and:
A. M. Glazer, *Acta Crystallogr.*, **B28**, 3384–3392 (1972); Acta Crystallog., **A31**, 756–762 (1972).
C. J. Howard, H. T. Stokes, *Acta Crystallogr.*, **B54**, 782–789 (1998).
M. O'Keeffe, B. G. Hyde, *Acta Crystallogr.*, **B33**, 3802–3813 (1977).
P. M. Woodward, *Acta Crystallogr.*, **B53**, 32–66, 44–66 (1997).

For information on the Jahn–Teller distortion in $KCuF_3$, see:
J.-S. Zhou *et al.*, *J. Fluor. Chem.*, **132**, 1117–1121 (2011).

For a review of the $LnScO_3$ phases with the $GdFeO_3$ orthorhombic structure, see:
R. P. Liferovich, R. H. Mitchell, *J. Solid State Chem.*, **177**, 2188–2197 (2004)

For symmetry and group theory, see:
C. J. Howard, H. T. Stokes, *Acta Crystallogr.*, **B54**, 782–787 (1998).
C. J. Howard, H. T. Stokes, *Acta Crystallogr.*, **A61**, 93–111 (2005).
C. J. Howard, B. J. Kennedy, P. M. Woodward, *Acta Crystallogr.*, **B59**, 463–471 (2003).
M. C. Knapp, P. M. Woodward, *J. Solid State Chem.*, **179**, 1076–1085 (2006).
U. Müller, *Symmetry Relationships between Crystal Structures*, Oxford University Press, Oxford (2013).

For a review of antiperovskites, see:
R. Niewa, *Z. Anorg. Allg. Chem.*, **639**, 1699–1715 (2013).

For theoretical calculations see:
R. LeSar, *Introduction to Computational Materials Science*, Materials Research Society, Cambridge University Press, Cambridge (2013).
R. Søndenå *et al.*, *Phys. Rev.*, **B75**, 184105 (2007).

2

ABX_3–Related Structures

Here the narrow prescription of Chapter 1 is widened to deal with more chemically complex phases, in which the materials may contain mixtures of A, B and X ions as well as chemical defects. In these cases, using an ionic model, it is only necessary that the nominal charges balance to obtain a viable perovskite composition. In many instances these ions are distributed at random over the available sites, but for some simple ratios they can order to form phases with double or triple perovskite-type unit cells. The distribution and valence of these ordered or partly ordered cations and anions are often not totally apparent from diffraction studies, and they are often clarified by use of the bond valence sums derived from experimentally determined bond distances. Information on the bond valence method is given in Appendix A for readers unfamiliar with it. Point defects also become significant in these materials. The standard Kroger–Vink notation, used for labelling these defects, is outlined in Appendix B.

2.1 Double Perovskites and Related Ordered Structures

2.1.1 Rock-Salt Ordered Double Perovskites

When several different ionic species are present in which the charges or the ionic sizes differ enough, lattice energy can often be reduced by ordering some or all of the ions over the available crystallographic sites. The most widely encountered

---

Perovskites: Structure–Property Relationships, First Edition. Richard J. D. Tilley.

ordering pattern is that found in the double perovskites with a formula $A_2(BB')O_6$, in which there is a B-site cation substitution (Table 2.1). The charge balance must be the same as in ordinary oxide perovskites, so that

$$q_A + q_B + q_{B'} = -6$$

Frequently encountered (but not exclusive) combinations are

$$q_X = -2,\ q_A = 2,\ q_B + q_{B'} = 8; (2,6), (3,5), (4,4);$$

$$q_X = -2,\ q_A = 3,\ q_B + q_{B'} = 6; (1,5), (2,4), (3,3);$$

The commonest ordering pattern encountered corresponds to the two different cations, B and B′, arranged in a chessboard pattern, often called rock-salt ordering (Figure 2.1).

Table 2.1 *Phases with the rock-salt-type double perovskite structure*

Phase	Space group	*a* (nm)	*b* (nm)	*c* (nm)	β (°)
Ca_2MgOsO_6	M, $P2_1/n$ (14)	0.54097	0.55403	0.76991	90.05
Ca_2CoNbO_6	M, $P2_1/n$ (14)	0.57664	0.57521	0.81790	89.74
Ca_2FeOsO_6	M, $P2_1/n$ (14)	0.54004	0.54957	0.76820	90.03
Ca_2FeReO_6	M, $P2_1/n$ (14)	0.54002	0.55251	0.76826	90.07
Ca_2MnTiO_6	T, $P4_2mc$ (105)	0.75376		0.76002	
Ca_2YRuO_6	M, $P2_1/n$ (14)	0.56820	0.57507	0.80715	90.25
Sr_2MgOsO_6	T, $I4/m$ (87)	0.55606		0.79212	
Sr_2CrMoO_6	T, $I4/m$ (87)	0.55335		0.78251	
Sr_2CrReO_6	T, $I4/m$ (87)	0.55206		0.78023	
Sr_2MnMoO_6	C, $Fm\bar{3}m$ (225)	0.80056			
Sr_2FeMoO_6	T, $I4/m$ (87)	0.55705		0.79253	
Sr_2FeReO_6	T, $I4/m$ (87)	0.5561		0.79008	
Sr_2CoReO_6	T, $I4/m$ (87)	0.55659		0.79508	
Sr_2EuMoO_6	M, $P2_1/n$ (14)	0.58206	0.58639	0.82498	90.23
Sr_2GdMoO_6	M, $P2_1/n$ (14)	0.58189	0.58621	0.82489	90.21
Sr_2YRuO_6	M, $P2_1/n$ (14)	0.57686	0.57826	0.81688	90.21
Sr_2YNbO_6	M, $P2_1/n$ (14)	0.57902	0.58175	0.82009	90.11
Sr_2YTaO_6	M, $P2_1/n$ (14)	0.57915	0.58205	0.82030	90.11
Ba_2FeMoO_6	C, $Fm\bar{3}m$ (225)	0.81021			
Ba_2MnReO_6	C, $Fm\bar{3}m$ (225)	0.81865			
Ba_2FeReO_6	C, $Fm\bar{3}m$ (225)	0.80513			
Ba_2MnMoO_6	C, $Fm\bar{3}m$ (225)	0.81817			
Ba_2FeWO_6	T, $I4/m$ (87)	0.57446		0.80199	
$BaPrCoNbO_6$	T, $I4/mmm$ (139)	0.56829		0.8063	
Dy_2NiMoO_6	M, $P2_1/n$ (14)	0.52425	0.55469	0.75026	90.04

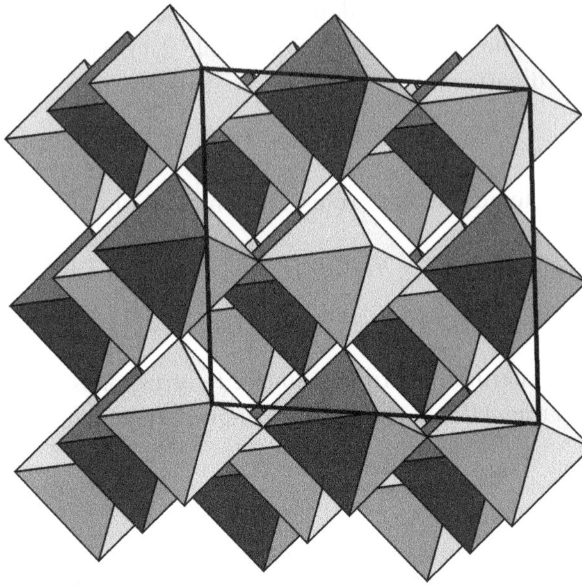

Figure 2.1 The octahedral framework of the rock-salt ordered double perovskite structure. The A-site cations are omitted for clarity

The symmetry of ordered rock-salt $A_2(BB')O_6$ double perovskites, taking into account octahedral tilting, has been systematically derived via group theoretical analysis. If the structure is cubic, with a tilt system $a^0a^0a^0$, the space group is $Fm\bar{3}m$ (225), the new axes run parallel to the simple cubic perovskite axes, and the lattice parameter is double the simple cubic lattice parameter a_p. Thus, in the case of Ba_2FeMoO_6, $a = 0.81865$ nm ($\approx 2a_p$). Relatively small A cations usually lead to octahedral tilting, as in simple perovskites, and this leads to a reduction in symmetry and a realignment of the unit cell axes. For example, the tilt system $a^0a^0c^-$ leads to tetragonal symmetry, space group *I*4/*m* (87). The new tetragonal unit cell **a**- and **b**-axes are parallel to the cubic double perovskite face diagonals [110], [$\bar{1}$10], and the new **c**-axis is parallel to the cubic double perovskite **c**-axis [001]. The new *a*- and *b*-lattice parameters are thus related to the perovskite cubic cell by the relationships: $a = b \approx \sqrt{2}a_p$, $c \approx 2a_p$, where a_p is the ideal cubic perovskite cell; for example, Sr_2FeMoO_6 has $a = 0.55705$ nm, $c = 0.79253$. Frequently the cell is slightly distorted to monoclinic, space group $P2_1/c$ (14), with similar lattice parameters, $a \approx b \approx \sqrt{2}a_p$, $c \approx 2a_p$, $\beta \approx 90°$. Similarly, the tilt system $a^0a^0c^+$ leads to tetragonal symmetry, space group *P*4/*mnc* (128). The new tetragonal unit cell **a**- and **b**-axes are parallel to the old double perovskite cubic face cell diagonals [110], [$\bar{1}$10], and the new **c**-axis is parallel to the cubic double perovskite **c**-axis [001]. The new *a*- and *b*-parameters are related to the double perovskite cubic cell by the same relationships as the $a^0a^0c^-$ tilt system: $a = b \approx a_d/\sqrt{2}$, $c \approx a_d$.

Bond valence sums are often calculated for these phases to clarify aspects of valence and structure – for example, bond valence sums for the double perovskite Ca_2MgOsO_6 are: Ca—O, 2.03; Mg—O, 2.15; and Os—O, 6.49. These suggest that assigning the ionic states Mg^{2+} and Ca^{2+} is reasonable, but the Os cation is significantly overbonded, suggesting that there may be a considerable departure from the formal Os^{6+} state in the OsO_6 units.

The commonest form of disorder among the B-site cations results in the B- and B′-cations swapping positions, an example of antisite disorder. Antisite disorder is temperature sensitive and is often influenced by the preparation methods employed and the particle size of the product, with nanoparticles often having different surface ordering than bulk samples. These defects have consequences for the electronic and magnetic properties of the phase.

Phase transitions as a function of temperature are commonplace and can be illustrated by the perovskite phase Ba_2BiIrO_6. At 500 K the structure is a classical cubic rock-salt-type double perovskite: space group $Fm\bar{3}m$ (225), a=0.85178 nm. This transforms to a rhombohedral phase at approximately 450 K. The (hexagonal) lattice parameters at 300 K are: space group $R\bar{3}c$ (167), a=0.60006 nm, c=1.47497 nm, γ=120°. Below 160 K, down to 140 K, the rhombohedral phase is in contact with a monoclinic phase with space group $C2/m$ (12). The monoclinic phase has not been prepared pure and no detailed structure is available. Below 140 K the two-phase region becomes single phase triclinic: space group $P\bar{1}$ (2), a=0.59932 nm, b=0.59918 nm, c=0.84438 nm, α=90.14°, β=90.49°, γ=89.81° at 80 K. Despite the low symmetry, this cell can be seen to be approximately tetragonal, closely related in size to the cubic high-temperature phase.

2.1.2 Other Ordered Perovskites

Although the A cations can be thought of as somewhat passive in these phases, they can affect the B-site ordering patterns. The series $La_{2-x}Sr_xCoRuO_6$ shows rock-salt ordering of the B-site cations Co and Ru over the composition range La_2CoRuO_6–$La_{0.4}Sr_{1.6}CoRuO_6$, while the B cations are disordered for the composition range $La_{0.2}Sr_{1.8}CoRuO_6$–Sr_2CoRuO_6. A similar feature occurs in the series $La_xCa_{2-x}FeReO_6$, and disorder on the B-sites increases with La incorporation. The phase β-Ca_3UO_6 has a rock-salt arrangement of Ca^{2+} and U^{6+} octahedra, with the remaining Ca^{2+} ions occupying the A-sites.

Apart from the rock-salt arrangement of B and B′ cations, the BO_6 and $B'O_6$ octahedra can be arranged in sheets, perpendicular to a cubic or pseudocubic axis, or at a diagonal to these axes, although such structures are relatively rare compared to the rock-salt type. The phase $Ba_4LiSb_3O_{12}$ has a cubic 2×2×2 unit cell and a 1 : 3 ordering on the B-sites. In a different ordering pattern, $Ba_3ZnTa_2O_9$ has ordered

(111) layers, in which one B layer is followed by one B′ layer. Examples of a quadruple unit cell with a stacking sequence along the **c**-axis of BB′B′B′ is given by the phases $Sr_2La_2CuTi_3O_{12}$ (a=0.39098 nm, c=1.5794 nm) and $Ca_2La_2CuTi_3O_{12}$ (a=0.38729 nm, c=1.5687 nm) (Figure 2.2). If either of these two latter phases is heated to temperatures of the order of 1100°C, the B and B′ cations disorder and the structure becomes equivalent to a simple cubic perovskite.

A different ordering pattern in the phases $K_4CaU_3O_{12}$ and $K_4SrU_3O_{12}$ leads to the cubic structures: space group $Im\bar{3}m$ (229), a=0.8483 nm (Ca^{2+}) and a=0.8582 nm (Sr^{2+}). In these latter phases the U^{6+} cation shares the octahedral B-sites with Ca^{2+} or Sr^{2+}.

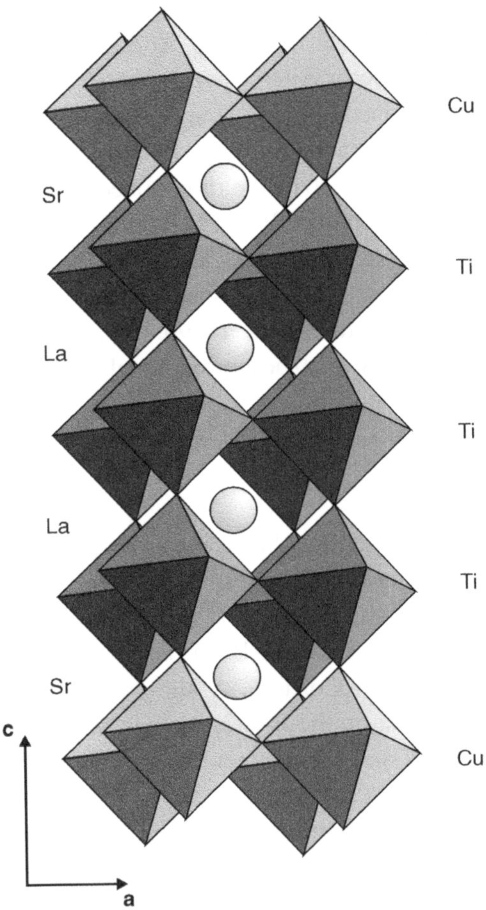

Figure 2.2 *The idealised structure of $Sr_2La_2CuTi_3O_{12}$. Both the A- and B-site cations are ordered*

As distinct from B, B′ ordering, ordering of A, A′ cations frequently takes the form of layers … A, A′, A, A′…. These structures tend to occur in non-stoichiometric oxygen-deficient phases and are especially associated with the perovskite-related layer structures, including the perovskite-related superconductors, that are described later (Chapter 4). The double perovskite $BaYMn_2O_{5+\delta}$ is similar to these layer structures. The material can take in varying amounts of oxygen, associated with a phase range from O_5 to O_6, corresponding to the nominal ionic forms $BaYMn^{2+}Mn^{3+}O_5$, $BaYMn^{3+}Mn^{3+}O_{5.5}$ and $BaYMn^{3+}Mn^{4+}O_6$. Both the Ba^{2+} and Y^{3+} cations are layered here, as are the oxygen vacancies.

Other instances of A, A′ ordering are less common. One example is provided by the double ordered $NaLaMgWO_6$ phase. Here the Mg and W cations are ordered in a rock-salt arrangement, and the Na and La cations are ordered in layers (Figure 2.3). The ordering pattern gives rise to a monoclinic unit cell: space group *C2/m* (12),

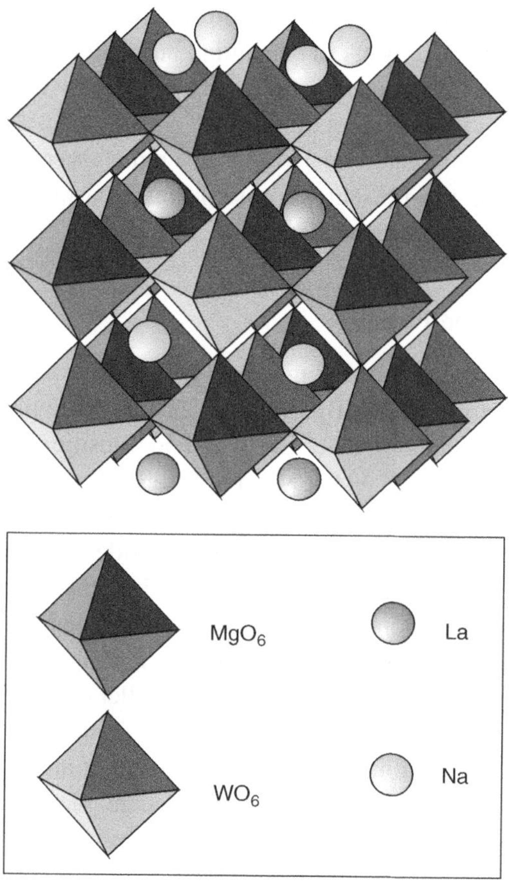

Figure 2.3 *The perovskite* $NaLaMgWO_6$ *with doubly ordered A-site and B-site cations*

$a = 0.78074$ nm, $b = 0.78158$ nm, $c = 0.78977$ nm, $\beta = 90.14°$, although the cell can be treated as approximately cubic for a discussion of many properties. The structure has two different anion sites with respect to the A and A′ cations, one surrounded by $4Na^+$ cations and one surrounded by $4La^{3+}$ cations. Bond valence sums suggest that the lattice strain imposed by this ordering is relieved in part by displacement of the W^{6+} cations to compensate for the overbonding and underbonding that this generates.

The superconducting double perovskite $(Na_{0.25}K_{0.45})Ba_3Bi_4O_{12}$ has a complex ordering of A-site cations. The structure is cubic: space group $Im\bar{3}m$ (229), $a = 0.8550$ nm. The A-sites are subdivided so that one set is occupied by $(Na_{0.25}K_{0.45}V)$, where V represents an A-site vacancy, while the other (A′-sites) are fully occupied by Ba. The Bi cations occupy the B-site octahedra.

In the unusual double perovskite $CaMnTi_2O_6$, the Mn^{2+} cations occupy the A-sites. The Mn^{2+} and Ca^{2+} are ordered in columns along the **c**-axis, with structural distortions allowing the Mn^{2+} to take tetrahedral and square planar positions in the cage sites. The phase $CaFeTi_2O_6$ appears to be similar, but with Fe only occupying tetrahedral positions. These phases are structurally related to the $AA'_3B_4O_{12}$-related phases described in the following section.

Note that other ordered structures can occur. The minerals cryolite Na_3AlF_6 (better written Na_2NaAlF_6 for the present purposes) and elpasolite K_2NaAlF_6 can be regarded as perovskite superstructures.

2.1.3 $AA'_3B_4O_{12}$-Related Phases

A surprising example of A-site ordering is provided by a group of cubic perovskite oxides with a general formula $AA'_3B_4O_{12}$ (Table 2.2). In these phases, A is an alkali metal, alkaline earth, lanthanoid, Pb or Bi; A′ is a 3d transition metal ion; and B can be a transition metal such as Ti, V, Cr, Mn, Fe, Ru, Rh, Ir or Pt, as well as the non-transition metals Ga, Ge, Sb or Sn. Ideally the ionic charge states are $A^{2+}A'^{2+}_3B^{4+}_4O^{2-}_{12}$ or $A^{3+}A'^{2+}_3B^{3+}B^{4+}_3O^{2-}_{12}$ and the cubic unit cell is of side approximately $2a_p$.

The assimilation of an A′ 3d transition metal ion into the large cage sites is unusual because these are commonly found as high-charge medium-sized cations that prefer octahedral coordination. In order to utilise a cage site, the A′ transition metal cations must show a strong expression of the Jahn–Teller effect (Section 1.6). So far the phases synthesised have been limited to the Jahn–Teller ions Mn^{3+} and Cu^{2+}. In these, the Jahn–Teller distortion usually takes the form of an elongated octahedron or, when the distortion is extreme, as here, a square planar configuration. The formation of the square planar groups is achieved by extreme B-site octahedral tilting. In particular the tilt system $a^+a^+a^+$, when combined with large tilt angles, can generate the requisite square planar geometry for the Cu^{2+} and Mn^{3+} cations.

The structure of these phases shows a framework of corner-linked BO_6 octahedra, together with normal A-site cations at the unit cell corner and centre, and square planar coordinated sites formed from the apical and equatorial oxygen anions of the BO_6 octahedra (Figure 2.4).

Table 2.2 *Compounds with the cubic $AA'_3B_4O_{12}$ structure*[a]

Compound	Unit cell (nm)
$NaMn_3V_4O_{12}$	0.73551
$CaCu_3Co_4O_{12}$	0.71226
$CaCu_3Ti_4O_{12}$	0.73935
$CaCu_3Rh_4O_{12}$	0.73937
$CaCu_3Fe_4O_{12}$	0.72955
$CaMn_3V_4O_{12}$	0.74070
$SrCu_3Fe_4O_{12}$	0.73492
$BiCu_3Fe_4O_{12}$	0.74332
$LaCu_3Fe_4O_{12}$	0.74287
$LaMn_3V_4O_{12}$	0.74849
$LaCu_3Mn_4O_{12}$	0.73272
$LaCu_3Ir_4O_{12}$	0.75242
$PrCu_3Fe_4O_{12}$	0.74036
$NdCu_3Fe_4O_{12}$	0.73478
$YCu_3Co_4O_{12}$	0.71195

[a] All compounds have a cubic unit cell: space group $Im\bar{3}$ (204).

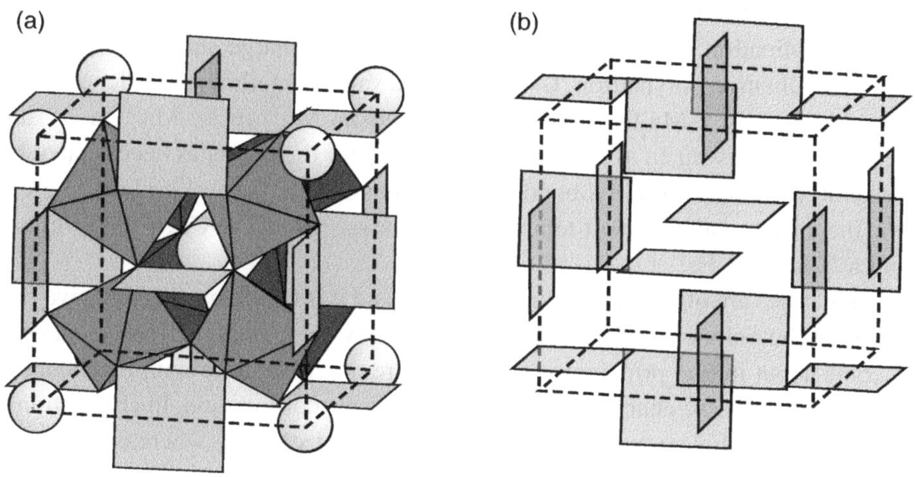

Figure 2.4 *(a) The $AA'_3B_4O_{12}$ structure; the squares represent square planar $A'O_4$ coordination polyhedra. (b) The distribution of the $A'O_4$ polyhedra within a unit cell*

The $AA'_3B_4O_{12}$ structures all show remarkable physical properties. The phase $SrCu_3Fe_4O_{12}$ has an extraordinary large thermal contraction (negative thermal expansion), $CaCu_3Ti_4O_{12}$ has an enormous relative permittivity, and $LaCu_3Mn_4O_{12}$ (note Mn^{4+} in B-sites) shows colossal magnetoresistance. In addition, a number of complex minerals are known with variants of this structure, for example, $LaCu_{2.5}(Mn_{3.91}Fe_{0.6})O_{11.4}$. However, the most surprising of this group of phases is the high-pressure phase ζ-Mn_2O_3, which is metastable at normal pressures. Here A is Mn^{2+} in cage sites, A′ is Mn^{3+} in square planar sites, and B is Mn^{3+} plus Mn^{4+} in octahedral sites, so that the phase can be written as $Mn^{2+}(Mn_3^{3+})(Mn_3^{3+}Mn^{4+})$. The almost cubic unit cell is triclinic: space group $F\bar{1}$ (2) (non-standard setting), $a = 1.46985$ nm, $b = 1.46482$ nm, $c = 1.46705$, $\alpha = 89.211°$, $\beta = 89.203°$, $\gamma = 89.196°$.

The charge states in these perovskites are often expressed in terms of noninteger values, such as $Mn^{2+}(Mn_3^{3+})(Mn_4^{3.25+})O_{12}$, $Ln^{3+}Cu_3^{2+}Fe_4^{3.75+}O_{12}$, $Ca^{2+}Cu_3^{2.8+}Rh_4^{3.4+}O_{12}$ and $CaCu_3^{3+}Co_4^{3.25+}O_{12}$. The noninteger charge states are associated with mixed cationic valence which often include unusually high valence states: Cu^{3+} (d^8), Fe^{4+} (d^4), Fe^{5+} (d^3), Co^{4+} (d^5) and Mn^{4+} (d^3). The noninteger values represent varying populations of these cations, for example, $(Fe^{3.75+})_4$ being equal to $(Fe_3^{4+}Fe^{3+})$ and $(Co^{3.25+})_4$ being equal to $(Co_3^{3+}Co^{4+})$. The higher charge states are sometimes described in terms of the surrounding oxygen coordination polyhedron. For example, Fe^{4+} is sometimes considered to consist of the octahedral unit $[FeO_6]^{11-}$, also written as $Fe^{3+}L$ or $Fe^{III}L$, where L is taken to mean a ligand oxygen hole. The Kroger–Vink notation for this defect would be (Fe_{Fe} $h^•$) or $Fe_{Fe}^•$, depending upon the chemistry involved (see Appendix B for a summary of the Kroger–Vink notation for point defects).

The utilisation of bond valence sums is widely used in these rather complex structures for either the allocation of an ion with a known nominal valence to the correct coordination polyhedron or the allocation of a charge state to an ion in a known coordination polyhedron. Using the bond valence method, the valence states of the Mn ions in ζ-Mn_2O_3 were determined to be Mn^{2+} in A, Mn^{3+} in A′ and $Mn_3^{3+}Mn^{4+}$ in B, giving an average valence of Mn^{3+} for the phase, as necessitated by the overall formula. Similarly, bond valence sums for the perovskite $LaCu_3Fe_4O_{12}$ allow the charge state at room temperature to be assigned $La^{3+}Cu_3^{3+}Fe_4^{3+}O_{12}$ rather than $La^{3+}Cu_3^{2+}(Fe^{3+}Fe_3^{4+})O_{12}$.

Charge states are often temperature sensitive, and it is these changes that give rise to some of the unique physical properties shown by these phases. Charge changes can come about in two principal ways, which are described here with the example of $ACu_3Fe_4O_{12}$ phases: charge transfer or charge disproportionation. In charge transfer, electrons (or holes) are moved between different cations, whereas in charge disproportionation a single cation type is involved. For example, charge transfer between A′-site Cu ions and B-site Fe cations occurs in $LaCu_3Fe_4O_{12}[La^{3+}Cu_3^{2+}(Fe_3^{4+}Fe^{3+})O_{12}]$ in the following way. Each Cu^{2+} cation loses an electron to form

Cu^{3+}, and these electrons are transferred to the $3Fe^{4+}$ cations to form Fe^{3+} cations. Schematically,

$$3Cu^{2+} \rightarrow 3Cu^{3+} + 3e'$$

$$3Fe^{4+} + 3e' \rightarrow 4Fe^{3+}$$

Charge transfer can also be viewed in terms of the ligand hole concept. The coordination polyhedra involved in charge transfer between Cu^{3+} and Fe^{4+} are linked by an oxygen shared between an FeO_6 octahedron and a Cu square planar group. The simultaneous valence change is mediated by double charge transfer between Cu and O to fill the ligand hole and immediately recreate it by transfer from O to Fe. Charge disproportionation between B-site cations in $CaCu_3Fe_4O_{12}$ ($Ca^{2+}Cu_3^{2+}Fe_4^{4+}O_{12}$) simply involves a redistribution of charge between the Fe cations:

$$4Fe^{4+} \rightarrow 2Fe^{3+} + 2Fe^{5+}$$

Charge transfer and charge disproportionation need not be regarded as discontinuous step-like alterations but can take place gradually as the temperature changes. For instance, at 300 K, the charge situation in $SrCu_3Fe_4O_{12}$ is believed to be Sr^{2+}-$Cu_3^{\sim 2.4+}Fe_4^{\sim 3.7+}O_{12}$. As the temperature is lowered, *charge transfer* takes place continuously, so that at 200 K the situation is represented as $Sr^{2+}Cu_3^{\sim 2.8+}Fe_4^{\sim 3.4+}O_{12}$. The transformation can be denoted by

$$3Cu^{\sim 2.8+} + 4Fe^{\sim 3.4} \rightarrow 3Cu^{\sim 2.4} + 4Fe^{\sim 3.7}$$

Further cooling leads to *charge disproportionation* between the Fe cations, so that at 4 K the situation can be written as $Sr^{2+}Cu_3^{\sim 2.8+}Fe_{\sim 3.2}^{3+}Fe_{\sim 0.8}^{5+}O_{12}$.

2.2 Anion Substituted Perovskites

2.2.1 Nitrides and Oxynitrides

Only a few nominally ionic nitride perovskites are known, of which the best characterised is $ThTaN_3$. This has the ideal cubic structure:

$ThTaN_3$: cubic; $a = 0.40205$ nm, $Z = 1$; space group, $Pm\bar{3}m$ (221);

Atom positions: Th : 1 (*a*) 0,0,0;
Ta : 1 (*b*) ½,½,½;
N : 3 (*c*) ½,½,0; ½,0,½; 0,½,½;

The Th^{4+} ions lie at the corners of the unit cell. The Ta^{5+} ions lie at the cell centre and are surrounded by a regular octahedron of N^{3-} ions.

Table 2.3 *Oxynitride and oxyhalide perovskite phases*

Phase	Space group	*a* (nm)	*b* (nm)	*c* (nm)
$CaTaO_2N$	O, *Pmna* (62)	0.56134	0.78891	0.55472
$SrTaO_2N$	T, *I4/mcm* (140)	0.57025		0.80542
$CeTiO_2N$	O, *Pmna* (62)	0.55580	0.78369	0.55830
$PrTiO_2N$	O, *Pmna* (62)	0.55468	0.78142	0.55514
$NdTiO_2N$	O, *Pmna* (62)	0.55492	0.78017	0.55290
$LaZrO_2N$	O, *Pmna* (62)	0.58725	0.82503	0.581011
$EuNbO_2N$	C, $Pm\bar{3}m$ (221)	0.40324		
$EuTaO_2N$	C, $Pm\bar{3}m$ (221)	0.40195		
$EuTaO_2N$ (annealed)	T, I4/mcm (140)	0.56829		0.80620
$EuWO_{0.84}N_{2.16}$	C, $Pm\bar{3}m$ (221)	0.39472		
$EuWO_{1.17}N_{1.83}$	C, $Pm\bar{3}m$ (221)	0.39621		
$EuWO_{1.46}N_{1.54}$	C, $Pm\bar{3}m$ (221)	0.39779		
$KTiO_2F$	C, $Pm\bar{3}m$ (221)	0.39607		
$SrFeO_2F$	C, $Pm\bar{3}m$ (221)	0.39529		
$BaFeO_2F$	C, $Pm\bar{3}m$ (221)	0.40603		
$Sr_{0.5}La_{0.5}FeO_{2.5}F_{0.5}$	O, *Pmna* (62)	0.55586	0.78746	0.55892

Many more perovskite phases are known in which both oxygen and nitrogen occupy the anion positions to form the oxynitride perovskites, with a formula ABO_2N, typified by $NdTiO_2N$ (Table 2.3). These materials are generally approximated as ionic compounds, so that

$$q_A + q_B = (2q_O + q_N) = 7$$

Ensuring charge balance is maintained gives a number of possibilities:

$$(q_A, q_B) = (3, 4); \text{for example } LaZrO_2N;$$

$$(q_A, q_B) = (2, 5); \text{for example } SrTaO_2N;$$

Broadly speaking, in the compounds that have been studied structurally, it appears that the O^{2-} and N^{3-} anions occupy the X-sites at random, although there is some evidence to suggest that local ordering of the anions may occur at a microdomain level.

The oxynitride perovskites exhibit the same octahedral distortions and tilting as the oxide counterparts described earlier. For example, $NdTiO_2N$ and $LaZrO_2N$ both adopt the $GdFeO_3$ structure, tilt system $a^+b^-b^-$ (10), with the anion sites filled with O and N distributed at random. $CeTiO_2N$ and $PrTiO_2N$ are similar, but there is evidence for short-range ordering of the O and N which suggests that a microdomain structure might exist that is unresolved by conventional X-ray and neutron diffraction. $LaTiO_2N$ adopts the octahedral tilt system $a^-b^-c^-$ (12) with a triclinic unit cell: $a = 0.56097$ nm, $b = 0.78719$ nm, $c = 0.55752$ nm $\alpha = 90.20°$, $\beta = 90.15°$, $\gamma = 89.99°$.

A number of oxynitrides are known in which the O^{2-} and N^{3-} amounts differ from those described. The rock-salt ordered double perovskites $LaMg_{0.5}Ta_{0.5}O_{2.5}N_{0.5}$ and $LaMg_{0.33}Ta_{0.66}O_2N$ have ordering of the B-site Mg^{2+} and Ta^{5+} cations. Non-stoichiometric variability in the O^{2-} and N^{3-} content is also known in a number of $ABO_{2-x}N_{1+x}$ phases in which A is an alkaline earth or lanthanoid cation and B a transition metal cation Ti, Zr, V, Nb, Ta, Mo and W. In these, the variable charge on the anion complement is balanced by variable valence of the cations. For example, in $LaWO_{0.6}N_{2.4}$, the nominal valence states are $La^{3+}W^{6+}_{0.4}W^{5+}_{0.6}O_{0.6}N_{2.4}$ and in $SrWO_{1.7}N_{1.3}$ the nominal charge states are $Sr^{2+}W^{6+}_{0.3}W^{5+}_{0.7}O_{17}N_{13}$. The cubic oxynitride phases $EuWO_{1+x}N_{2-x}$ ($-0.16<x<0.46$) derived from the phase $Eu^{2+}W^{6+}ON_2$ are of interest in this respect because the nominal oxidation state of both the Eu and the W can be altered, thus changing the magnetic properties considerably. Negative values of x correspond to oxidation of Eu^{2+} to give a nominal composition of $Eu^{2+}_{1+x}Eu^{3+}_{-x}W^{6+}O_{1+x}N_{2-x}$. For positive values of x, the W^{6+} is reduced to give a nominal formula of $Eu^{2+}W^{6+}_{1-x}W^{5+}_{x}O_{1+x}N_{2-x}$.

2.2.2 Oxyfluorides

Oxyhalides have a halide ion substituted for an oxygen ion in the perovskite structure. As with the oxynitrides, addition of a halogen requires charge compensation, typically via a change in the nominal cation valence. Oxyhalides with a formula ABO_2F are generally approximated as ionic compounds, so that

$$q_A + q_B = (2q_O + q_F) = 5$$

Ensuring charge balance is maintained gives the possibilities:

$$(q_A, q_B) = (1,4) \text{ or } (2,3);$$

By far the commonest of these phases contain Ca, Sr or Ba as the A-site ion and F as the halide (Table 2.3), for example, $SrFeO_2F$. As with the oxynitrides, it appears that the O^{2-} and F^- anions occupy the X-sites at random to give a cubic perovskite unit cell, although short-range order is not ruled out in these phases.

A number of oxyfluorides are known that have substitution on the A-sites as well as the X-sites. Thus the phases $Sr_xLa_{1-x}FeO_{3-x}F$, which represents a solid solution between $SrFeO_2F$ and $LaFeO_3$, nominally contain only Fe^{3+}. Over most of the composition range, $0<x<0.9$, the phases are orthorhombic, with the $GdFeO_3$ structure, although the octahedral tilting decreases as x increases. For x values of up to 0.6, the X-sites are filled at random with O^{2-} and F^- anions. However, in the composition range between x values of 0.6 and 1.0, oxygen loss is common and the compounds show up to 4% vacancies, also distributed at random. This remains true for greater oxygen deficits in many oxyhalide phases and even considerable oxygen loss, as in

$Sr_{0.8}Ba_{0.2}FeO_{2.48}F_{0.06}$, with a composition close to the brownmillerite structure type (Section 2.4) still retains the disordered perovskite structure rather than the ordered variant.

2.3 A-Site-Deficient Perovskite Structures

2.3.1 ReO_3, WO_3 and Related Structures

Many perovskites are known with deficiency in the A-site populations. In the extreme case the A-site can be empty if the B-site cation has a charge of 6+. The simplest of these perovskite-related structures is that of rhenium trioxide, ReO_3, a metallic conductor, with an apex-sharing array of ReO_6 octahedra identical in disposition to the TiO_6 octahedra in $SrTiO_3$ (Figure 2.5).

ReO_3: cubic; $a = 0.37518$ nm, $Z = 1$; space group, $Pm\bar{3}m$ (No. 221);

$$\text{Atom positions:} \quad \text{Re}: 1\,(a)\,0,0,0;$$
$$\text{O}: 3\,(d)\,½,0,0;\ 0,½,0;\ 0,0,½;$$

The oxyfluorides NbO_2F ($a = 0.38854$ nm), TaO_2F ($a = 0.3896$ nm) and $TiOF_2$ ($a = 0.38102$) are similar. In these cases the O and F atoms are placed at random over the oxygen sites. Both ReO_3 and NbO_2F display octahedral tilting and a drop in symmetry when subjected to high pressure.

Tungsten trioxide, WO_3, is similar to ReO_3, but the octahedra now exhibit cation displacement and octahedral tilting so that a variety of temperature-sensitive lower symmetry forms occur. Below −43°C the ε-structure is monoclinic, with an $a^-b^-c^-$ tilt system coupled with displacement of the W^{6+} cations away from the centres of the octahedra. Between −43 and +17°C the δ-form has a triclinic structure with the

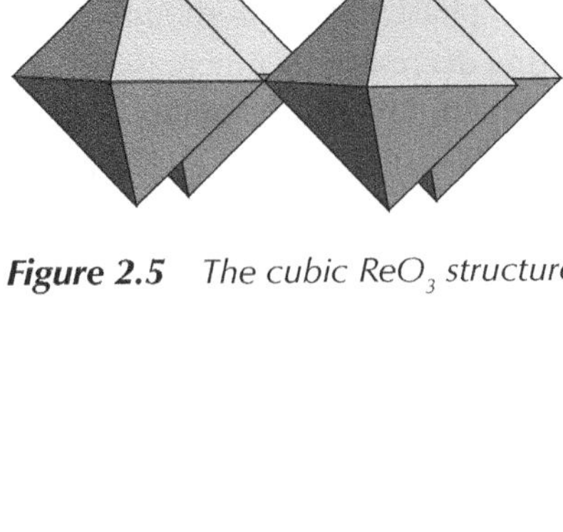

Figure 2.5 *The cubic ReO_3 structure*

same $a^-b^-c^-$ tilt system but different cation displacements. Between approximately 17 and 350°C monoclinic γ-WO_3 occurs with $a^-b^+c^-$ tilting. Above this temperature the structure becomes first orthorhombic β-WO_3, with $a^0b^+c^-$ tilting and then tetragonal, $a^0a^0c^-$.

2.3.2 Perovskite Tungsten Bronzes

The compounds known as tungsten 'bronzes', with a general formula A_xWO_3, are so called because, when they were first discovered by Wohler in 1837, their metallic lustre and high electrical conductivity lead him to believe that he had made alloys of tungsten rather than new oxides. The tungsten bronzes adopt several structures, depending upon the foreign metal and the overall composition of the phase.

The perovskite tungsten bronzes have a structure in which the foreign atoms occupy the empty A-sites in the WO_3 structure at random to form A-site-deficient perovskite phases A_xWO_3. Although many metals seem to be able to form these phases over a small range of x values, those with the greatest composition range are Li_xWO_3 (x=0–0.5), Na_xWO_3 (x=0–0.11 and 0.41–0.95), Ca_xWO_3 (x=0–0.15), Cd_xWO_3 (x=0–0.115), La_xWO_3 (x=0–0.25), Ce_xWO_3 (x=0–0.30), Nd_xWO_3 (x=0–0.40), Tb_xWO_3 (x=0–0.30) and Dy_xWO_3 (x=0–0.25). The sodium tungsten bronze, Na_xWO_3, has two composition ranges associated with the perovskite bronze structure, WO_3 to $Na_{0.11}WO_3$ and $Na_{0.41}WO_3$ to $Na_{0.95}WO_3$.

The structures of many of these perovskite bronzes are not known in detail, but at the lowest degrees of interpolation of the A cations, the structure is monoclinic and identical or similar to the room temperature WO_3 structure. As the amount of A cation increases, the structure changes from monoclinic through orthorhombic, tetragonal and ultimately cubic. For example, $Ca_{0.01}WO_3$ is orthorhombic, with lattice parameters a=0.74890 nm, b=0.73507 nm and c=0.38535 nm ($a \approx b \approx 2a_p$, $c \approx a_p$). In the composition range x=0.03–0.11, the phase is tetragonal, with lattice parameters for $Ca_{0.03}WO_3$ of a=0.52436 nm, c=0.38735 nm ($a = b \approx \sqrt{2}a_p$, $c \approx a_p$). For compositions in the range 0.12–0.15, the phase is cubic, with a lattice parameter for $Ca_{0.12}WO_3$ of a=0.37963 nm.

As expected, these structures are also sensitive to temperature and pressure as well as composition.

2.3.3 A-Site-Deficient Titanates, Niobates and Tantalates

A number of perovskite phases are able to support considerable populations of A-site vacancies provided that charge neutrality is maintained. The best known of these are the (3, 4) perovskites $La_{2/3}TiO_3$, which has a vacancy population on 1/3 of the A-sites, and the superionic conductors $La_{(2/3)-x}Li_{3x}V_{(1/3)-2x}TiO_3$ ($0<x<0.167$). These latter phases are described in Section 2.6.

In $La_{2/3}TiO_3$ and oxygen-deficient $La_{2/3}TiO_{3-\delta}$, with δ taking values from approximately 0.007–0.079, the A-site vacancies are generally considered to be ordered, with a doubled cell of approximate dimensions $a \approx a_p = b_p$, $c \approx 2c_p$, with respect to an idealised perovskite cubic unit cell. The crystal structure of slightly oxygen-deficient $La_{2/3}TiO_{2.970}$ has been described as orthorhombic: space group *Pmmm* (47), $a=0.38789$ nm, $b=0.38668$ nm, $c=0.77866$ nm. The more oxygen-deficient $La_{2/3}TiO_{2.870}$ is reported as tetragonal: space group *P4/mmm* (123), $a=0.38980$ nm, $c=0.77949$ nm. The oxygen loss in these materials appears to involve a charge compensation in which Ti^{4+} ions are reduced to Ti^{3+}. To maintain charge balance, $2Ti^{3+}$ cations (Ti'_{Ti} defects) are formed for each oxygen lost to the bulk:

$$\mathrm{nul} \rightarrow \delta V_O^{2\bullet} + 2\delta e' + \tfrac{1}{2}\delta O_2(g) \rightarrow \delta V_O^{2\bullet} + 2\delta Ti'_{Ti} + \tfrac{1}{2}\delta O_2(g)$$

Nul represents the perfect structure with no defects present at all. The defects formed from this ideal state must be balanced in terms of charge and atom numbers as the equation shows.

Additional La^{3+} can also be accommodated in the structure to partially occupy the remaining A-site vacancies. In this case the substitution may be balanced by the creation of further Ti^{3+} or oxygen ion vacancies. For example, the phase $La_{5/6}TiO_{2.982}$ has a nominal formula $La_{5/6}Ti^{3+}_{0.536}Ti^{4+}_{0.464}O_{2.982}$.

A number of lanthanoid niobates and tantalates exist which are analogous to the titanates, with a formula of $Ln_{1/3}NbO_3$ and $Ln_{1/3}TaO_3$, in which 2/3 of the A-sites are vacant. Not all the structures of these phases are completely resolved although some have been shown to have ordering of the Ln^{3+} cations. The compound $Ce_{1/3}NbO_3$ has the Ce^{3+} ions ordered onto alternate (010) planes, at $y=0$ (and empty at $y=1/2$), with the occupied A-sites 2/3 occupied (Figure 2.6) to give a monoclinic cell: space group *P2/m* (10), $a=0.55267$ nm, $b=0.78824$ nm, $c=0.55245$ nm, $\beta=90.29°$, approximately

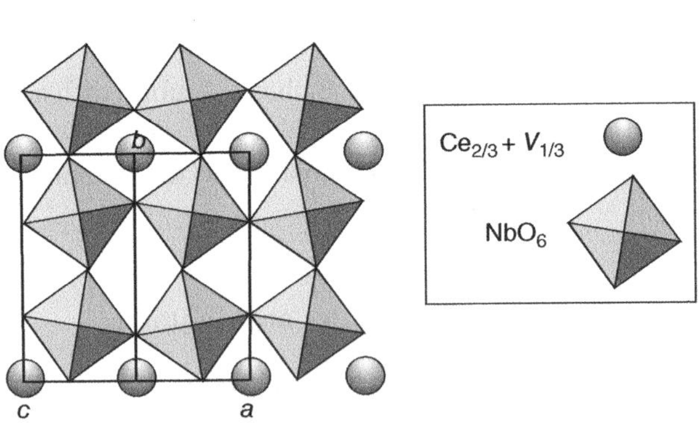

Figure 2.6 *The structure of $Ce_{1/3}NbO_3$ projected down [101]*

$\sqrt{2}a_p \times 2a_p \times \sqrt{2}a_p$. The high-temperature form of this phase appears to be orthorhombic: space group *Pmmm* (47), $a=0.3899$ nm, $b=0.3917$ nm, $c=0.7881$ nm, approximately $a_p \times a_p \times 2a_p$ with a different pattern of vacancy ordering than in the low-temperature form.

A number of species populate the A-sites to give ordered structures of the cubic $AA'_3B_4O_{12}$ type (Section 2.1.3). If the A^{2+} or B^{4+} ions are replaced with cations of higher charge, the structure still persists, but now the charge excess is corrected by a vacancy population on the A-sites. For example, $La_{2/3}Cu_3Ti_4O_{12}$, $Bi_{2/3}Cu_3Ti_4O_{12}$ and $Y_{2/3}Cu_3Ti_4O_{12}$ all have 1/3 of the A-sites vacant, while in $Cu_3Ti_2Ta_2O_{12}$, all of the A-sites are vacant.

This latter phase can take in further Ta content to form the perovskite $Cu_{2+x}Ta_4O_{12+\delta}$, in which not only are the A-sites vacant but also a considerable percentage of the A′-sites. The composition varies widely, from $Cu_{2.125}Ta_4O_{12+\delta}$ to $Cu_{2.5}Ta_4O_{12+\delta}$. Polycrystalline samples that are rapidly cooled in air from the synthesis temperature of 1025°C are found to have a cubic unit cell: space group $Pm\overline{3}$ (200), $a=0.75272$ nm for $Cu_{2.5}Ta_4O_{12+\delta}$, with a statistical distribution of vacancies. Slow cooled samples have a tetragonal or orthorhombic unit cell, for compositions between $Cu_{2.125}Ta_4O_{12+\delta}$ and $Cu_{2.425}Ta_4O_{12+\delta}$. The oxygen content of these samples is higher than the rapidly cooled phases. The vacancies in the Cu sites for slow cooled samples close to $Cu_{2.125}Ta_4O_{12+\delta}$ in composition are ordered into alternately completely filled or 1/3 filled layers, similar to the ordering patterns in Figure 2.6. Ordering for other compositions has not been clarified.

2.4 Anion-Deficient Phases Containing Tetrahedra

2.4.1 Brownmillerites

Brownmillerite is a mineral of approximate composition Ca_2FeAlO_5. Mineral specimens vary in real composition with isomorphous substitution of both the Fe and Al components being commonplace. A large number of brownmillerite phases are now known, the simplest having a formula $A_2B_2O_5$, typified by $Ba_2In_2O_5$, but many others, involving partial substitution of the A cations to form $A_{2-x}A'_xB_2O_5$, the B cations to form $A_2B_{2-x}B'_xO_5$ (as in brownmillerite itself) or both A and B cations, have been synthesised (Table 2.4).

Brownmillerite phases form an important constituent of cement powder. Portland cement, one of the most widely used types, is made by heating mixed limestone and clay in a kiln. This results in multiple reactions together with partial melting and sintering of the mixture, producing a dense mass called clinker, which is ground to produce cement powder. The resulting complex material contains four major

Table 2.4 *Oxides with the brownmillerite structure*

Compound	Space group	*a* (nm)	*b* (nm)	*c* (nm)
$Ca_2Fe_2O_5$	O, *Pmna* (62)	0.54258	1.47658	0.55974
$Ca_2Fe_{0.922}Al_{1.078}O_5$	O, *I2bm* (46)	0.53269	1.44687	0.55432
Ca_2FeCoO_5	O, *Pbcm* (57)	0.53652	1.10995	1.47982
Ca_2FeMnO_5	O, *Pmna* (62)	0.53251	1.53865	0.54787
$Ca_2Fe_{1.5}Mn_{0.5}O_5$	O, *Pmna* (62)	0.53622	1.50508	0.55284
$Ca_2Fe_{1.33}Mn_{0.67}O_5$	O, *Pmna* (62)	0.53385	1.51540	0.55009
$Ca_2Fe_{1.5}Mn_{0.5}O_{\sim5.5}$	O, *Icmm* (74)	0.54728	1.48948	0.53842
Ca_2MnAlO_5	O, *I2mb* (46)	0.52313	1.49533	0.54626
$Ca_2MnGaO_{5.045}$	O, *Pmna* (62)	0.52685	1.5301	0.54686
$Sr_2Co_2O_5$	O, *Imma* (74)	0.54639	1.56487	0.55667
$Sr_2Co_2O_5$	O, *Iam2* (46)	1.57450	0.55739	0.54697
Sr_2GaMnO_5	O, *Icmm* (74)	0.55033	1.6234	0.53717
Sr_2CoFeO_5	O, *Icmm* (74)	0.56243	1.56515	0.55017
Sr_2GaMnO_5	O, *Icmm* (74)	0.55033	1.6234	0.53717
$Sr_2Al_{1.07}Mn_{0.93}O_5$	O, *Imma* (74)	0.54358	1.56230	0.56075
$Sr_2Fe_2O_5$	O, *Icmm* (74)	0.56721	1.5592	0.55271
Sr_2GaMnO_5	O, *Imma* (74)	0.535450	1.62256	0.54888
$CaSrFe_{1.5}Mn_{0.5}O_5$	O, *Icmm* (74)	0.55630	1.54625	0.54172
$CaSrFe_{1.33}Mn_{0.67}O_5$	O, *Icmm* (74)	0.55344	1.55129	0.54093
$CaSrFe_{1.5}Mn_{0.5}O_{\sim5.5}$	C, *Pm$\bar{3}$m* (221)	0.38281		
$CaSrFe_{1.33}Mn_{0.67}O_{\sim5.5}$	C, *Pm$\bar{3}$m* (221)	0.38219		
$CaSrGaMnO_{5.035}$	O, *I2mb* (46)	0.530562	1.57705	0.54809
Ba_2GaInO_5	O, *Ibm2* (46)	0.61124	1.55649	0.59221
$Ba_2In_2O_5$	O, *Icmm* (74)	0.60961	1.68676	0.59962
$Ca_{0.5}La_{1.5}Mn_2O_5$	O, *Pmna* (62)	0.54901	1.6140	0.56477
$Sr_{0.8}La_{1.2}Mn_2O_5$	O, *Pcmb* (57)	0.54190	1.66207	1.10320
$Ba_{0.4}La_{1.6}Mn_2O_5$	O, *I2mb* (46)	0.55591	1.63997	0.56666

constituents, of which one, generally called ferrite[1] or referred to as C_2F or C_4AF, where C stands for CaO, A stands for Al_2O_3 and F stands for Fe_2O_3, has the brownmillerite structure. The chemical formula of ferrite in cement powder is variously written as $Ca_2Fe_2O_5$, $Ca_4Al_2Fe_2O_{10}$ or $Ca_6Al_3Fe_3O_{15}$. Ferrite has an important role to play in the hardening of cement power as the brownmillerite structure absorbs water to give a hydrated product which is beneficial to the ultimate strength of the material.

The brownmillerite structure is closely related to perovskite, and it is convenient to describe it with respect to an ideal perovskite parent (Figure 2.7a). The oxygen stoichiometry $A_2B_2O_5$ is achieved, in principle, by arranging rows of oxygen

[1] More properly tetracalcium aluminoferrite. This must be differentiated from ferrites, which are spinels of general formula AFe_2O_4, and from the hexagonal ferrites, that adopt structures related to that of magnetoplumbite, $PbFe_{12}O_{19}$.

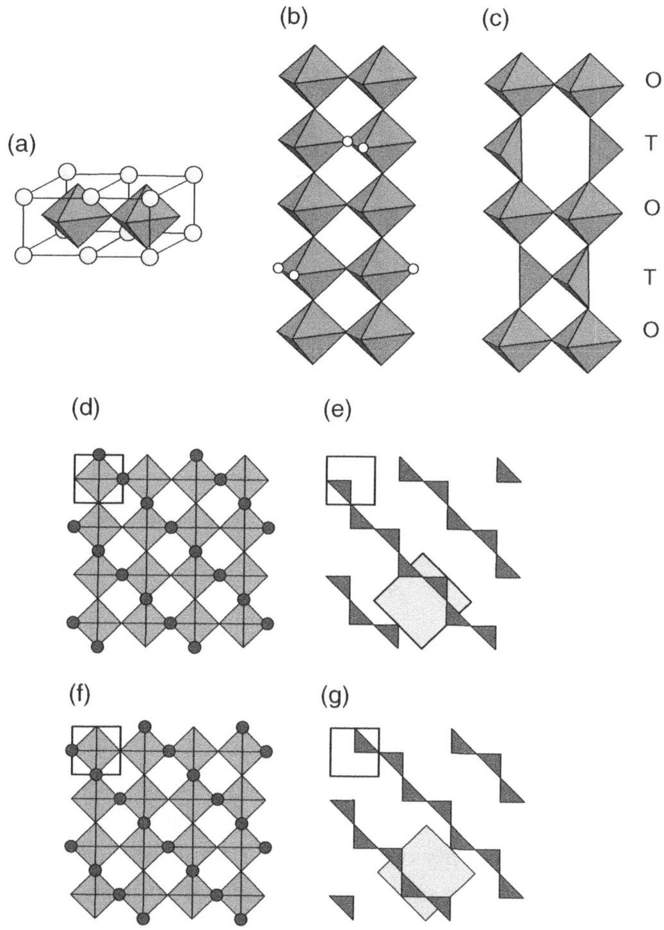

Figure 2.7 *The $A_2B_2O_5$ brownmillerite structure: (a) two unit cells of the ideal perovskite structure ABO_3; (b) the octahedral framework of ideal perovskite, with A cations omitted, showing the location of oxygen vacancies required to generate the brownmillerite structure; (c) the arrangement of octahedra and tetrahedra in the brownmillerite structure; (d and e) the arrangement of oxygen vacancies and chains of tetrahedra at y = ¼; (f and g) the arrangement of oxygen vacancies and chains of tetrahedra at y = ¾. The brownmillerite and perovskite cells are outlined in (c–f)*

vacancies along [110] directions in an ordered fashion in ideal perovskite (Figure 2.7b). In this configuration, each BO_6 octahedron loses two equatorial oxygen atoms, and with a slight displacement of the B cation, the remaining four form a tetrahedron. Thus each alternate sheet of BO_6 octahedra in perovskite is transformed into a sheet containing only apex-linked BO_4 tetrahedra in brownmillerite (Figure 2.7c). The tetrahedra are linked into chains which run along [110] of the

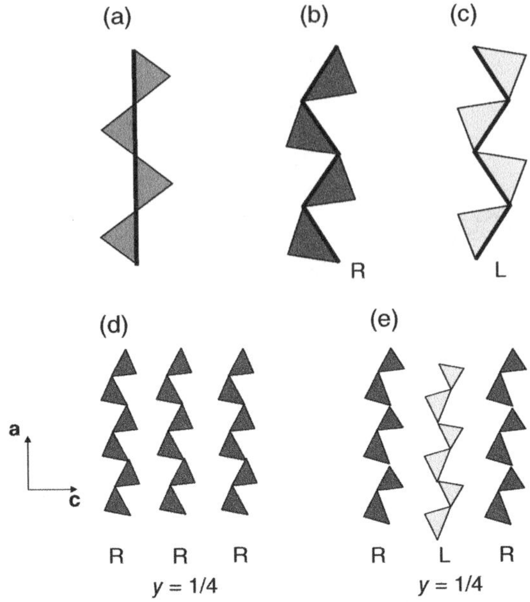

Figure 2.8 *Tetrahedral chain configurations: (a) regular; (b) R conformation; (c) L conformation; (d) all conformations the same within a layer at* y = ¼ *or* ¾*; (e) conformations alternating within the same layer at* y = ¼ *or* ¾

cubic perovskite structure (Figure 2.7d–g). Brownmillerites can be represented as a stacking of layers of corner-linked polyhedra in the sequence ...OTOT... where O stands for a sheet containing only octahedra and T for a sheet containing only tetrahedra. The ideal brownmillerite unit cell parameters are usually written in terms of the ideal perovskite cell as $a_{bm} \sim \sqrt{2}a_p$, $b_{bm} \sim 4a_p$, $c_{bm} \sim \sqrt{2}a_p$, the ...OTOT... sequence running parallel to the long (b_{bm}) axis, so that the T sheets are at $y = ¼$ and ¾.

The real structure of brownmillerite phases is more difficult to resolve because the apex-linked chains of tetrahedra are not regular (Figure 2.8a) but display a corrugation compared to the ideal motif. There are two possibilities for this chain deformation, either right (R) or left (L) handed (Figure 2.8b and c). The organisation of these chains has repercussions for the space group and unit cell axes of the brownmillerite phase:

1. If all chains have the same handedness (Figures 2.8d and 2.9a), the structure is most often (but not always) described within space group *I2mb* (46), with $b > c > a$.
2. If chains with the same handedness in layer 1 are paired with chains of the alternate handedness in layer 2, that is, alternating *interlayer* ordering, the space group *Pnma* (62) with $b > c > a$ is usually (but not always) chosen (Figure 2.9b).

3. If the R and L chains are disordered throughout the structure, the space group *Imma* (74) with $b>c>a$ is mostly (but not always) used.
4. If the chains alternate in handedness within a layer (Figure 2.8e), there is a doubling of the lattice parameter in a direction normal to the chains. In this case there are two ways of stacking the chains, (a) with a stacking vector ½ [011], called A-type stacking, and (b) with a stacking vector ½ [01$\bar{1}$], called B-type stacking (Figure 2.10). A large number of structures can now be derived in terms of various A and B sequences. The simplest of these are the uniform stacking ... AAA... described with space group *C2/c* (15) and the alternating stacking ... ABAB... described by space group *Pcbm* (57) with $b>c>a$.

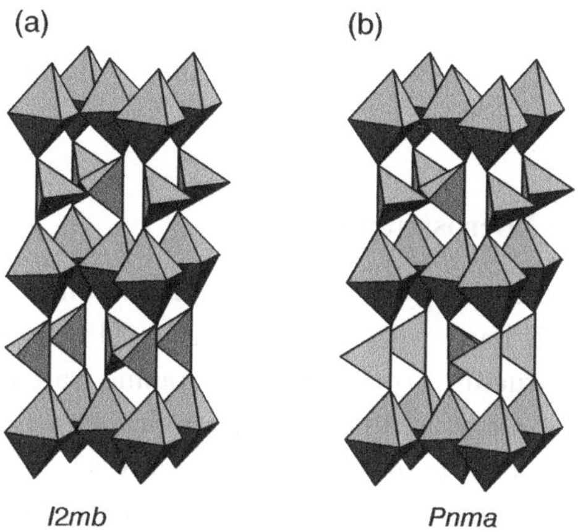

Figure 2.9 *The brownmillerite structure: (a) identical layers of tetrahedra, space group* I2bm *(46); (b) alternating layers of tetrahedra, space group* Pnma *(62)*

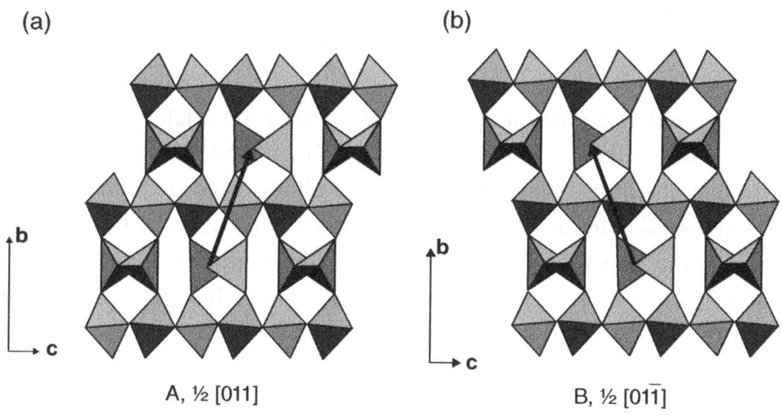

Figure 2.10 *Two ways of stacking layers of R and L alternating rows of tetrahedra in brownmillerites: (a) A-type, stacking vector ½ [011]; (b) B-type, stacking vector ½ [01$\bar{1}$]*

In addition to these conventional structures, in at least one case the tetrahedral chains have been described in terms of a modulation in which the deformations take on a wave-like periodicity that can be commensurate or incommensurate with the underlying octahedral framework. These modulated structures are described in terms of a four-dimensional superspace description rather than by a conventional space group. (For further information on modulated structures, see Sections 2.6 and 3.2.)

The R and L chain handedness in brownmillerites can be modified by changes in composition or temperature. For example, the room temperature structure of $Ca_2Fe_2O_5$ is described with the space group *Pmna* (62) with interlayer ordering of R and L layers, but the space group reverts to *I2bm* (46) at 1073 K in air as the chains become similar. In the brownmillerite series $Ca_2Fe_{2-x}Al_xO_5$, for x in the range 0–0.65 the space group is *Pnma* (62) with interlayer ordering of R and L conformations. When x is above 0.65, the symmetry changes to *I2bm* (46) and the chain orientations become the same.

2.4.2 Brownmillerite Microstructures

Transmission electron microscopy shows that there is often a complex microdomain structure at the unit cell level within these phases. Microdomains are nanoscale volumes of modified structure, often crystallographic variants, coherently intergrown within the matrix of a phase that gives the impression of being a single crystal at a macroscopic scale. The microdomains generally show a variety of ordering patterns related by symmetry to that of the parent matrix but may also consist of small volumes of disorder. Microdomain texture is difficult to identify using X-ray and neutron diffraction, as these techniques tend to yield the average atom positions in relatively large volumes of material. The preparation conditions, especially rates of cooling, can modify the microdomain structures so that a crystal may well contain a plethora of local structures. These are not easily described in terms of conventional crystallography.

It has been found that at this level $Sr_2Co_2O_5$ shows a symmetry consistent with an orthorhombic unit cell: space group *Pcmb*, (57) $a = 0.54639$ nm, $b = 1.56486$ nm, $c = 1.1340$ nm, with intra-layer RLRL ordered chains, and a doubled **c**-axis, the chains running parallel to the **a**-axis. However, on a scale which averages over many microdomains, the apparent structure satisfies the requirements of an alternative orthorhombic cell: space group *Imma*, (74) $a = 0.54639$ nm, $b = 1.56487$ nm, $c = 0.55667$ nm, indicative of disordered chains.

Microdomain texture is very dependent upon the preparation method employed as well as composition and the thermal history of the sample. For example, the oxide Ca_2FeMnO_5 is a brownmillerite with well-ordered chains of tetrahedra. Replacement of half of the Ca by Sr to form $CaSrFeMnO_5$ results in a phase with short-range order within the chains of tetrahedra, resulting in a crystal microstructure composed of microdomains of the brownmillerite type approximately 1.6 nm in size, but arranged randomly with respect to the chain directions. Complete replacement of the Ca by Sr to form Sr_2FeMnO_5 destroys the tetrahedral chains and any remnants of order, so that the resulting phase shows a disordered cubic perovskite structure.

2.4.3 Temperature Variation and Disorder

The brownmillerite ordered structure is generally a lower-temperature form, stable at room temperature and up to approximately 700°C. Above this temperature the O-site vacancies begin to disorder, with the result that the tetrahedral coordination polyhedra are destroyed and a cubic perovskite with a population of disordered oxygen vacancies is formed. The oxide $Ba_2In_2O_5$ is a well-studied phase that illustrates this general trend. At room temperature it adopts the brownmillerite structure. It appears that the structure begins to disorder above a temperature of approximately 900°C and ultimately transforms to a cubic perovskite phase: space group $Pm\bar{3}m$ (221), $a = 0.42743$ nm, with disordered O-site vacancies above approximately 1040°C. An intermediate tetragonal phase – space group $I4cm$ (108), $a = 0.60384$ nm, $c = 1.70688$ nm – has also been reported to form within the temperature range 925–1040°C, lying between the low-temperature brownmillerite form and the high-temperature disordered perovskite, although not all studies agree on this point.

In general, an ordered brownmillerite phase transforms to the disordered state via a two-phase region in which the ordered and disordered forms coexist. The metal atom positions are approximately unchanged on passing from the brownmillerite structure to the disordered perovskite structure and the transformation can be viewed as the creation of defect clusters which gradually expand to form a recognisable unit of a new structure coherently embedded in the surrounding matrix in the form of microdomains of a disordered perovskite matrix randomly distributed throughout a defect-free matrix. The size of the microdomains encountered depends upon the composition and the thermal history of the samples. These microdomains, which can generally be revealed by transmission electron microscopy, are a contributing factor in the difficultly of accurate structure determination at intermediate temperatures.

The thermal history of the sample may be important. The orthorhombic brownmillerite phase $Sr_2Co_2O_5$ (which has also been reported to be Co deficient, as $Sr_6Co_5O_{15}$) is only obtained when samples are heated in air and quenched rapidly to room temperature from above 900°C. If this quenched material is heated, it transforms at 653°C to the hexagonal $BaNiO_3$ structure (Section 3.1) and then at between 920 and 940°C to a disordered cubic phase: space group $Pm\bar{3}m$, $a = 0.395286$ nm.

The $Ba_2In_2O_5$–$BaCoO_3$ ($Ba_2In_{1-x}Co_xO_{5+\delta}$) system yields a cubic disordered perovskite for the composition range x of between 0 and 0.8 when the samples are quenched from high temperatures, while annealing the 1:1 phase $Ba_2InCoO_{5+\delta}$ at 850°C produces an orthorhombic structure brownmillerite with $a = 0.59433$ nm, $b = 1.5810$ nm, $c = 0.59872$ nm.

2.4.4 B-Site Doped Brownmillerite Phases

As already indicated, B-site substitution in brownmillerites is readily achieved and manipulation of the B-site constituents is widely used to modify the chemical and physical properties of the phases. Order or partial order between two B-site cations is common, but not universal. In Ba_2GaInO_5 the Ga^{3+} ions are confined to the tetrahedra and the In^{3+} ions confined to the octahedra. In Ca_2FeCoO_5, part of the Fe^{3+} and Co^{3+} cations each occupy one set of tetrahedral chains in an ordered fashion leading to the doubling of one of the short axes, while the remainder of the Fe^{3+} and Co^{3+} are distributed at random over the octahedral sites. Similarly, in the complex oxides $CaSrFe_{1.5}Mn_{0.5}O_5$ and $CaSrFe_{1.33}Mn_{0.67}O_5$ part of the Fe^{3+} exclusively occupies all of the tetrahedral sites, with Mn^{3+} and the remainder of the Fe^{3+} occupying the octahedral positions. In the solid solution series $Ca_2Fe_{2-x}Al_xO_5$, at values of x below approximately 0.9, the Al^{3+} preferentially enters the tetrahedral sites, although this segregation is not perfect. Above a value of x of 0.9, the tetrahedral sites are about 2/3 filled by Al^{3+} and additional Al^{3+} is allocated to octahedral and tetrahedral sites equally.

Doping on the B-sites can significantly alter the defect populations of brownmillerite structure phases. In the simplest cases of isovalent substitution, the main result is to form a disordered cubic perovskite phase in place of the ordered brownmillerite. For instance, the phases $Ba_2In_{2-x}Ga_xO_5$, for the range of doping values $0.4 < x < 0.9$, are all cubic perovskites with disordered O-site vacancies. The phases $Sr_2Co_{2-x}Al_xO_5$ ($0.3 < x < 0.5$) are similar.

In the brownmillerite $Ba_2In_2O_5$, insertion of Si^{4+} at In^{3+} sites leads to an oxygen increase, expressed as $Ba_2In_{2-2x}Si_{2x}O_{3+x}$. The Si^{4+} occupies the tetrahedral sites, and the additional oxygen occupies some of the available O-site brownmillerite 'vacancies' converting them to octahedra. A similar oxygen increase can occur in the phases $Sr_2Fe_{2-x}V_xO_{5+x}$ and $Ca_2Mn_{2-x}Fe_xO_{6-x}$. These form cubic perovskites with disordered O-site vacancies. However, transmission electron microscopy shows that the materials frequently consist of microdomains (~4 nm diameter) of a brownmillerite structure that are unresolved using conventional X-ray and neutron diffraction.

In the well-studied system $Ba_2In_{2-2x}Ti_{2x}O_{5+x}$, which can be considered to form within the system $Ba_2In_2O_5$ brownmillerite and $BaTiO_3$ ($Ba_2Ti_2O_6$) perovskite, the brownmillerite phase persists only up to $0 < x < 0.075$. Between the composition limits of $0.075 < x < 0.15$, the structure is tetragonal and composed of interpenetrating microdomains. In the range $0.15<x<1.0$ the structure becomes cubic. In the $BaIn_xZr_{1-x}O_{3+x/2}$ system that is formed between $Ba_2In_2O_5$ brownmillerite and $BaZrO_3$ ($Ba_2Zr_2O_6$) perovskite, brownmillerite formation is also suppressed and only the disordered cubic perovskite phase forms for x values between 0.1 and 0.9.

2.4.5 B-Site Doping and Oxygen Pressure

When B-site ions can take on a variety of oxidation states, a change in the external oxygen partial pressure can lead to an oxygen composition range for the brownmillerite phase. This additional oxygen has a number of structural consequences. The oxygen can occupy the 'vacancies' in the brownmillerite phase and lead to new ordered phases, new coordination polyhedra such as trigonal prisms or square pyramids, discussed below (Section 2.5), or result in a disordered cubic perovskite phase.

Oxygen partial pressure is important in the phases formed when Fe is replaced by Co in $SrCo_{0.8}Fe_{0.2}O_{3-\delta}$. On preparation in pure oxygen at 1 atm, the phase $SrCo_{0.8}Fe_{0.2}O_{2.53}$ forms in the temperature range 600–800°C. The additional oxygen disorders the structure and a cubic perovskite phase forms. At lower oxygen partial pressures, for example, 0.1 atm, a two-phase mixture of cubic perovskite and virtually stoichiometric brownmillerite exists between 600 and 649°C. At an oxygen partial pressure of 10^{-2} atm, additional oxygen is not incorporated into the phase and the brownmillerite structure is present between the temperatures of 600 and 743°C. Above these upper temperature limits, the brownmillerite disorders into a cubic perovskite due to thermal effects.

A similar disordering effect occurs when the brownmillerite structure oxide $CaSrFe_{1.5}Mn_{0.5}O_5$ prepared in argon is heated in air, when it will oxidise to a composition of approximately $CaSrFe_{1.5}Mn_{0.5}0_{5.5}$, corresponding to oxidation of all of the Mn^{3+} cations to Mn^{4+}. This material, unlike the starting phase, is no longer ordered but shows a disordered cubic perovskite structure: space group $Pm\bar{3}m$, $a=0.38281$ nm.

The brownmillerite phase $Sr_2MnGaO_{5+\delta}$ which exists over the composition range of δ between −0.03 and 0.13 is transformed to an ordered perovskite-like structure for compositions above this range. In these, the Ga coordination polyhedra are converted from tetrahedral to trigonal bipyramidal, these being arranged in an ordered fashion in the phases $Sr_2MnGaO_{5.41}$ and $Sr_2MnGaO_{5.46}$.

2.4.6 A-Site Doped Brownmillerite Phases

A-site doping can often be written in terms of a reaction between a brownmillerite and a perovskite, $A_2B_2O_5$–$A'BO_3$ ($A'_2B_2O_6$). For the $Ba_2In_2O_5$–$LaInO_3$ ($La_2In_2O_6$) system, $Ba_{2-x}La_{2x}In_2O_{5+x}$, a brownmillerite structure, forms when x is less than 0.1.

Above this value the structure is that of a disordered perovskite, initially with a tetragonal cell and then a cubic cell.

An extensively studied series of brownmillerite phases can be prepared from the cubic perovskite series $La_{1-x}A_xMnO_3$, where A is Ca, Sr or Ba. The oxidised cubic phases contain Mn^{3+} and Mn^{4+}, the ratio dependent upon the value of x. The cubic form can be reduced to brownmillerite structures $La_{1-x}A_xMnO_{2.5}$, where all the B-site cations are now Mn^{2+} and Mn^{3+}. The approximate phase ranges, x, over which the brownmillerite structure exists are as follows: Ca, 0.225–0.5; Sr, 0.155–0.5; and Ba, 0.11–0.325.

2.4.7 Brownmillerite-Related Phases

The ideal brownmillerite structure can be described by the stacking sequence, along the (usually) $\mathbf{b}_{bm}$-axis of ...OTOT.... However, as outlined, the structure can accommodate additional oxygen fairly easily if the cation valence permits. If this is added in an ordered fashion, a number of new structures can result. One series of phases is formed if the oxygen is accommodated solely in the T layers, which are then converted into O layers. These phases can be considered to be members of the series $A_nB_nO_{3n-1}$, of which the brownmillerite structure type is the $n=2$ member and perovskite the $n=\infty$ member. The $n=3$ phases, also known as Grenier phases, with the stacking ...OOTOOT... are exemplified by $Ca_2LaFe_3O_8$ and $Ca_2NdFe_3O_8$. They can be considered to be 1:1 intergrowths between the perovskite $LaFeO_3$ or $NdFeO_3$ and brownmillerite $Ca_2Fe_2O_5$.

The phase diagram for $LaFeO_3$–$CaFeO_{2.5}$ (Figure 2.11) shows that $Ca_2LaFe_3O_8$ exists at a precise composition of $Ca_{0.67}La_{0.33}FeO_{2.67}$. At the $LaFeO_3$ side of the phase range, the perovskite structure, which has a change of symmetry at temperatures between 600 and 1000°C, exists up to a composition of $La_{0.62}Ca_{0.38}FeO_{2.81}$. At normal oxygen pressures, incorporation of $2Ca^{2+}$ cations onto the La^{3+} sites is balanced by the formation of one oxygen vacancy and the oxidation state of the Fe^{3+} remains unchanged. An increase in the $Ca_2Fe_2O_5$ content results in a two-phase region in which the perovskite and $Ca_2LaFe_3O_8$ phases coexist. The ordered $Ca_2LaFe_3O_8$ structure shows no composition range at all. At the $Ca_2Fe_2O_5$ end of the phase range, the brownmillerite structure also has no composition range at all. Incorporation of oxygen results in the formation of $Ca_2LaFe_3O_8$ with the additional oxygen filling the tetrahedral sites selectively, until pure $Ca_2LaFe_3O_8$ is found. There is evidence to suggest that at high oxygen pressures the oxygen vacancies in all of these phases fill, and charge balance is achieved by the formation of a proportional number of Fe^{4+} cations.

The $n=4$ phases of the $A_nB_nO_{3n-1}$ series $Ca_4Ti_2Fe_2O_{11}$ (Fe^{3+}), $Ca_4Mn_2Ga_2O_{11}$ (Mn^{4+}) and $La_4Mn_4O_{11}$ ($2Mn^{2+}+2Mn^{3+}$) adopt a stacking sequence ...OOOTOOOT....

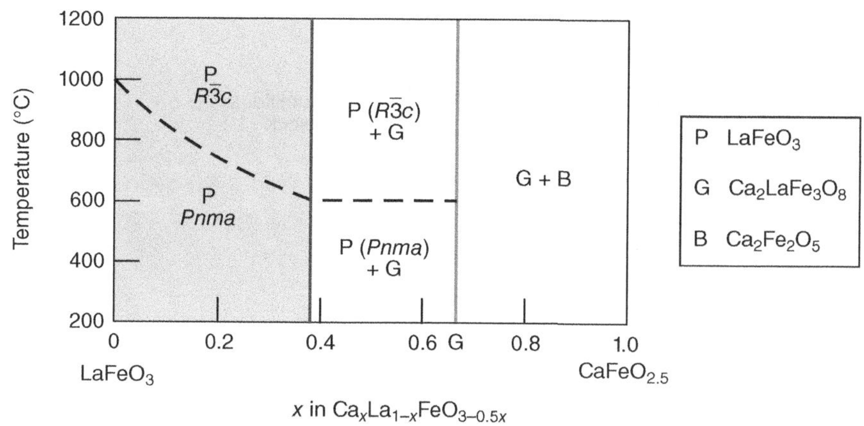

Figure 2.11 *Simplified phase diagram for the $LaFeO_3$ (perovskite)–$CaFeO_{2.5}$ ($Ca_2Fe_2O_5$ brownmillerite) phase region. The perovskite phase has a symmetry change from orthorhombic* Pnma *at lower temperatures to trigonal $R\bar{3}c$ at higher temperatures*

Electron microscopy has shown that under normal preparation conditions, rapidly cooled samples are rarely monophasic but contain intergrowths between the $n=2$, 3, 4 and higher members of the series. Once again, preparations made under high oxygen pressures result in the filling of oxygen vacancies which is balanced by a variation in the valence of the Fe and Mn cations.

A layered structure is adopted by $Sr_3NdFe_3O_9$, $Ca_4Fe_2Mn_{0.5}Ti_{0.5}O_9$ and a number of other $Ca_4B_2B'O_9$ compounds where B is Al, Fe or a mixture of both and B′ is Ti, Mn or a mixture of both. In these materials, alternate sheets of tetrahedra are empty of B cations and the phase is built up from slabs of brownmillerite type, each consisting of three layers, OTO (Figure 2.12). (These are analogous to the hexagonal perovskites related to $La_4Ti_3O_{12}$ and similar phases, Chapter 3, which have layers of empty octahedra.) The structure can be considered to consist of brownmillerite slabs separated by rock-salt structure interfaces or as intergrowths related to the K_2NiF_4 structure (Section 4.1). The *average* structure of $Ca_4Fe_2Mn_{0.5}Ti_{0.5}O_9$, in which all the slabs are identical, has an orthorhombic unit cell: space group *Amma* (63), $a=0.53510$ nm, $b=2.6669$ nm, $c=0.54914$ nm. However, the real structure is more complex and crystals frequently show one-dimensional planar disorder in a direction parallel to the **b**-axis. This more complex structure has alternating R and L tetrahedral chains within each T layer, necessitating a doubling of the c-parameter. In addition, the OTO blocks can now stack in two different ways, as in the case of the brownmillerite phases described previously, with stacking vectors ½ [011] and ½ [01$\bar{1}$] (Figure 2.13). This stacking can be disordered or quasi-ordered, in which case the real structure is then described in terms of a four-dimensional superspace group.

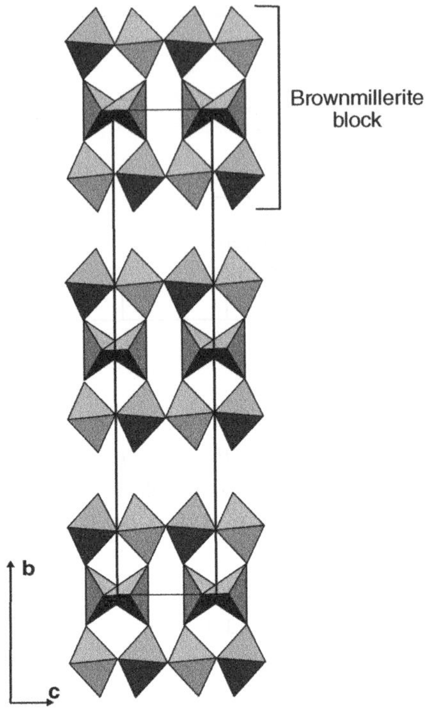

Figure 2.12 Average structure of the layered brownmillerite $Ca_4Fe_2Mn_{0.5}Ti_{0.5}O_9$

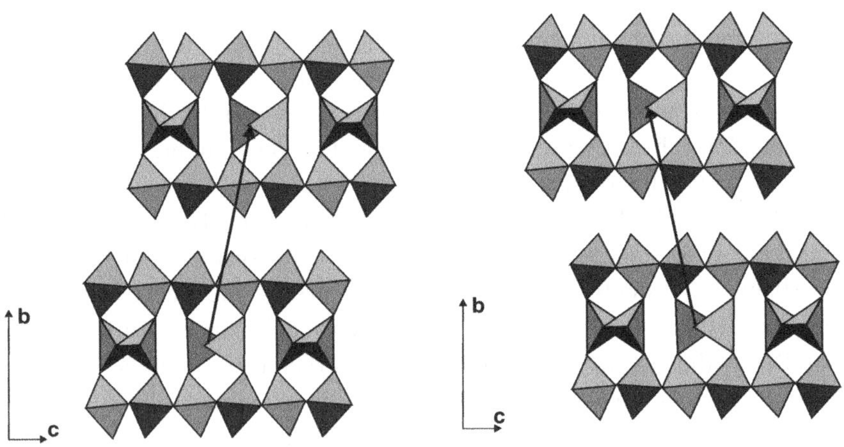

Figure 2.13 The two ways of stacking OTO blocks when R and L tetrahedral chain configurations exist within a T layer

2.5 Anion-Deficient Phases Containing Square Pyramids

The ordering pattern of oxygen vacancies found in brownmillerite, resulting in chains of apex-linked tetrahedra, can be supplemented by arrangements in which the coordination polyhedra formed are BO_5 square pyramids. There are large numbers of these structures known within the manganites, ferrites and cobaltites, and only representative examples are illustrated here.

As with brownmillerite the formation of the square pyramidal groups can readily be visualised in terms of an ideal cubic perovskite parent. In the structures here, ordered oxygen vacancies occupy single sites on some of the BO_6 octahedra rather than a pair of sites as in brownmillerite, so that the octahedra are converted into chains of square pyramidal polyhedra rather than tetrahedra (Figure 2.14a and b). The different phases known, some of which are described in the following, contain ordered variants of these octahedral and square pyramidal chains.

2.5.1 Manganites

The phase $Sr_2Mn_2O_5$ (also described as $SrMnO_{2.5}$, $Sr_4Mn_4O_{10}$ or $Sr_5Mn_5O_{12}$) is built entirely of square pyramids, although the composition is identical to that of the brownmillerites (Figure 2.15a). If the square pyramids are mentally reconverted to octahedra, the relationship to perovskite is apparent, with $a \approx \sqrt{2}a_p$, $b \approx 2\sqrt{2}a_p$, $c \approx a_p$. The Mn^{3+} cations all occupy square pyramidal sites, with the Sr^{2+} cations occupying the now enlarged perovskite cage sites. The bonds in the square pyramids consist of one long bond to the apical oxygen, three equatorial bonds of intermediate length and one shorter bond, linked to a long bond in the adjoining square pyramid. This is taken as strong evidence for orbital ordering, with the half-filled d_{z^2} orbital aligned along the long bond direction. This arrangement can be associated with the Jahn–Teller distortion found in Mn^{3+} octahedral sites in which the four equatorial bonds are shorter and the two apical bonds longer, corresponding to the half-filled d_{z^2} orbital pointing along the apical (long bond) axis (also see Section 8.6).

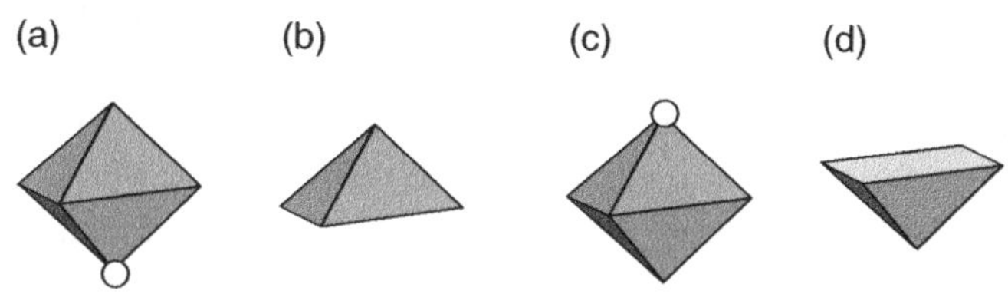

Figure 2.14 (a–c) An oxygen vacancy (open circle) (a and c) converts an octahedron into a square pyramid (b and d)

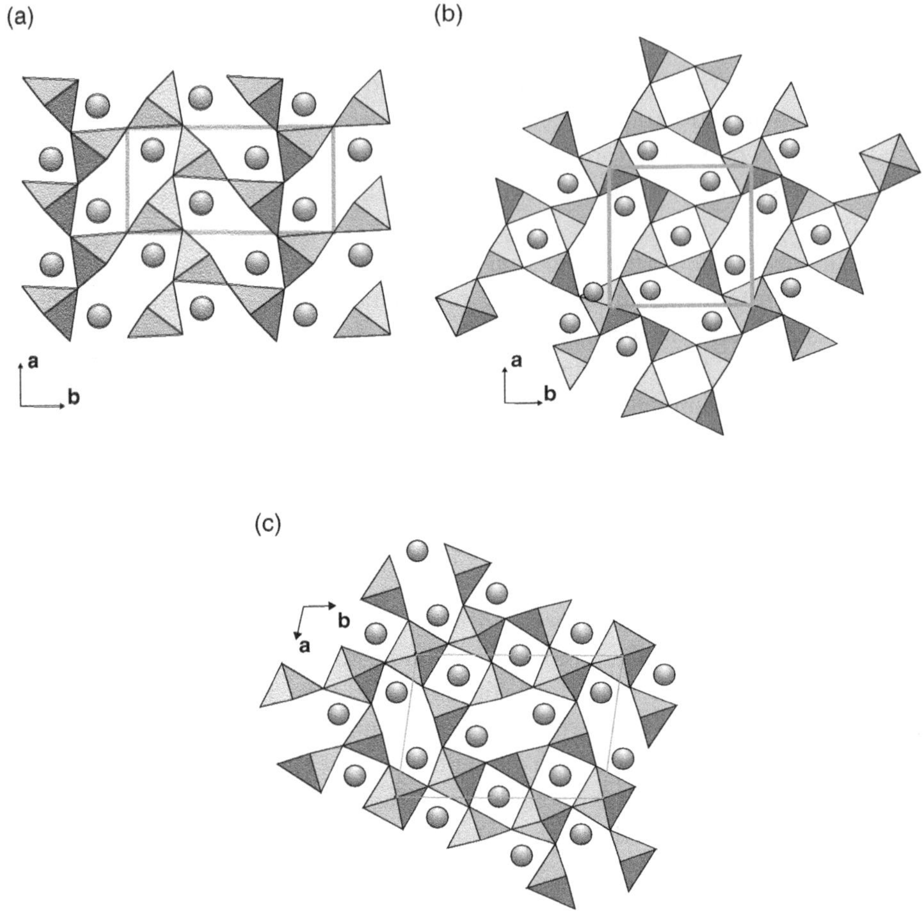

Figure 2.15 *The slightly idealised structures of (a) $Sr_2Mn_2O_5$; (b) $Sr_5Mn_5O_{13}$; (c) $Sr_7Mn_7O_{19}$, viewed down [001]*

As the oxygen content increases, two very similar phases are found, $Sr_5Mn_5O_{13}$ (Figure 2.15b) and $Sr_7Mn_7O_{19}$ (Figure 2.15c). These are built from a linking of square pyramidal and octahedral polyhedra. Once again, charge and orbital ordering is present, with the Mn^{4+} cations preferring the octahedra and the Jahn–Teller prone Mn^{3+} ions confined to the square pyramids. The formula of $Sr_5Mn_5O_{13}$ can be written $Sr_5Mn^{4+}(Mn^{3+})_4O_{13}$ and the d_{z^2} orbitals are arranged along the apical directions of the square pyramids as before. The idealised unit cell has dimensions $a \approx \sqrt{5}a_p$, $b \approx \sqrt{5}a_p$, $c \approx a_p$. The formula of $Sr_7Mn_7O_{19}$ can be written $Sr_7(Mn^{4+})_3(Mn^{3+})_4O_{13}$ and the d_{z^2} orbitals are arranged as before, along the apical directions of the square pyramids. The idealised monoclinic unit cell has dimensions $a \approx \sqrt{5}a_p$, $b \approx \sqrt{10}a_p$, $c \approx a_p$, $\gamma \approx 98.2°$, taking the (non-conventional) **c**-axis as unique.

These phases can be considered to form part of a homologous series of structures with a formula $Sr_{n+4}Mn_4^{3+}Mn_n^{4+}O_{10+3n}$, with $Sr_2Mn_2O_5$ ($Sr_4Mn_4O_{10}$) being the $n=0$ phase, $Sr_5Mn_5O_{13}$ being the $n=1$ phase and $Sr_7Mn_7O_{19}$ being the $n=3$ phase. The index n corresponds to the number of apex-linked octahedra present, none in $Sr_4Mn_4O_{10}$, isolated octahedra in $Sr_5Mn_5O_{13}$ and chains of three in $Sr_7Mn_7O_{19}$. The $n=2$ compound $Sr_6Mn_6O_{16}$ with linked pairs of octahedra has yet to be synthesised.

Other examples of these structures are known, especially several containing copper. Phases isostructural with $Sr_4Mn_4O_{10}$ include $Ca_4Mn_4O_{10}$, $Sr_4Cu_4O_{10}$, $La_4Cu_4O_{10}$ and $Nd_4Cu_4O_{10}$. Phases isostructural with $Sr_5Mn_5O_{13}$ include $La_5Cu_5O_{13}$, $Nd_5Cu_5O_{13}$ and $La_4BaCu_5O_{13}$, if, in this latter structure, the ordering of the La and Ba cations is ignored. The charge and orbital distribution in these Cu-containing phases mirrors that in the manganites. Divalent Cu^{2+} is a Jahn–Teller ion as is Mn^{3+} and Cu^{3+} corresponds to the non-Jahn–Teller cation Mn^{4+}.

2.5.2 $SrFeO_{2.5}$ and Related Phases

One of the most investigated systems that make use of square pyramidal coordination, at least in part, is $SrFeO_{2.5+\delta}$, which exists over the composition range from $SrFeO_{2.5}$ to $SrFeO_3$. The stoichiometric phase $SrFeO_3$, stable only when prepared under high oxygen pressures, adopts the cubic perovskite structure and contains only Fe^{4+} located in octahedral coordination. The other end member of the series $Sr_2Fe_2O_5$ ($SrFeO_{2.5}$) adopts the brownmillerite structure, containing solely Fe^{3+}, in both tetrahedra and octahedra.

The oxygen content of these oxides is dependent upon the ambient oxygen partial pressure during preparation. Nominally $SrFeO_3$ samples prepared under 1 atm partial pressure of oxygen have a composition of approximately $SrFeO_{2.86}$. The incorporation of one oxygen vacancy results in the concurrent replacement of $2Fe^{4+}$ species with $2Fe^{3+}$ species. As the composition falls towards $Sr_2Fe_2O_5$, the vacancy content increases in concert with an increase in Fe^{3+} content. At high temperatures these vacancy-rich materials adopt a cubic perovskite structure across the whole composition range, with a varying population of disordered oxygen vacancies distributed over the O-sites. Below a composition-dependent transition temperature, of the order of 900°C, these oxygen vacancies order along alternate chains of octahedra, to convert them into sequences of pairs of apex-sharing square pyramids (Figure 2.16a and b). Two phases, $Sr_4Fe_4O_{11}$ ($SrFeO_{2.75}$) and $Sr_8Fe_8O_{23}$ ($SrFeO_{2.875}$), have been characterised. These belong to a homologous series $Sr_nFe_nO_{3n-1}$, with $n=4$ and $n=8$. In this terminology $SrFeO_3$ is the $n=\infty$ phase and $Sr_2Fe_2O_5$ the $n=2$ phase.

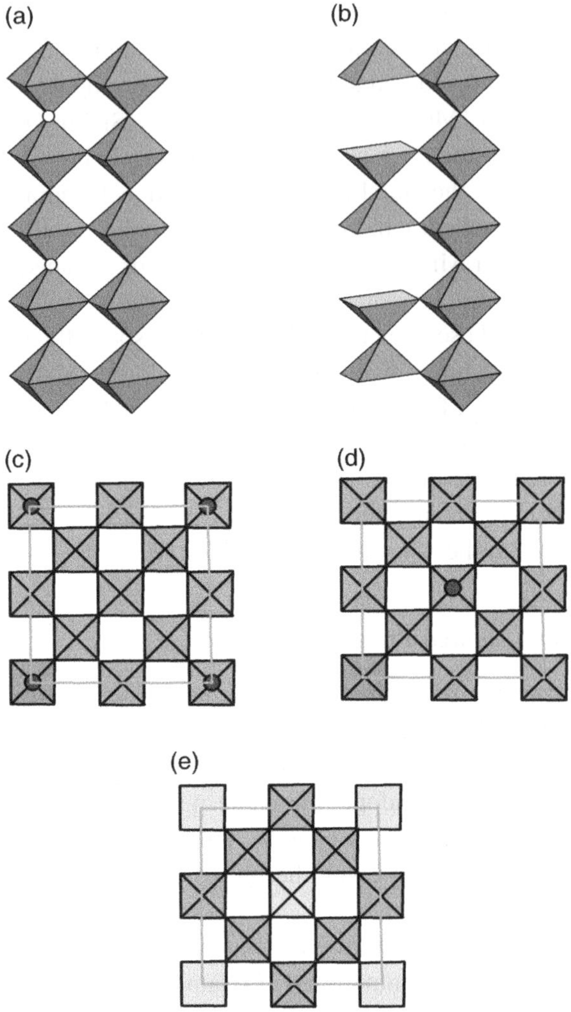

Figure 2.16 The ideal structure of $Sr_8Fe_8O_{23}$ ($SrFeO_{2.875}$): (a) ordered arrangement of oxygen vacancies in the ideal perovskite structure; (b) the formation of chains of apex-sharing square pyramids; (c and d) the arrangement of oxygen vacancies at z = ¼ (c) and z = ½ (d) in the idealised perovskite structure; (e) arrangement of chains of square pyramids in a unit cell. In (c, d) and e) the unit cell is outlined and projected down [001]

As mentioned, the phase $Sr_8Fe_8O_{23}$ is the phase normally synthesised under oxygen partial pressures of 1 atm instead of stoichiometric $SrFeO_3$. The tetragonal unit cell – space group *I4/mmm* (139), $a = 1.0929$ nm, $c = 0.7698$ nm – contains a chain containing the square pyramidal units at each corner and another, out of step by one octahedron, at the centre (Figure 2.16c–e). The Fe^{4+} cations preferentially occupy square pyramids, with the Fe^{3+} and residue of the Fe^{4+} cations in the octahedra. The phase $Sr_4Fe_4O_{11}$ is somewhat similar, again with chains of apex-linked square pyramidal FeO_5 coordination polyhedra and FeO_6 octahedra and an orthorhombic unit cell: space group *Cmmm* (65), a = 1.0974 nm, b = 0.7702 nm, c = 0.5473 nm.

The related phase $BaFeO_{2.5}$ has an extremely complex structure of ordered oxygen vacancies, with a monoclinic unit cell: space group $P2_1/c$ (14), a = 0.69753 nm, b = 1.17281 nm, c = 2.34509 nm, β = 98.81°. The Fe^{3+} cations are found in several different coordination polyhedra, octahedra (1/7), square pyramidal (2/7) and tetrahedral (4/7).

2.5.3 Cobaltite-Related Phases

The cobaltite phases show a similar complexity to the manganites and ferrites described previously. $Sr_2Co_2O_5$ is a brownmillerite with Co^{3+} in both octahedral and tetrahedral coordination. Careful oxidation of the brownmillerite allows a series $Sr_nCo_nO_{3n-1}$ to form analogous to those in the $Sr_2Fe_2O_{5+\delta}$ system. Brownmillerite represents the $n = 2$ phase and $SrCoO_3$ the $n = \infty$ form. Series members with $n = 4$, 5, 6, 7 and 8 have been reported.

The cobaltites $LnBaCo_2O_{5.50+\delta}$, with Ln being Y, La, Pr, Nd, Sm, Eu, Gd, Tb and Dy, have been widely studied. The phases have structures similar to that of orthorhombic $PrBaCo_2O_5$: space group *Pmmm* (47), a = 0.39049 nm, b = 0.78737 nm, c = 0.76084 nm, with idealised unit cell of dimensions $a \approx a_p$, $b \approx 2a_p$, $c \approx 2a_p$. In common with the majority of perovskite-related phases, most show structural changes at higher temperatures or with changing oxygen stoichiometry.

The basic structure of this material is very similar to that of the ferrites in the previous section. Oxygen vacancies are ordered in the Ln planes to create chains of pairs of square pyramids apex linked to chains of octahedra. The ordering pattern, though, is different from the iron equivalents (Figure 2.17). The value of the oxygen

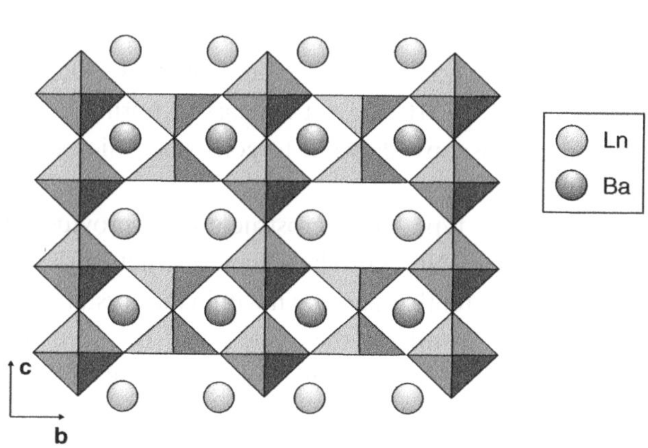

Figure 2.17 Idealised structure of the $LnBaCo_2O_{5.50}$ phases

non-stoichiometry, δ, nominally lies in the range –0.5 to +0.5. When δ is 0, the phase $LnBa_2Co_2O_{5.5}$ contains only Co^{3+}. When δ becomes negative, the material contains both Co^{2+} and Co^{3+}, and when δ becomes positive, the material contains both Co^{3+} and Co^{4+}, these changes leading to variation in the magnetic and electronic properties of the solid.

2.6 Point Defects, Microdomains and Modulated Phases

The compositional changes in perovskites are most often described in terms of disordered, or occasionally ordered, point defects, especially vacancies. This approach can be exemplified by $SrFeO_3$, which, when prepared in air at high temperatures and cooled slowly, has a composition of approximately $SrFeO_{2.875}$ with a $GdFeO_3$-type structure. The oxygen loss is described in terms of changed valence on the cations ($Fe^{4+}_{0.75}Fe^{3+}_{0.25}$) and random oxygen anion vacancies.

However, the defect structures of many stoichiometric and non-stoichiometric perovskites are far more complex than this simple model permits. As was outlined earlier with respect to the brownmillerites (Section 2.4), microdomains (also called nanodomains or nanoregions) are often present. These are small volumes of modified structure, coherently intergrown within the matrix of an apparent phase, giving the impression of being a single crystal at a macroscopic scale. The structures of many perovskites with mixed A-site and B-site cations, together with vacancies, often show such microstructures. In these phases, there are a number of competing factors that must be accommodated during crystallisation. These include the tilting of (approximately rigid) octahedra so as to accommodate A—X/B—X bond mismatch, A-site displacement, B-site displacement, Jahn–Teller distortions and A-site, B-site and X-site vacancy ordering. The conflicts between these differing priorities often result in a micro- or nanostructure containing a variety of different microdomains.

This feature is frequently observed in brownmillerites and related phases (Sections 2.4.2, 2.4.3 and 2.4.5). Non-stoichiometric $SrCoO_{2.5}$ can lose oxygen readily. In the range $SrCoO_{2.42}$–$SrCoO_{2.29}$ it appears to normal X-ray diffraction to be cubic, but electron microscopy shows it to consist of microdomains of brownmillerite-type material.

Another example of this type of microstructure is provided by the relaxor ferroelectrics related to $PbMg_{1/3}Nb_{2/3}O_3$. These have a structure described as consisting of polar nanoregions, coherently intergrown within the crystal matrix and sensitive to temperature (Section 6.7). Similarly, the oxide $Sm_{1.875}Ba_{3.125}Fe_5O_{14.65}$ is nominally cubic with respect to X-ray diffraction, with a cubic unit cell of a=0.3934 nm. However, electron microscopy shows the presence of twinned microdomains with a tetrahedral symmetry, $a \times a \times 5a$ in extent, which contains A cation ordering: Sm – Ba – (Sm + Ba) – (Sm + Ba) – Ba – Sm, that is not detected by powder X-ray diffraction.

The superionic conductors $La_{(2/3)-x}Li_{3x}V_{(1/3)-2x}TiO_3$ ($0<x<0.167$) have uncertain structures as there is considerable difficulty in locating the positions of both the Li atoms and the distribution of the vacancies *V*. This is exacerbated by the fact that the structures are sensitive to the preparation method and subsequent thermal history of the sample. In addition, octahedral tilting and Ti cation displacement are both temperature sensitive. Electron microscopy has revealed that rapidly cooled samples are composed of microdomains of order which vary from region to region. One such example is the orthorhombic unit cell of $La_{0.55}Li_{0.35}TiO_3$: space group *Pmma* (51), $a=0.54353$ nm, $b=0.76942$ nm, $c=0.54366$ nm. The tilt system is described as $a^0b^-b^-$ and the unit cell is approximately given by $\sqrt{2}a_p \times 2a_p \times \sqrt{2}a_p$. In this structure, the A-sites are occupied in an ordered fashion. Every other layer is almost completely occupied by La^{3+}. Alternating with these filled layers are layers containing the residue of the La^{3+} cations, the Li^+ cations and residual vacancies, *V*, with a composition of $La_{0.1}Li_{0.7}V_{0.2}$ (Figure 2.18). Small changes in octahedral tilt or the relative distributions of the La^{3+}, Li^+ and *V* among the A-sites lead to the many alternative unit cells described in the literature.

Although the microdomain description is adequate for many compounds, for some phases a more accurate picture is obtained if they are described in terms of two (or more) mutually interacting subsystems. The distribution of atom/vacancies, octahedral tilt, distortion or similar features may not follow a strict crystallographic repeat distance but may be modulated into a wave-like repetition, the wavelength of which is a function of composition, preparation conditions and so on. As an example, suppose that the unit cell for a perovskite has instead of octahedra with alternate tilts $+\varphi$ and $-\varphi$ so that the repetition along the chain is equal to the pseudocubic parameter a_p (Figure 2.19a), a single wave-like modulation imposed such that the octahedral tilt

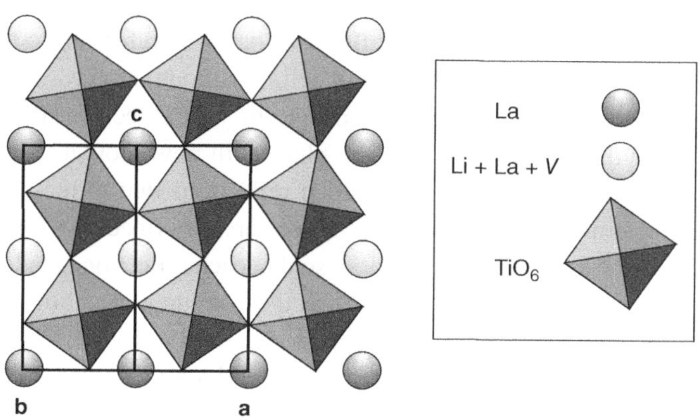

Figure 2.18 *Schematic structure of the orthorhombic $La_{(2/3)-x}Li_{3x}V_{(1/3)-2x}TiO_3$ superionic conducting phases*

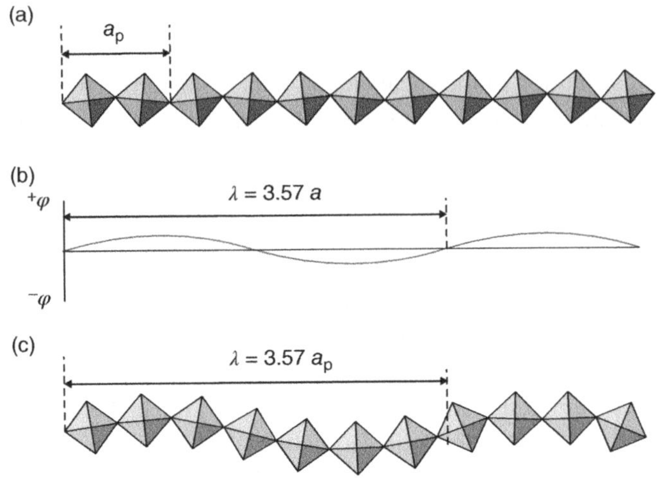

Figure 2.19 *Modulated structures, schematic: (a) chain of unmodulated BO_6 octahedra, with a repeat a_p; (b) modulation of octahedral tilt angle φ, with modulation wavelength 3.57 a_p; (c) resultant octahedral tilt wave (exaggerated)*

angle varies between $+\varphi$ and $-\varphi$ with a wave-like repetition (Figure 2.19b and c). The resultant true unit cell will now depend upon the relative values of the modulation wavelength λ to that of the basic lattice parameter a_p. In cases where the modulation wavelength comes into perfect register after small distances, the structure is described as a *commensurate modulated composite structure*. In other cases the wavelength of the modulation does not come into register at all or at least not over small distances, in which case the structures are called *incommensurate modulated composite structures*. The modulation wavelength in Figure 2.19b is 3.57 a_p, so that the true repeat along the **a**-axis is approximately 25 a_p.

The wave-like modulations are generally described with reference to the reciprocal lattice of the phase rather than the direct unit cell. The modulation wave is specified in terms of a wave vector **q**, which is defined in terms of the reciprocal lattice vectors **a***, **b*** or **c***. The diffraction pattern of the phase will now show a set of reflections corresponding to the subcell plus superlattice reflections due to the modulation. In cases with a single modulation wave, the indices are described, not by conventional Miller indices but by a four-index extension, so that each reflection is indexed as $ha^* + kb^* + lc^* + mq$. The value of q is obtained directly from the diffraction pattern. When there are two distinct modulations, the diffraction pattern reflections must be indexed on a five-index system $ha^* + kb^* + lc^* + mq_1 + mq_2$, and for the system with modulation in three independent directs a six index notation is needed (also see Section 3.2.2).

The $(1-x)BiFeO_3$–$xLaFeO_3$ system shows a one-dimensional modulated structure over the composition range $x = 0.19$–0.3 at 1075°C. The basic unit cell of this structure is orthorhombic: $a \approx \sqrt{2}a_p = 0.55977$ nm, $b \approx 2a_p = 0.78172$ nm, $c \approx \sqrt{2}\,a_p = 0.56206$ nm. The one-dimensional modulation vector is 0.4855 **c*** and the modulated subsystem primarily consists of wave-like displacement of Bi and O atoms in the (Bi, La) – O layers accompanied by a change in local octahedral tilt.

The double perovskites $NaLaMgWO_6$, $KLaMgWO_6$, $KLaMnWO_6$, $NaCeMnWO_6$ and $NaPrMnWO_6$ show rock-salt ordering of the B, B′ cations and layer ordering of the A, A′ cations, both of which are accompanied by octahedral tilting. The interaction between these ordering modes is somewhat in conflict and gives rise to a modulated structure. In $NaLaMgWO_6$ the modulation is one-dimensional and along either the **a**-axis or the **b**-axis. The underlying tetragonal unit cell has dimensions: space group *P4/nmm*, $a = b = 0.55160\,\text{nm} \approx \sqrt{2}a_p$, $c = 0.78763\,\text{nm} \approx 2a_p$. A modulation along the **a**-axis gives rise to a supercell of approximate dimensions: $a = 12a_p$, $b = 2a_p$, $c = 2a_p$. However, the equivalence of the **a**- and **b**-axes in the tetragonal parent structure means that the modulation can be along either of the basic perovskite $[100]_p$ and $[010]_p$ directions. Thus the crystal will show a twinned internal structure in addition to the modulation.

The other double perovskite phases all have two-dimensional modulations which run along $[100]_p$ and $[010]_p$, the modulation vector **q** being the same in each direction. In $NaCeMnWO_6$ the modulation vector $\mathbf{q} \approx 0.067\,\mathbf{a}^*$ (along both the $[100]_p$ and $[010]_p$ directions) gives a unit cell repeat of $\approx 14.9\,a_p \times 14.9\,a_p$, which produces a chessboard-like microstructure of modulated cells when viewed down $[001]_p$. In $NaPrMnWO_6$ the modulation vector $\mathbf{q} \approx 0.046\mathbf{a}^*$, again running along both the $[100]_p$ and $[010]_p$ directions, gives a unit cell repeat of $\approx 21.7\,a_p \times 21.7\,a_p$.

The system $Li_{3x}Nd_{2/3-x}TiO_3$ is also incommensurate in two directions. The phase $Li_{0.15}Nd_{0.6167}TiO_3$ has a unit cell: $a = 0.383105$ nm, $b = 0.382798$ nm, $c = 0.772436$ nm, $\mathbf{q}_1 = 0.45131\ \mathbf{a}^* + ½\ \mathbf{b}^*$, $\mathbf{q}_2 = ½\ \mathbf{a}^* + 0.41923\ \mathbf{b}^*$. It appears that the modulation is principally related to the tilting of the octahedra in the structure.

A number of incommensurate structures occur in the $LnBaCo_2O_{5+\delta}$ phases. In these phases the nominal valence of Co varies from +2.5 ($\delta = 0$) to +3.5 ($\delta = ½$). The inter-perovskite layers are ordered in the sequence $BaO–CoO_2–LnO_\delta–CoO_2$. The coordination polyhedra of the Co atoms can be square pyramidal or octahedral, depending upon the oxygen content, and the Co can take 2+, 3+ and 4+, low-spin, intermediate-spin and high-spin states (see Chapter 7). The prototype structure has a unit cell of approximately $a_p \times a_p \times 2a_p$ due to the ordering of the Ba- and Ln-containing layers and is generally adopted by phases in which $\delta = 0$. At lower temperatures the systems are able to take up oxygen, and the structures become more complex and varied as the oxygen content increases, due to ordering of the nominal oxygen vacancies. The room temperature form of $GdBaCo_2O_{5+\delta}$ in which $\delta \approx 0.39$ has an incommensurate structure with a modulation along two directions, giving an approximate unit cell $3a_p \times 3a_p \times 2a_p$. The incommensurate structure evolves reversibly, with no large change in oxygen content, as the temperature increases. At a temperature of approximately 380 K, the structure has a conventionally commensurate $1a_p \times 2a_p \times 2a_p$ orthorhombic unit cell: *Pmmm* (47), $a = 0.38764$ nm, $b = 0.78337$ nm, $c = 0.75594$ nm.

Further Reading

For a review of double perovskites, especially with respect to magnetic properties, see:
D. Serrate, J. M. DeTeresa, M. R. Ibarra, *J. Phys: Condens. Matter*, **19**, 023201 (2007).

An extensive list of double perovskites with single magnetic sublattices is given by:
Y. Yuan *et al.*, *Inorg. Chem.*, **54**, 3422–3431 (2015).

For information on ζ-Mn_2O_3 and $AA'_3B_4O_{12}$ perovskites see:
S. V. Ovsyannikov *et al.*, *Angew. Chem. Int. Ed.*, **52**, 1494–1498 (2013).

Recent reviews of oxynitride perovskites are:
S. G. Ebbinghaus *et al.*, *Prog. Solid State Chem.*, **37**, 173–205 (2003).
A. Fuertes, *J. Mater. Chem.*, **22**, 3293–3299 (2012).

To start on the extensive literature of substituted perovskite phases including brownmillerites, consult the references in:
C. Didier, J. Claridge, M. Rosseinsky, *J. Solid State Chem.*, **218**, 38–43 (2014)
T. G. Parsons, H. D'Hondt, J. Hadermann, M. A. Hayward, *Chem. Mater.*, **21**, 5527–5578 (2009).
F. Ramezanipour *et al.*, *J. Am. Chem. Soc.*, **134**, 3215–3227 (2012).

For details of the $Sr_{4+n}Mn_{4+n}O_{10+3n}$ homologous series of manganites see:
L. Suescun, B. Dabrowski, *Acta Crystallogr.*, **B64**, 177–186 (2008).

For details of the crystal chemistry of especially Cu, Mn and Co containing perovskites see:
B. Raveau, *Ang. Chem. Int. Ed.* **52**, 167–175 (2013) and references therein.

3

Hexagonal Perovskite–Related Structures

This chapter deals with structures that, instead of cubic close packing of AX_3 layers, have hexagonal close packing of AX_3 layers, or sequences of hexagonal and cubic AX_3 close-packed layers. These structures go under the general title of hexagonal perovskites. The large number of known hexagonal perovskites makes a classification in terms of composition difficult. In this chapter the phases are organised mainly in terms of structures, but unifying aspects of stoichiometry are included as often as possible.

3.1 The $BaNiO_3$ Structure

The $BaNiO_3$ structure (often called the 2H—$BaNiO_3$ or just the 2H structure) is the hexagonal analogue of the ideal cubic perovskite structure (Section 1.2).

$BaNiO_3$: hexagonal; $a = 0.5629$ nm, $c = 0.4811$ nm; $Z = 2$; space group, $P6_3/mmc$ (194);

$$\begin{aligned}\text{Atom positions: } & \text{Ba}: 2(d)\tfrac{1}{3},\tfrac{2}{3},\tfrac{3}{4};\tfrac{2}{3},\tfrac{1}{3},\tfrac{1}{4};\\ & \text{Ni}: 2(a)0,0,0;0,0,\tfrac{1}{2};\\ & \text{O}: 6(h)x,2x,\tfrac{1}{4};-2x,-x,\tfrac{1}{4};x,-x,\tfrac{1}{4};\\ & \quad -x,-2x,\tfrac{3}{4};2x,x,\tfrac{3}{4};-x,x,\tfrac{3}{4};x=0.1462\end{aligned}$$

Perovskites: Structure–Property Relationships, First Edition. Richard J. D. Tilley.

The (unusual) Ni^{4+} ions lie at the corners of the unit cell and are surrounded by an octahedron of O^{2-} ions (Figure 3.1a). The structure can thus be envisaged as columns of *face-sharing* BX_6 octahedra interleaved with chains of Ba^{2+} ions, both running parallel to the **c**-axis (Figure 3.1b–d).

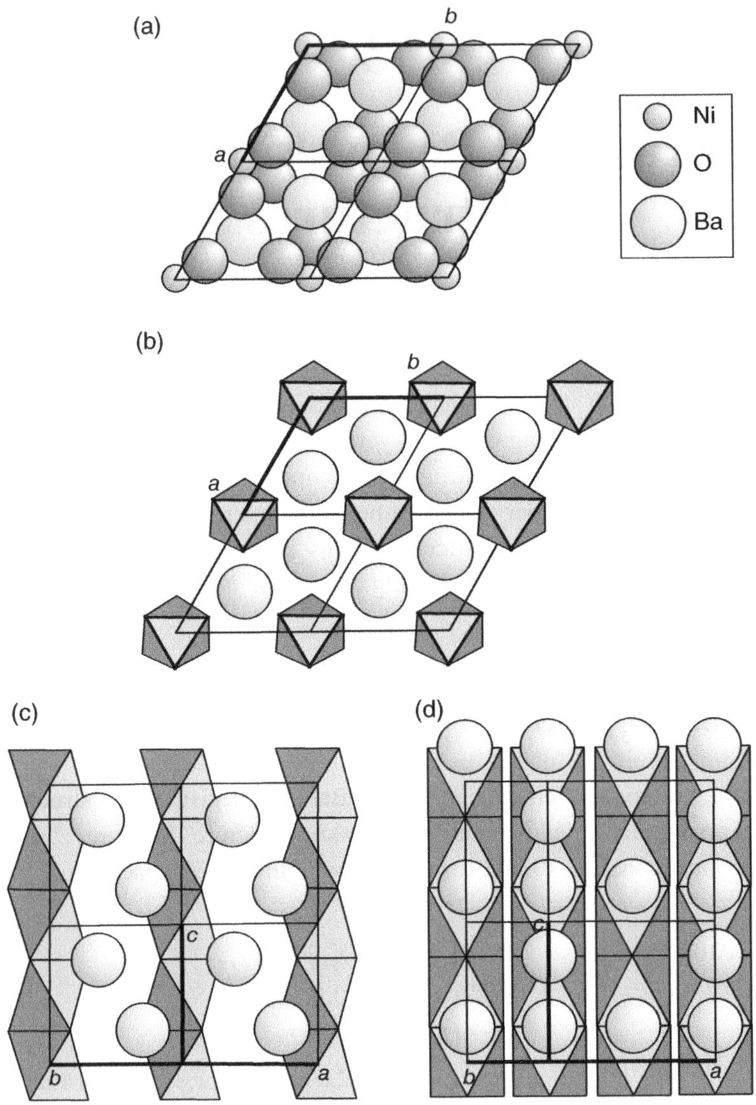

Figure 3.1 *The 2H–$BaNiO_3$ structure: (a) 2×2 unit cells projected down [001]; (b) as (a), emphasising the columns of NiO_6 face-sharing octahedra; (c) projection down [110]; (d) projection down [210]. In (c and d) the **a**- and **b**-axes do not lie in the plane of the page*

Table 3.1 *Phases with the 2H–$BaNiO_3$ structure*

Phase	Space group	*a* (nm)	*c* (nm)
$BaNiO_3$	H, $P6_3/mmc$ (194)	0.5629	0.4811
$SrCoO_3$	H, $P6_3/mmc$ (194)	0.5486	0.420
$Sr_6Co_5O_{15}$	Tr, $R32$ (135)	0.94808	1.24353
$BaCoO_3$	H, $P6_3/mmc$ (194)	0.56525	0.47629
$BaMnO_3$	H, $P6_3/mmc$ (194)	0.56991	0.48148
$SrTiS_3$	H, $P6_3/mmc$ (194)	0.6577	0.5696
$BaTiS_3$	H, $P6_3/mmc$ (194)	0.6743	0.5832
$BaTaS_3$	H, $P6_3/mmc$ (194)	0.6826	0.5776
$BaTiSe_3$	H, $P6_3/mmc$ (194)	0.7054	0.6033
$KNiCl_3$	H, $P6_3cm$ (185)	1.1795	0.5926
$RbCoCl_3$	H, $P6_3/mmc$ (194)	0.6999	0.5996
$CsMgCl_3$	H, $P6_3/mmc$ (194)	0.7269	0.6187
$RbNiBr_3$	H, $P6_3/mmc$ (194)	0.7268	0.6208

Viewed as a layer stacking, the Ba^{2+} and O^{2-} positions in the $BaNiO_3$ structure are identical to that of the Au and Cu positions in the alloy Cu_3Au and the SrO_3 layers in $SrTiO_3$ (Figure 1.2a and b). In $BaNiO_3$ these layers are stacked in a two-layer hexagonal close-packed sequence ...ABAB..., also designated as h (Sections 3.3 and 3.4). As with the cubic perovskite phase, the charge balance needed to maintain neutrality in this skeleton structure is provided by an ordered distribution of the Ni^{4+} ions in the available octahedral interstices that are bounded by O^{2-} ions only (Figure 1.3).

The occurrence of oxide $BaNiO_3$-type structures can be interpreted via the tolerance factor. When $t = 1.0$ the ideal cubic perovskite structure is stable, and as t drops below 1.0 the octahedral framework distorts, but the apex-sharing arrangement of BX_6 octahedra remains, as in $CaMnO_3$, $t = 0.987$, with a distorted cubic structure. For values of t above 1 the hexagonal $BaNiO_3$ structure type is preferred, in which the apex-sharing arrangement is exchanged for face sharing, exemplified by $BaMnO_3$, $t = 1.089$, with the $BaNiO_3$ structure (Table 3.1). Most phases are stoichiometric, but $Sr_6Co_5O_{15}$, formed when nominally $SrCoO_3$ with the 2H–$BaNiO_3$ structure is slow cooled to room temperature, has a defect 2H- structure with a Co deficit and a tripled **c**-axis.

3.2 $BaNiO_3$-Related Phases Containing Trigonal Prisms

3.2.1 Commensurate Structures

A number of phases are known that are closely related to the 2H–$BaNiO_3$ structure in that they are built from columns of approximate composition BX_3 and chains of A atoms, both running parallel to the **c**-axis of the structure. Ideally, all of these phases can be described in terms of a hexagonal unit cell with *a* and *b* lattice parameters of approximately $\sqrt{3} \times a_{2H}$ (0.97 nm) in length, where a_{2H} is the hexagonal 2H—$BaNiO_3$ lattice parameter. Each unit cell contains three columns and six chains of A atoms (Figure 3.2). The *c*-parameter mainly depends upon the arrangement of polyhedra within and between the columns (Table 3.2).

The structures can be described in terms of *two* types of hexagonally close-packed layers with ...AB... stacking that lie normal to the **c**-axis. The first of these has a composition (A_3O_9), with A mainly an alkaline earth metal cation (Ca^{2+}, Sr^{2+} or Ba^{2+}) and is similar to the BaO_3 layers that make up 2H—$BaNiO_3$ (Figure 3.3a). The second type, which leads to anion deficiency compared to 2H—$BaNiO_3$, is formed from the first by replacing three oxygen ions by an alternative cation A′ to form ($A_3A'O_6$) layers where A′ can come from across the periodic table (Figure 3.3b).

The stacking of (A_3O_9) layers in hexagonal close packing creates octahedral interstices between each pair of layers (Figure 3.4a and b), which, when occupied by B cations, generate the 2H—$BaNiO_3$-type structure, with columns composed only of face-sharing octahedra ····o-o-o-o····. Stacking ($A_3A'O_6$) layers in a similar way produces trigonal prisms, which are occupied by the A′ cations (Figure 3.4c and d), and octahedral interstices, which, when occupied by B cations, give rise to columns consisting of alternating face-sharing BO_6 octahedra and $A'O_6$ trigonal prisms in the sequence ····o-tp-o-tp-o-tp-o-tp··· (Figure 3.5a). The formula of the phases is $A_3A'BO_6$. This arrangement occurs in the K_4CdCl_6 structure, where K occupies the A and A′ sites and is adopted by a number of phases (Table 3.2) including Sr_3ZnCoO_6, in which A′ is Zn^{2+} occupying trigonal prisms and B is Co^{4+} occupying octahedra and $Ca_3Co_2O_6$, in which Co^{3+} takes the role of both A′ and B (Figure 3.5b). Note that although the sequence of octahedra and trigonal prisms is the same in each column, there is an offset between columns in the unit cell.

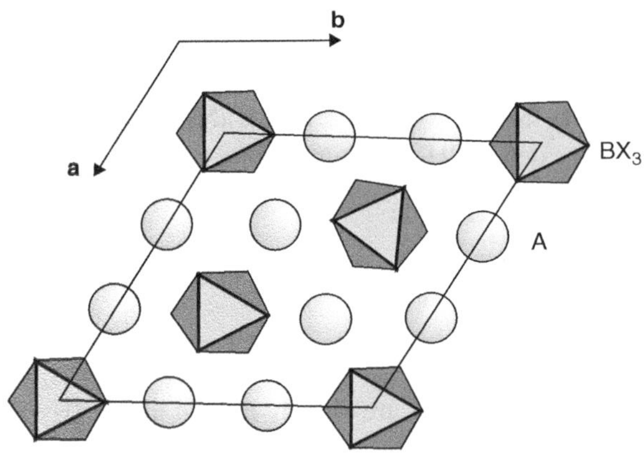

Figure 3.2 *The idealised structure of a number of 2H—$BaNiO_3$-related phases projected down [001]*

Table 3.2 *$BaNiO_3$-related structures containing trigonal prisms*

Phase	Space group	a/nm	c/nm
Commensurate structures			
$n'=1$, -o-tp-, ($m=0$, $n=1$), $A_3A'BO_6^a$			
K_4CdCl_6	Tr, $R\bar{3}c$ (167)	1.2105	1.4909
Ca_3NiMnO_6	Tr, $R\bar{3}c$ (167)	0.90958	1.05671
Ca_3FeRhO_6	Tr, $R\bar{3}c$ (167)	0.91884	1.07750
$Ca_4IrO_6^b$	Tr, $R\bar{3}c$ (167)	0.9303	1.10864
$Ca_3Co_2O_6^c$	Tr, $R\bar{3}c$ (167)	0.90793	1.0381
$Ca_4PtO_6^b$	Tr, $R\bar{3}c$ (167)	0.9332	1.1264
Ca_3NaIrO_6	Tr, $R\bar{3}c$ (167)	0.9273	1.1214
$Sr_4IrO_6^b$	Tr, $R\bar{3}c$ (167)	0.97344	1.1892
$Sr_4PtO_6^b$	Tr, $R\bar{3}c$ (167)	0.97477	1.18791
Sr_3ZnCoO_6	Tr, $R\bar{3}c$ (167)	0.955191	1.088790
Sr_3NiRhO_6	Tr, $R\bar{3}c$ (167)	0.95951	1.1062
Sr_3GdRhO_6	Tr, $R\text{-}3c$ (167)	0.97840	1.14196
$Sr_3Co_2O_6^c$	Tr, $R\bar{3}c$ (167)	0.961107	1.0701
Sr_3HoCrO_6	Tr, $R\bar{3}c$ (167)	0.97802	1.13063
Sr_3NiIrO_6	Tr, $R\bar{3}c$ (167)	0.95783	1.11323
Sr_3NaIrO_6	Tr, $R\bar{3}c$ (167)	0.96383	1.15852
$Ba_4PtO_6^b$	Tr, $R\bar{3}c$ (167)	1.02101	1.26172
Ba_3NaIrO_6	Tr, $R\bar{3}c$ (167)	1.01282	1.1905
Ba_4USe_6	Tr, $R\bar{3}c$ (167)	1.225997	1.56067
Ba_3UFeSe_6	Tr, $R\bar{3}c$ (167)	1.24842	1.39258
Ba_3UMnSe_6	Tr, $R\bar{3}c$ (167)	1.24561	1.41354
Ba_3UMnS_6	Tr, $R\bar{3}c$ (167)	1.20220	1.36410
Ba_3PrInS_6	Tr, $R\bar{3}c$ (167)	1.21196	1.39803
Ba_3SmInS_6	Tr, $R\bar{3}c$ (167)	1.2063	1.3853
Ba_3GdInS_6	Tr, $R\bar{3}c$ (167)	1.20718	1.3816
Ba_3YbInS_6	Tr, $R\bar{3}c$ (167)	1.20565	1.3713
$n'=2$, -o-o-tp-, ($m=1$, $n=3$), $A_4A'B_2O_9$			
$Sr_4Ni_3O_9^c$	Tr, $P321$ (150)	0.9447	0.7825
$Sr_4NiMn_2O_9$	Tr, $P321$ (150)	0.96007	0.77646
$Sr_4CuMn_2O_9$	Tr, $P321$ (150)	0.95918	0.78114
$Sr_4ZnMn_2O_9$	Tr, $P321$ (150)	0.95679	0.7851
$Ba_4UCr_2O_9$	Tr, $P321$ (150)	1.201	0.9644
$n'=3$, -o-o-o-tp-, ($m=2$, $n=3$), $A_5A'B_3O_{12}$			
$(Sr_{0.8}Ca_{0.2})_5Co_4O_{12}^c$	Tr, $P\bar{3}c1$ (165)	0.94196	1.99857
$Ba_5CuIr_3O_{12}$	Tr, $P3c1$ (158)	1.01406	2.16599
$Ba_5CuIr_3O_{12}$	Tr, $P321$ (150)	1.014382	2.16553
$n'=4$, -o-o-o-o-tp-, ($m=1$, $n=1$), $A_6A'B_4O_{15}$			
$Sr_6Co_5O_{15}^c$	Tr, $R32$ (155)	0.9497	1.23956
$Sr_6Co_5O_{14.70}^c$	Tr, $R\bar{3}$ (148)	0.9459	1.2469
$Sr_6Co_{4.9}Ni_{0.1}O_{14.36}^c$	Tr, $R\bar{3}$ (148)	0.9440	1.2476
$Ba_6CuIr_4O_{15}$	Tr, $R32$ (155)	1.01196	1.34097
$Ba_6CuMn_4O_{15}$	Tr, $R32$ (155)	1.00211	1.28241
$Ba_6Ni_5O_{15}^c$	Tr, $R32$ (155)	0.98890	1.2867
$Ba_6MgMn_4O_{15}$	Tr, $R32$ (155)	1.0042	1.2954
$Ba_6NiMn_4O_{15}$	Tr, $R32$ (155)	1.0044	1.2924
$Ba_6ZnMn_4O_{15}$	Tr, $R32$ (155)	1.00498	1.28304

(continued)

Table 3.2 (continued)

Phase	Space group	a/nm	c/nm
$n'=7$, -o-o-o-o-o-o-o-tp-, ($m=2$, $n=1$), $A_9A'B_7O_{24}$			
$Ba_9Rh_8O_{24}^c$	Tr, $R\bar{3}c$ (167)	1.00899	4.1462
$n'=2,3$ intergrowth, -o-o-tp-o-o-o-tp- ($m=1$, $n=2$), $A_9A'_2B_5O_{21}$			
$Sr_9Ni_7O_{21}^c$	Tr, $R\bar{3}c$ (167)	0.9524	3.6008
$n'=2,3,3$ intergrowth, -o-o-tp-o-o-o-tp-o-o-o-tp-, ($m=5$, $n=9$)			
$Ba_{14}Cu_3Ir_8O_{33}$	Tr, $P321$ (150)	1.014585	2.99574
$n'=3,3,4$ intergrowth, -o-o-o-tp-o-o-o-o-tp-o-o-o-o-tp-, ($m=7$, $n=9$), $A_{16}A'_3B_{10}O_{39}$			
$Ba_{16}Cu_3Ir_{10}O_{39}$	Tr, $P321$ (150)	1.013442	3.50564
Modulated structures			
Approximately $n'=2$, o-o-tp-, ($m=1$, $n=3$), $A_4A'B_2O_9$			
$Ca_4CuMn_2O_9$		0.9195	$c_1=0.2580$, $c_2=0.3744$
$Sr_4CuMn_2O_9$		0.9602	$c_1=0.2593$, $c_2=0.3940$
Approximately $n'=4$, -o-o-o-o-tp-, ($m=1$, $n=1$), $A_6A'B_4O_{15}$			
$SrMn_{0.6}Co_{0.4}O_{3-y}$		0.9576	$c_1=0.2548$, $c_2=0.4032$
$Ba_6ZrIr_4O_{15}$		1.01228	$c_1=0.26982$, $c_2=0.44099$
$Ba_6MgMn_4O_{15}$		1.0042	$c_1=0.2565$, $c_2=0.4318$
$Ba_6NiMn_4O_{15}$		1.0044	$c_1=0.2551$, $c_2=0.4308$
$Ba_6Mn_4PdO_{15}$		1.0030	$c_1=0.2616$, $c_2=0.4251$
Approximately $n'=2,3$ intergrowth, -o-o-tp-o-o-o-tp-, ($m=1$, $n=2$), A9A'2B5O21			
$Sr_{14/11}CoO_3$		0.9508	$c_1=0.2534$, $c_2=0.3982$
$Sr_{1.2872}NiO_3$		0.95177	$c_1=0.2574$, $c_2=0.3999$
$Ba_8CoRh_6O_{21}$		1.00431	$c_1=0.25946$, $c_2=0.45405$

Notes: [a] n, m refer to equation 3.1, n' refers to equation 3.6; [b] $A'=A$; [c] $A'=B$; $c_1=MX_3$ column repeat, c_2=A chain repeat

A number of Cu containing phases Ca_3CuIrO_6, Ca_3CuRhO_6 and Sr_3CuRhO_6 have distorted versions of the structure with monoclinic unit cells. This change is the result of the displacement of the Cu^{2+} A′ cations from the centres of the trigonal pyramidal sites to an approximately square trigonal prism face. This allows Cu^{2+} to adopt square pyramidal coordination, also favoured in the $AA'_3B_4O_{12}$ perovskite phases (Section 2.1.3).

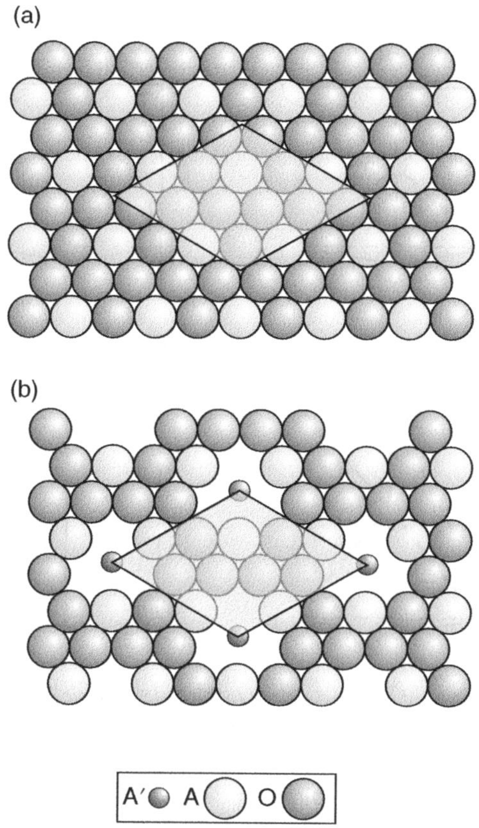

Figure 3.3 *The hexagonal layer packing in a number of 2H–$BaNiO_3$-related phases: (a) composition (A_3O_9); (b) composition ($A_3A'O_6$)*

Other structures can be constructed by stacking m (A_3O_9) sheets with n ($A_3A'O_6$) layers in a hexagonal close-packed arrangement, with A′ cations occupying the trigonal prisms and B cations occupying the octahedral sites that are generated (Table 3.2). The series of phases so produced have compositions given by

$$A_{3m+3n}A'_{n}B_{3m+n}O_{9m+6n} \tag{3.1}$$

In this formula A is mainly an alkaline earth metal cation, Ca^{2+}, Sr^{2+} or Ba^{2+}, while A′ and B can be taken from a wide range of metals. The columns in these phases consist of face-sharing octahedra and trigonal prisms and are separated by chains of A cations. The ratio of the trigonal prismatic sites to octahedral sites, tp/o, is given by the ratio of A′ to B, so that

$$\frac{\text{tp}}{\text{o}} = \frac{n}{(3m+n)} \tag{3.2}$$

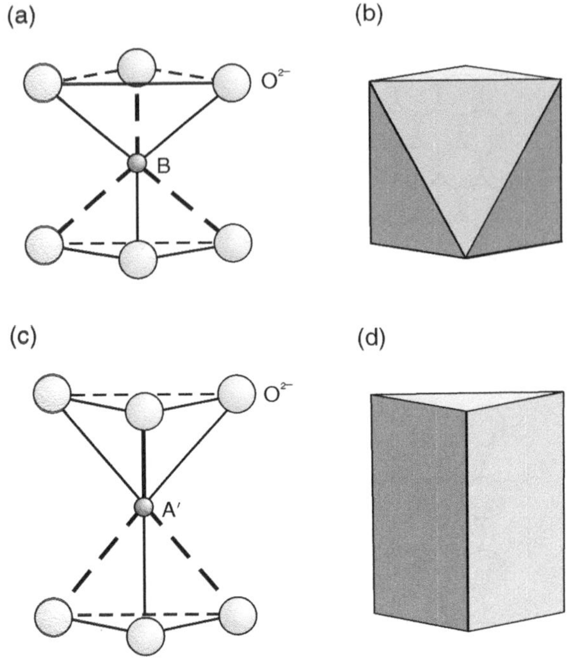

Figure 3.4 *The creation of (a and b) octahedral interstices and (c and d) trigonal prismatic interstices; by stacking of hexagonal close-packed layers of anions*

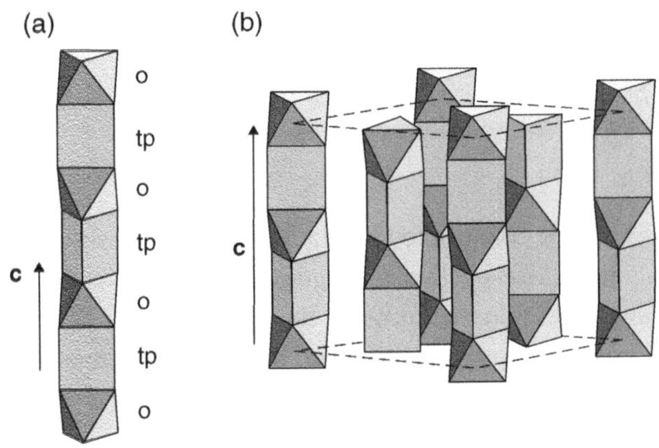

Figure 3.5 *The idealised $A_3A'BO_6$ (K_4CdCl_6) structure: (a) skeleton of columns of octahedra and trigonal prisms; (b) arrangement of columns within a unit cell. The A cations are omitted for clarity*

As in the $A_3A'BO_6$ structures, all columns are taken to have the same sequence of octahedra and trigonal prisms, but there is an offset between the different columns in the unit cell. Note, though, that although the ratio of trigonal prisms to octahedra is easily calculated, the arrangement of these polyhedra along the columns is not intuitively obvious, particularly for higher values of m and n.

In terms of Equation (3.1) all the structures can be regarded as lying between ($m=1$, $n=0$) 2H—$BaNiO_3$ structure type, with tp/o = 0, and ($m=0$, $n=1$) $A_3A'BO_6$ (K_4CdCl_6) structure type, with tp/o = 1. Other structures can then be described:

1. The structure generated by ($m=1$, $n=1$) has an overall composition of $A_6A'B_4O_{15}$. The columns have tp/o = ¼, giving a likely column sequence of ···-o-o-o-o-tp-o-o-o-o-tp-··· (Figure 3.6a), and is adopted by $Ba_6MgMn_4O_{15}$ and $Ba_6CuIr_4O_{15}$.
2. The structure generated by ($m=1$, $n=2$) has a composition $A_9A'_2B_5O_{21}$ and tp/o = ⅖. The most likely sequence of octahedra and trigonal prisms

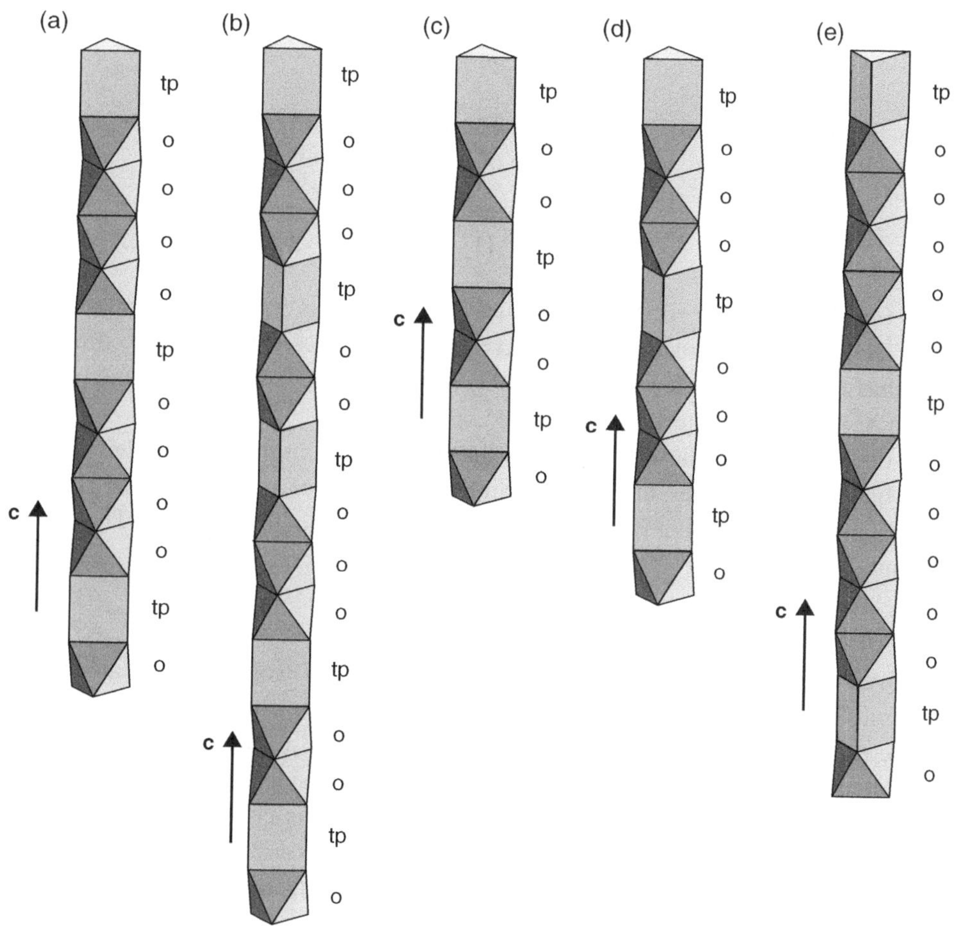

Figure 3.6 *Structures composed of columns of trigonal prisms (tp) and octahedra (o): (a) $A_6A'B_4O_{15}$; (b) $A_9A'_2B_5O_{21}$; (c) $A_4A'B_2O_9$; (d) $A_9A'B_3O_{12}$; (e) $A_7A'B_5O_{18}$*

is ···tp-o-o-tp-o-o-o-tp··· with alternating blocks of two and three octahedra (Figure 3.6b). This structure is found in $Sr_9Ni_7O_{21}$, where both A′ and B are Ni.

3. The phases generated by ($m=1, n=3$) have a composition $A_4A'B_2O_9$ and tp/o = ½, giving a likely column structure of ···tp-o-o-tp-o-o-tp··· (Figure 3.6c). This arrangement is adopted by the phases $Sr_4NiMn_2O_9$ and $Ba_4Cr_2US_9$.

The series formula Equation (3.1) can also be rewritten in terms of the conventional perovskite formula as

$$A_{1+x}\left(A'_xB_{1-x}\right)O_3 \tag{3.3}$$

where

$$x = \frac{n}{\left(3m+2n\right)} \tag{3.4}$$

and can take values between 0 and ½. The ratio of the trigonal prismatic sites to octahedral sites, tp/o, given by the ratio of A′ to B, is

$$\frac{\text{tp}}{\text{o}} = \frac{x}{\left(1-x\right)} \tag{3.5}$$

The ($m=1$, $n=0$) 2H–$BaNiO_3$ structure type then has $x=0$, the ($m=0$, $n=1$) $A_3A'BO_6$ (K_4CdCl_6) structure type has $x=½$, the ($m=1$, $n=1$) $A_6A'B_4O_{15}$ structure type has $x=⅕$, the ($m=1$, $n=2$) $A_9A'_2B_5O_{21}$ structure type has $x=2/7$, and the ($m=1$, $n=3$) $A_9A'_2B_5O_{21}$ structure type has $x=3/9$.

A disadvantage of these formulae is that there is no intuitively simple correlation between the values of m and n and the number of octahedra and trigonal prisms in each column, or the arrangement of these within the columns. The series formula

$$A_{n'+2}A'B_{n'}O_{3n'+3} \tag{3.6}$$

resolves this for cases in which the columns contain isolated trigonal prisms separated by n' adjacent face-sharing octahedral sites containing a B cation. (Note that n′ is *not* identical to n in Equation (3.1) and following equations.) In this alternative description, the various phases are considered to lie between the $n'=1$ $A_3A'BO_6$ (K_4CdCl_6) structure type with columns composed of alternate octahedra and trigonal prisms (Figure 3.5) and the $n'=\infty$ 2H–$BaNiO_3$ structure type with columns exclusively made up of face-sharing octahedra (Figure 3.1c and d).

Broadly speaking, the phases which form depend upon the size of the alkaline earth cation: Ca^{2+} only forms $n'=1$ structures, Sr^{2+} and Ba^{2+} form $n'=1$, 2, 3 and 4 structures, and in addition Ba^{2+} is also able to form phases with n' greater than 4. The simpler compounds that can be described in this way are:

1. $n'=1$. These phases, with a composition $A_3A'BO_6$ and column sequence ···tp-o-tp-o-tp-o-tp-···, are adopted by Ca_3NiMnO_6 (K_4CdCl_6) analogues, described earlier (Figure 3.5). The simplest lattice parameters of these phases are all close to $a \approx 0.97$ nm, $c \approx 1.13$ nm.
2. $n'=2$. These phases contain columns of alternating trigonal prisms and pairs of octahedra, sharing end faces in the sequence ···tp-o-o-tp-o-o-tp-··· (Figure 3.6c). The composition of the phases is $A_4A'B_2O_9$, typified by $Sr_4ZnMn_2O_9$. The ideal lattice parameters of these compounds are all close to $a \approx 0.97$ nm, $c \approx 0.78$ nm.
3. $n'=3$. These phases contain columns of alternating trigonal prisms and triplets of octahedra, sharing end faces in the sequence ···tp-o-o-o-tp-o-o-o-tp-··· (Figure 3.6d). The composition of the phases is $A_5A'B_3O_{12}$, represented by $Ba_5PdMn_3O_{12}$. The ideal lattice parameters of these phases are all close to $a \approx 0.97$ nm, $c \approx 2.16$ nm.
4. $n'=4$. These phases contain columns consisting of four face-sharing octahedra separated by a trigonal prism in the sequence ···tp-o-o-o-o-tp-o-o-o-o-tp··· (Figure 3.6a). The composition of the phases is $A_6A'B_4O_{15}$, represented by $Ba_6MgMn_4O_{15}$. The ideal lattice parameters of these phases are all close to $a \approx 0.97$ nm, $c \approx 1.25$ nm.
5. $n'=5$. These phases contain columns consisting of five face-sharing octahedra separated by a trigonal prism in the sequence ···tp-o-o-o-o-o-tp-o-o-o-o-o-tp-··· (Figure 3.6e). The composition is $A_7A'B_5O_{18}$, represented by $Ba_7PdMn_5O_{18}$. The ideal lattice parameters of these phases are all close to $a \approx 0.97$ nm, $c \approx 2.80$ nm.

It is also possible to envisage intergrowths between any two adjacent n' values. For example, a 1 : 1 intergrowth between an $n'=2$ and $n'=3$ structure results in columns built of a stacking of three face-sharing octahedra, one trigonal prism, and two face-sharing octahedra, one trigonal prism in the sequence ···tp-o-o-tp-o-o-o-tp-···(Figure 3.6b). The composition of the phases is $A_9A'_2B_5O_{21}$, typified by $Sr_9Ni_7O_{21}$, where A′ and B both are Ni. The ideal lattice parameters of these phases are all close to $a \approx 0.97$ nm, $c \approx 3.60$ nm.

3.2.2 Modulated Structures

For a number of 2H—$BaNiO_3$-related phases an accurate structural picture is only obtained if they are described in terms of two mutually interacting subsystems. This also applies to some non-stoichiometric phases such as $Sr_{1.2872}NiO_3$ and the $Sr_xTiS_{3-\delta}$ sulphides, which superficially appear to be normal 2H—$BaNiO_3$-type structures. In all of these materials the subsystems are (i) the $(A'B)O_3$ columns and (ii) the A chains. In the (001) plane normal to the **c**-axis, there is perfect crystallographic order (Figure 3.2). However, the spacing of the atoms in the columns and the chains parallel to the **c**-axis does not follow a strict crystallographic repeat distance, but is modulated into a wave-like repetition, the wavelength of which is a function of composition.

The two subsystems are conveniently described in terms of their individual *c*-parameters:

c_1: $(A'B)O_3$ columns, spacing approximately $c_{2H}/2 \approx 0.24$ nm
c_2: A chains, spacing approximately $c_{2H} \approx 0.48$ nm

where c_{2H} is the hexagonal 2H—$BaNiO_3$ lattice parameter. In these systems, both subsystems contribute to the reflections on a diffraction pattern. In such cases the indices are described not by conventional Miller indices, but by a four index extension, so that each reflection is indexed as $ha^* + kb^* + lc_1^* + mc_2^*$. The true values of c_1 and c_2 are obtained from diffraction patterns and can be written as

$$\gamma = \frac{c_2^*}{c_1^*} = \frac{c_1}{c_2} \tag{3.7}$$

where c_2^* and c_1^* are reciprocal distances measured directly from the diffraction patterns and *h*, *k*, *l* and *m* are integers. Moreover, the ratio γ can be directly related to the composition given by Equation (3.3): thus

$$\gamma = \frac{(1+x)}{2} \tag{3.8}$$

As *x* can take values of 0 – ½, γ can take values between ½ and ¾. Moreover, the ratio of trigonal prisms to octahedra in the columns, tp/o, given by Equation (3.5), can be written as

$$\frac{\text{t}}{\text{op}} = \frac{x}{(1-x)} = \frac{(2\gamma-1)}{2(1-\gamma)} \tag{3.9}$$

In cases where the two *c*-parameters come into perfect register after small distances, the structure is described as a *commensurate modulated composite structure*. In such circumstances, the value of γ is a rational fraction, *p*/*q* where *p* and *q* are (small) integers and the *c* lattice parameter of the phase is given by

$$c = qc_1 = pc_2$$

The structures described in Section 3.2.1 by and large fall into this group. As an example, the ideal $m=1, n=3$ (Equation 3.1) or $n'=2$ (Equation 3.6) $A_4A'B_2O_9$ phases are commensurate with $x=\frac{1}{3}$ (Equation 3.4) and $\gamma=\frac{2}{3}$ (Equation 3.8). Ideally

$$c = 3c_1 = 2c_2$$

In some structures, although the coincidence specified by ($qc_1 = pc_2$) is not strictly correct, it is possible to assume this is not too inaccurate an approximation, so that a conventional unit cell *c*-parameter can be described. This is the case with the phase $Sr_{1.273}CoO_3$, where the measured parameters are as follows: $a = 0.9508$ nm, c_1 (CoO_3) = 0.25343 nm, c_2 (Sr) = 0.3982 nm. The measured value of γ (c_1/c_2) is 0.6364, which is reasonably close to $7/11$, so that the phase can be approximated to a commensurate modulated structure, with a *c* lattice parameter given by $c = 11 \times 0.25343 \approx 7 \times 0.3982 = 2.79$ nm. As $\gamma = 7/11$, $(1+x) = 14/11$, and the formula of the phase can then be approximated as $Sr_{14/11}CoO_3$ or $Sr_{14}Co_{11}O_{33}$. The ratio of trigonal prisms to octahedra is given by Equation (3.9) to be, tp/o = $3/8$. In agreement with this, the sequence of the octahedra and trigonal prisms is found to be ···tp-o-o-o-tp-o-o-tp-o-o-o-tp···.

In cases where c_1 and c_2 fail to come into register at all, that is, when *p*/*q* is irrational, or at least when *q* is large, the structures are called *incommensurate modulated composite structures*. In order to analyse these incommensurate structures, a more complex *superspace* description needs to be adopted, which involves the introduction of a modulation vector $\mathbf{q} = \gamma\mathbf{c}$, where **c** is a vector corresponding to the basic repeat along the **c**-axis, corresponding to c_1.

One of the structural consequences of the modulations in incommensurate phases is to distort the octahedra and trigonal prisms along the $(A'B)O_3$ columns so that they cannot be accurately described either as an octahedron or a trigonal prism. These somewhat indeterminate structures are often called *intermediate polyhedra* (ip). The sulphide of composition $Sr_{1.125}TiS_3$ is an example. This phase has the following lattice parameters: $a = 1.1482$ nm, c_1 (TiS_3) = 0.29483 nm, c_2(Sr) = 0.53057 nm. This cannot be described in terms of a commensurate cell as $\gamma = c_1/c_2 = 0.555685 = 1427/2565$. The *c* lattice parameter is then approximately $c = 2565 \times c_1 = 1427 \times c_2 \approx 756$ nm. The only relatively simple conventional structure that can be assigned is to approximate $\gamma = 9/16$. If this gross approximation is made, $c = 16 \times c_1 = 9 \times c_2 \approx 4.78$ nm. Using Equations (3.3) and (3.8), $(1+x) = 2\gamma = 9/8$, which gives a phase of ideal formula $Sr_9Ti_8O_{24}$, with $a = 1.148$ nm, $c = 4.78$ nm. Ideally, it is possible to calculate the relative numbers of octahedra and trigonal prisms, via Equation (3.9), as tp/o = $1/7$. That is, the ideal chains are made up of seven octahedra and one trigonal prism ····o-o-o-o-o-o-o-tp-o-o-o-o-o-o-o-tp··· (Figure 3.7a). However, the modulations in the Sr chains have not been taken into account in this approximation, and when they are it is found that the trigonal prisms are distorted to the extent that the sequence is better given by five octahedra along the chain, separated by a triplet consisting of two intermediate polyhedra and one octahedron which is in a sideways on orientation (o′) compared with the others: thus ····o-o-o-o-o-ip-o′-ip-o-o-o-o-o-ip-o′-ip-··· (Figure 3.7b). Even this picture, though, is approximate, as the real repeat distance is closer to 756 nm than 4.78 nm, so that the spacing of the ip-o′-ip triplets varies from the five octahedra repetition suggested over a much longer cycle.

It can be noted that even small changes in stoichiometry can have significant effects upon these composite modulated structures. For example, although the

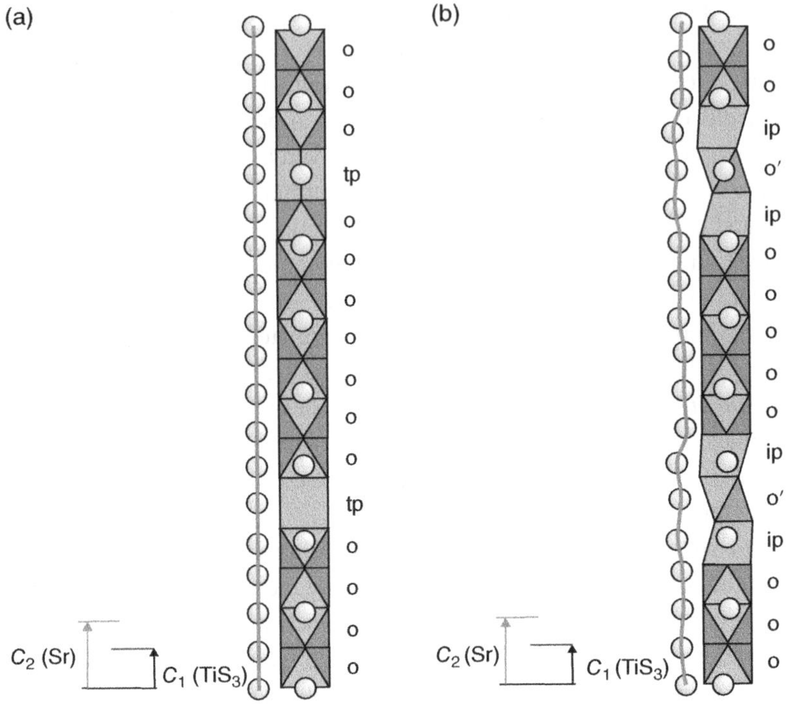

Figure 3.7 *Structures of $Sr_{1.125}TiS_3$: (a) ideal unmodulated $Sr_9Ti_8O_{24}$; (b) ideal modulated $Sr_9Ti_8O_{24}$*

commensurate $m=1$, $n=1$ (Equation 3.1) or $n'=4$ (Equation 3.6) phase $Sr_6Co_5O_{15}$ can be accurately described as having four octahedra followed by one trigonal prism, the oxygen-deficient counterpart $Sr_6Co_5O_{14.70}$ contains columns in which three octahedra are followed by two distorted intermediate polyhedra midway between an octahedron and a trigonal prism. The $A_3A'BO_6$ phases Sr_3NiRhO_6 and Sr_3CuRhO_6 are all able to gain oxygen on heating in air. These materials no longer show the conventional unit cell and symmetry of the stoichiometric parent, but incommensurate modulated structures. Similarly, small changes in the compositions of phases such as the strontium titanium sulphides described earlier can induce considerable changes in the column structure parallel to the **c**-axis.

3.3 Perovskites with Mixed Hexagonal/Cubic Packing: Nomenclature

Both the 2H—$BaNiO_3$ and $SrTiO_3$ structures can be described in terms of AX_3 close-packed layers. There are very small energetic differences between these two arrangements and crystals composed of ordered or disordered interleaved hexagonal and cubic packing of the layers readily form. Ideally such structures will have a hexagonal or rhombohedral symmetry, with the hexagonal **a**- and **b**-axes lying in the plane of the layers and the hexagonal **c**-axis lying normal to the planes.

In practice the structures are often complex when details are analysed, especially with respect to stoichiometric variations. In these cases, ordering of component atoms can skew the symmetry away from hexagonal or rhombohedral to orthorhombic or monoclinic. These forms are sensitive to pressure, temperature and composition, implying that complicated phase relationships are usual. In the descriptions that follow in this and the following section, these subtleties have largely been put to one side, and it has been assumed that the structures consist of ideal layer stacking.

The simplest of these can be regarded as intergrowths between the nominal parent structures, 2H—$BaNiO_3$ type and $SrTiO_3$ type. Both of these have the same ABX_3 stoichiometry, as will all of the intergrowth variants. Such phases, characterised by variable stacking sequences, but a single composition, are termed *polytypes*. In practice, many compounds are non-stoichiometric, and change in composition is a prime factor in stabilising one structure form over another. For instance, between $BaMnO_3$ and $BaMnO_{2.5}$, at least six different structures are found, all stabilised by changes in oxygen stoichiometry. Strictly speaking these non-stoichiometric phases should not be termed polytypes, as structure is linked to changing composition, but are frequently regarded as such. Both true polytypes and non-stoichiometric phases are described together in the following sections, in terms of a set of idealised arrangements, any one of which may tolerate a range of differing compositions.

These structures are conveniently depicted by the linkage of octahedra alone, with the large A ions generally omitted. The 2H—$BaNiO_3$ structure type is then represented by single columns of face-sharing octahedra (Figure 3.8a). Similarly the ideal cubic $SrTiO_3$ structure is depicted as an array of apex-shared octahedra, viewed down [110] (Figure 3.8b) and then rotated by 55° to make the (111) planes horizontal, to bring the orientation into coincidence with that of $BaNiO_3$ (Figure 3.8c). Intergrowths of cubic and hexagonal perovskite layers can then be drawn as a sequence of fragments of these two motifs, which emphasises the fact that the structures consist of slabs of ideal perovskite cut parallel to (111) interleaved with slabs of 2H—$BaNiO_3$ type cut parallel to (001). One of the simplest ordered arrangements is a 1 : 1 stacking of hexagonal and cubic layers, found in the high-temperature oxygen-deficient form of $BaMnO_3$ with a designation 4H—$BaMnO_3$ and a composition $BaMnO_{2.65}$. The structure consists of pairs of face-sharing octahedra linked to each other by apex sharing (Figure 3.8d).

In addition to polyhedral depictions, the stacking of the AX_3 (or non-stoichiometric $AX_{3-\delta}$) layers is often represented in terms of the usual close-packed A, B and C layer designation. Thus 2H—$BaNiO_3$ has the ···ABAB··· sequence (Figure 3.8a) and $SrTiO_3$ has the ···ABC··· arrangement (Figure 3.8b and c). Any hexagonal perovskite formed between these two parent stacking sequences can then be described as a

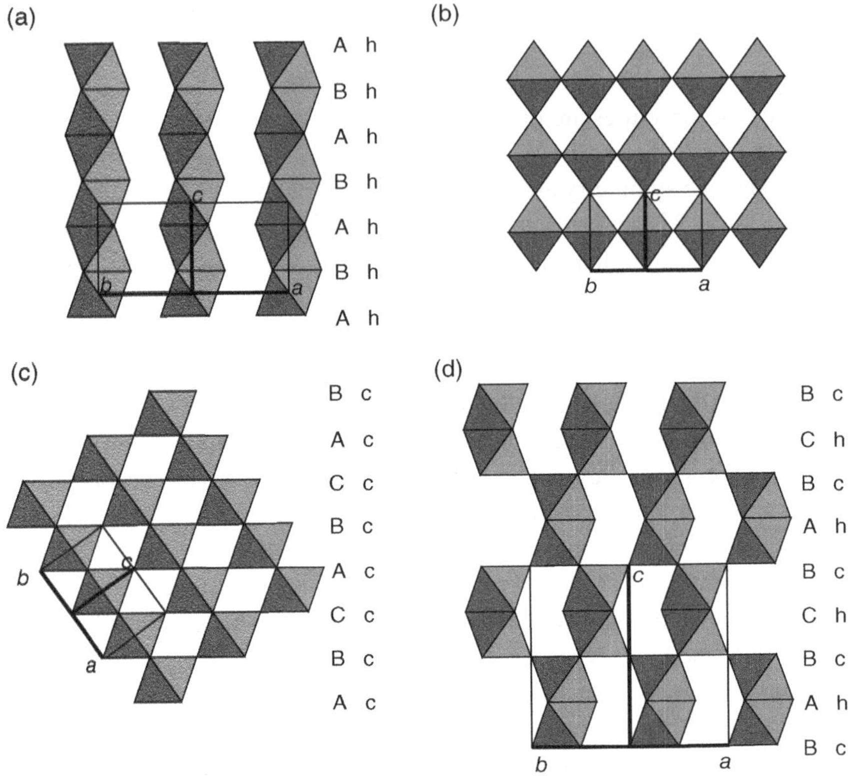

Figure 3.8 *Schematic structures: (a) 2H—BaNiO_3 projected down [110]; (b) SrTiO_3 projected down [110]; (c) as (b), rotated to bring the (111) planes parallel to the (001) planes in (a); (d) 4H—BaMn$O_{2.65}$ $(ch)_2$ projected down [110]. The BX_6 octahedral skeleton of the structures is drawn, with A ions and oxygen vacancies omitted*

series of A, B and C layers. The oxide 4H—BaMn$O_{2.65}$ has a stacking sequence ··· ABCB··· (Figure 3.8d).

The description of these frequently complex arrangements in terms of the ··· ABC··· notation is cumbersome for all except the simplest structures, and there are a number of compact forms of nomenclature in use. The most widely encountered of these is the *Ramsdell* notation, which gives the number of layers in the crystallographic repeat, together with the symmetry of the unit cell, specified by C for cubic, H for hexagonal and R for rhombohedral. Thus, as already employed, the BaNiO_3 structure type has the label 2H, as there are two layers in the hexagonal unit cell. Similarly the SrTiO_3 structure type is given the label 3C as there are three layers in the unit cell repeat distance, along [111]. The 1 : 1 packing arrangement of BaMn$O_{2.65}$ is labelled 4H, as described earlier.

In some publications this system of nomenclature has been modified so that if a structure is a slightly skewed version of an ideal type, say, with a monoclinic unit cell, the structure is labelled *n*M instead of *n*H or *n*R. In other descriptions the number of AX_3 layers in the repeat is simply specified, without recourse to the symmetry, as, for example, 12L, where L stands for layers.

A second widely used terminology relies upon specifying the relative position of an AX_3 layer in terms of its neighbours. A layer sandwiched between sheets of the same relative disposition, say, BAB or CAC, is labelled *h*, meaning that the middle layer is in hexagonal close packing with respect to its neighbours. Similarly, a layer sandwiched between sheets of different relative disposition, say, ABC or BCA, is labelled *c*, meaning that the middle layer is in cubic close packing with respect to its neighbours. The number of repeat layers in the unit cell is sometimes enclosed in parentheses to make things clearer, and in structures with repeating sequences these can be indicated by a subscript. Thus 2H—$BaNiO_3$ has the structure symbol (h) and 3C—$SrTiO_3$ (c). The oxide 4H—$BaMnO_{2.65}$ is designated as $(ch)_2$ where the subscript gives the repeat over the unit cell length.

The *Zhdanov* notation specifies the stacking sequence in yet another way. The translation between an A layer and a B layer or a B layer and a C layer can be represented by a rotation of +60° or a translation of $+\frac{1}{3}$. Reverse transformations, C to B or B to A, are then represented by −60° or $-\frac{1}{3}$. The stacking sequence of any polytype can then be written as a series of + and − signs. This, itself, does not result in a compact notation, and this is addressed by the Zhdanov symbol of the polytype, which gives the number of signs of the same type in the sequence. Thus the cubic 2C—$SrTiO_3$ perovskite structure has a stacking sequence ··· +++++ ···. The Zhdanov symbol is written as (1) rather than (∞). The hexagonal 2H—$BaNiO_3$ structure has a stacking sequence ··· +−+− ··· so the Zhdanov symbol is (11). Similarly, the high-temperature form 4H—$BaMnO_{2.65}$ has a stacking sequence ···++−−··· and has the Zhdanov symbol (22). Zhdanov symbols are particularly useful when depicting very long, complex polytypes, such as those found in the ZnS and SiC systems, but are less frequently encountered in descriptions of hexagonal perovskites.

3.4 Perovskites with Mixed Hexagonal/Cubic Packing: Stacking Sequences

In any group of stacked c and h planes given by a sequence $(c_p h_q)$, the number of arrangements of the symbols is given by *n* where

$$n = \frac{(p+q)!}{(p!q!)}$$

Thus for the sequence (cchh) there are 4!/(2!2!) or six configurations: (cchh), (chch), (hcch), (hhcc), (chhc) and (hchc). However, when each of these alternatives is concatenated with itself to form the stacking sequence appropriate to a crystal, only two distinct chains (or crystal structure types), *s*, are found: (cchh) and (chch). The number of possibilities of crystal structures for several values of *p* and *q* is given in Table 3.3. This shows that in any (ch_q) or (c_ph) sequence, regardless of the values of *p* or *q*, only one crystallographic repeat is possible but thereafter the number of non-identical stacking sequences (i.e. different crystal structures) that can arise in a stacking (c_ph_q) increases rapidly with both *p* and *q*. For example, when $p = q = 3$ there are four unique stacking sequences (i.e. structures): ccchhh, cchchh, cchhch, chchch. When $p = q = 4$ there are nine unique stacking sequences: ccccchhhh, ccchchhh, ccchhchh, ccchhhch, cchchchh, cchchhch, cchhchch, cchcchhh, cchhcchh. For $p = q = 5$ there are 26 sequences and when $p = q = 6$ there are 77.

Notice that many of these stacking sequences can be regarded as intergrowths between simpler configurations. For example, the stacking sequence (c_4h_2) has three unique structures: (ccccchh), (cchccch) and (chcccch). The structure (cchccch) is a 1 : 1 ordered intergrowth between (cch) and (ccch), and (chcccch) is a 1 : 1 ordered

Table 3.3 *Stacking sequences of c_ph_q layers*

h \ c	0	1	2	3	4	5
0		c: *n* = 1				
1	h: *n* = 1	ch: *n* = 1; *s* = 1 (ch)	c_2h: *n* = 1; *s* = 1 (cch)	c_3h: *n* = 1; *s* = 1 (ccch)	c_4h: *n* = 1; *s* = 1 (cccch)	c_5h: *n* = 1; *s* = 1 (ccccch)
2		ch_2: *n* = 1; *s* = 1 (chh)	c_2h_2: *n* = 6; *s* = 2 (cchh), (chch)	c_3h_2: *n* = 10; *s* = 2 (ccchh), (cchch)	c_4h_2: *n* = 15; *s* = 3 (ccccchh) ...	c_5h_2: *n* = 21; *s* = 3 (ccccchh) ...
3		ch_3: *n* = 1; *s* = 1 (chhh)	c_2h_3: *n* = 10; *s* = 2 (cchhh), (chchh)	c_3h_3: *n* = 20, *s* = 4 (ccchhh) ...	c_4h_3: *n* = 35; *s* = 5 (cccchhh) ...	c_5h_3: *n* = 56; *s* = 7 (ccccchhh) ...
4		ch_4: *n* = 1; *s* = 1 (chhhh)	c_2h_4: *n* = 15; *s* = 3 (cchhhh)...	c_3h_4: *n* = 35; *s* = 5 (ccchhhh)...	c_4h_4: *n* = 70; *s* = 9 (cccchhhh)...	c_5h_4: *n* = 126; *s* = 14 (ccccchhh)...
5		ch_5: *n* = 1; *s* = 1 (chhhh)	c_2h_5: *n* = 21; *s* = 3 (cchhhhh)...	c_3h_5: *n* = 56; *s* = 7 (ccchhhhh)...	c_4h_5: *n* = 126; *s* = 14 (cccchhhhh)...	c_5h_5: *n* = 252; *s* = 26 (ccccchhhhh)...

n, is the total number of arrangements of c and h; *s*, is the number of distinct crystal structures that can occur.

intergrowth between (ch) and (cccch). This description is often to be preferred when describing crystal structures.

In structural terms, h planes lie between a pair of face-sharing octahedra. A single h plane then is equivalent to a pair of face-sharing octahedra, hh is equivalent to three face-sharing octahedra, and h_q is equivalent to a column of $(q+1)$ face-sharing octahedra. Similarly, an isolated c layer between h layers means that the columns of face-shared octahedra are displaced with respect to each other and linked by apex sharing. Two successive c layers, cc, indicates that a complete layer of apex-sharing octahedra, that is, a slice of cubic perovskite structure type one octahedron in thickness, separates face-sharing columns and c_p represents a slab of p apex-sharing octahedra, equivalent to a slice of cubic perovskite structure $(p-1)$ octahedra in thickness, parallel to the cubic [111] direction.

The ideal *c*-parameter dimensions of these phases can be calculated by considering that an octahedron has an average thickness of 0.24 nm parallel to the **c**-axis, irrespective of whether the packing is c or h (Table 3.4). The *a*-parameter of all structures is similar to that of 2H—$BaNiO_3$, 0.56 nm.

Table 3.4 *Approximate c lattice parameters of $c_p h_q$ stacking sequences*

Layers	Representative stacking sequence	*c*-Parameter (nm)	c_{av} (expt) (nm)[a]
1	c or h	0.24	—
2	2H, hh	0.48	0.559
3	3C, ccc	0.72	0.676
4	4H, $(ch)_2$ (cchh)	0.96	0.929
5	5H, (chcch), (ccchh)	1.20	1.376
6	6H, $(cch)_2$, (chchhh)	1.40	1.416
7		1.68	
8	8H, $(chhh)_2$, $(ccch)_2$, (chchhhh)	1.92	1.895
9	9R, $(chh)_3$	2.16	2.126
10	10H, $(cccch)_2$, $(chchh)_2$	2.40	2.364
11		2.64	
12	12H, $(chhhhh)_2$, $(ccccch)_2$; 12R, $(cchh)_3$	2.88	
13		3.12	
14		3.36	
15	15R, $(chhhh)_3$, $(chcch)_3$	3.60	3.534
16	16H, $(ccccchhh)_2$	3.84	
17		4.08	
18	18R, $(ccchhh)_3$, $(cccchh)_3$	4.32	
19		4.56	
20		4.80	
21	21R, $(chhhhhh)_3$, $(ccccchh)_3$	5.04	

[a] Experimental c lattice parameters are for oxides and taken from the tables in the following sections.

3.5 Hexagonal Perovskites with ch_q and c_ph Stacking

3.5.1 (ch_q) Structures

Phases represented by a (ch_q) stacking symbol show only one unique stacking sequence (chh…h). The resulting materials (Table 3.5) are all composed of short columns of $(q+1)$ face-sharing octahedra separated by one layer of apex-shared octahedra, which contributes the isolated c motif.

The first member of this series of phases is the hexagonal $(ch)_2$ structure, typified by the high-temperature form 4H—$BaMnO_{2.65}$ (Figure 3.8d). This structure, which consists of pairs of face-sharing octahedra linked by apex sharing, is formed when room temperature 2H—$BaMnO_3$ is heated to high temperatures. (Although the change in structure is mediated by the loss of oxygen, the oxygen-deficient phase 4H—$BaMnO_{2.65}$ can be fully oxidised to 4H—$BaMnO_3$ by careful heat treatment.) The isostructural 4H—$SrMnO_3$ is also able to lose oxygen to form non-stoichiometric 4H—$SrMnO_{3-\delta}$. In this case, however, it seems that the oxygen loss is accommodated by the insertion of extra c layers rather than by disordered oxygen vacancies. Resulting sequences, such as (hcch) and (hcchcch), break up the regular $(ch)_2$ structure. The same feature is noted in reduced nanoparticles of 4H—$SrMnO_3$.

Table 3.5 *Phases with the (ch_q) stacking sequence*

Phase	Space group	*a* (nm)	*c* (nm)
$(ch)_2$, 4H			
$SrMnO_3$	H, $P6_3/mmc$ (194)	0.54432	0.90704
$BaMnO_{2.65}$	H, $P6_3/mmc$ (194)	0.56833	0.93556
$BaMnO_3$	H, $P6_3/mmc$ (194)	0.56376	0.92241
$Ba_{0.5}Sr_{0.5}MnO_{2.78}$	H, $P6_3/mmc$ (194)	0.55854	0.92123
$Ba_{0.5}Sr_{0.5}MnO_3$	H, $P6_3/mmc$ (194)	0.5528	0.91471
$BaRuO_3$	H, $P6_3/mmc$ (194)	0.5729	0.9500
$(chh)_3$, 9R			
$BaRuO_3$	Tr, $R\bar{3}m$ (166)	0.5747	2.1602
$BaIr_{0.3}Mn_{0.7}O_3$	Tr, $R\bar{3}m$ (166)	0.56993	2.12778
$Ba_{0.875}Sr_{0.125}MnO_3$	Tr, $R\bar{3}m$ (166)	0.56397	2.08973
$CsCoF_3$	Tr, $R\bar{3}m$ (166)	0.6194	2.261
$CsMgF_3$	Tr, $R\bar{3}m$ (166)	0.616	2.213
$CsNiF_3$	Tr, $R\bar{3}m$ (166)	0.6149	2.235
$(chhh)_2$, 8H			
$BaMnO_{2.875}$	H, $P6_3/mmc$ (194)	0.56736	1.87550
$(chhhh)_3$, 15R			
$BaMnO_{2.99}$	Tr, $R\bar{3}m$ (166)	0.56791	3.5345

The second member of the series has the stacking sequence $(chh)_3$, typified by 9R–$BaRuO_3$. It is built of triplets of face-sharing octahedra, linked by apex sharing (Figure 3.9a). The location of cations of variable valence within the groups of three has been clarified by bond valence methods (Appendix A). The stacking sequence $(chhh)_2$, adopted by 8H–$BaMnO_{2.875}$, contains columns of four face-shared octahedra linked by apex sharing (Figure 3.9b). The five face-shared octahedra analogue, with a packing sequence $(chhhh)_3$, is adopted by rhombohedral 15R–$BaMnO_{2.99}$ (Figure 3.9c). The $(chhhhh)_2$ stacking sequence gives a 12H polytype which has columns of six face-shared octahedra (Figure 3.9d). No examples of this structure appear to have been synthesised to date. The $(chhhhhh)_3$ stacking sequence is adopted by 21R–$BaMnO_{2.928}$ (Figure 3.9e).

3.5.2 (c_ph) Structures

Phases with (c_ph) stacking show only one unique stacking sequence (ccc…ch). The resulting compounds all contain pairs of face-sharing octahedra of the 2H–$BaNiO_3$ type, which contributes the isolated h motif, separated by $(p-1)$ layers of apex-shared octahedra. The nominal stoichiometry of all these phases is $A_nB_nO_{3n}$.

The simplest of these, the 4H–$BaMnO_{2.65}$ structure, with a stacking sequence $(ch)_2$ is described in Section 3.5.1 (Figure 3.8d). Thereafter we find $(cch)_2$, the hexagonal 6H–$BaTiO_3$ structure. This form occurs when cubic perovskite $BaTiO_3$ is heated to high temperatures and appears to be stabilised by B-site impurities or oxygen loss. The structure is made up of pairs of face-sharing octahedra separated by single sheets of apex-sharing cubic perovskite type (Figure 3.10a). There are large numbers of phases with this structure, mostly with a composition $A_3B_3O_9$, such as $Ba_3Ti_2VO_9$ and $Ba_3La^{3+}Ru_2^{4.5+}O_9$ (Table 3.6). The phase 6H–$BaFeO_2F$ has F ordered in the $BaFeX_3$ h layers.

There are several phases with a stacking sequence of $(ccch)_2$, including 8H–$Ba_4LiNb_{3-x}Ta_xO_{12}$ ($x=0$–3), which includes 8H–$Ba_4LiTa_3O_{12}$ and 8H–$Ba_4LiNb_3O_{12}$ as limiting structures (Figure 3.10b), and the phases $Ba_8Ga_{4-x}Ta_{4+0.6x}O_{24}$ ($1.8<x<3.2$). The phase 8H–$Ba_4LiTa_2SbO_{12}$ forms from a cubic perovskite structure at temperatures above 1450°C.

In addition to these nominally $A_4B_4O_{12}$ compositions, a group of cation-deficient phases with an $A_nB_{n-1}O_{3n}$ composition are also known, typified by 8H–$Ba_8Ti_3Nb_4O_{24}$. In these compounds the cation deficiency is distributed across the pairs of face-sharing octahedra, which therefore contain a complex ordered arrangement of B cations. For example, the phase 8H–$Ba_8Ti_3Ta_4O_{24}$ has an ordering of Ti, Ta and cation vacancies occupying these positions. Moreover, the phase has a considerable solid solution range, represented by $Ba_8Ti_{3-x}Ta_{4+0.8x}O_{24}$ ($0<x<0.8$), from

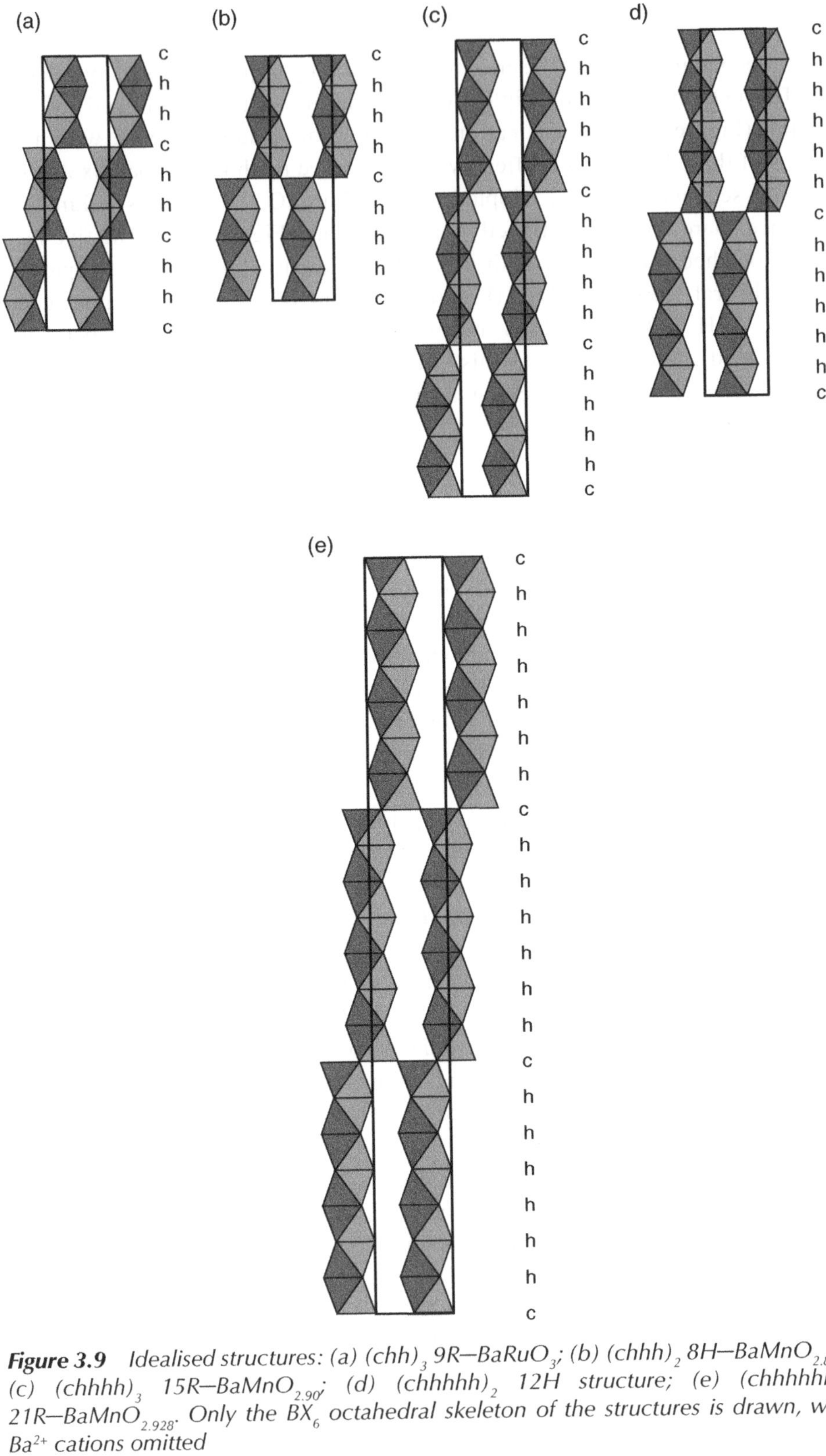

Figure 3.9 *Idealised structures: (a) $(chh)_3$ 9R—$BaRuO_3$; (b) $(chhh)_2$ 8H—$BaMnO_{2.875}$; (c) $(chhhh)_3$ 15R—$BaMnO_{2.90}$; (d) $(chhhhh)_2$ 12H structure; (e) $(chhhhhh)_3$ 21R—$BaMnO_{2.928}$. Only the BX_6 octahedral skeleton of the structures is drawn, with Ba^{2+} cations omitted*

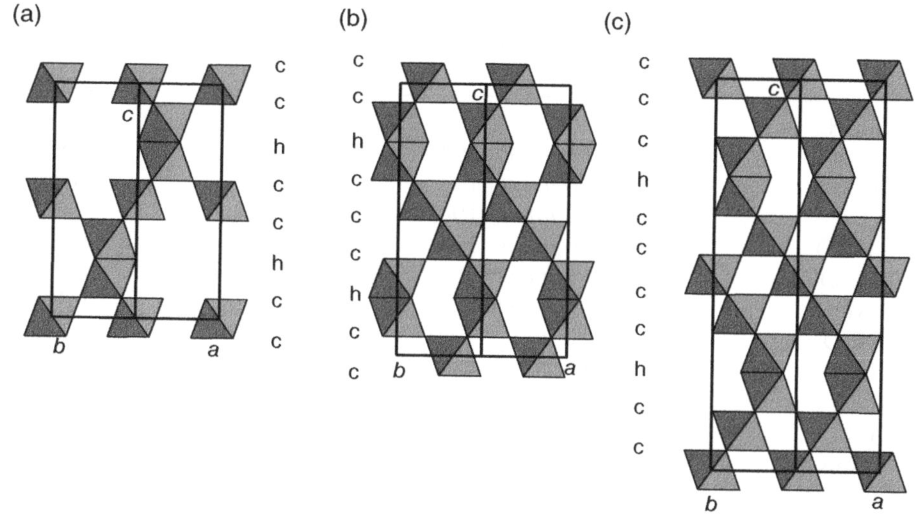

Figure 3.10 *Idealised structures: (a) $(cch)_2$ hexagonal 6H—$BaTiO_3$; (b) $(ccch)_2$ 8H—$Ba_4LiTa_3O_{12}$; (c) $(cccch)_2$ 10H—$Sr_5Mn_{3.6}Fe_{1.4}O_{14.95}$. Only the BX_6 octahedral skeleton of the structures is drawn, with Sr^{2+} or Ba^{2+} cations omitted*

$Ba_8Ti_3Ta_4O_{24}$ ($x=0$) to $Ba_8Ti_{2.2}Ta_{4.64}O_{24}$ ($x=0.8$). These different stoichiometries are accompanied by different ordering patterns of B-site cations and vacancies.

The structure of $Sr_5Mn_{3.6}Fe_{1.4}O_{14.95}$ has a layer stacking sequence of $(cccch)_2$, characteristic of a 10H- structure type (Figure 3.10c). The ideal composition of this and similar phases is $A_5B_5O_{15}$ (Table 3.6). The related cation-deficient phase 10H—$Ba_{10}Ta_{7.04}Ti_{2.2}O_{30}$ also shows a solid solution range represented by $Ba_{10}Ta_{8-0.8x}Ti_xO_{30}$ ($0.6<x<1.2$) in which the cation vacancies are segregated in the face-sharing octahedra, but in a disordered way.

A comparison of these three structures shows that the slabs of apex-sharing octahedra are in a *twinned* relationship to each other on passing from one slice to an adjacent one parallel to the **c**-axis (Figure 3.11). The twin plane is the h plane. The relationship of these c_ph 'twinned' structures is compared to the similar c_phh 'shifted' structures later (Section 3.6.4).

In addition to cation-deficit phases, a number of (c_ph) structures have an oxygen deficit. In the main the oxygen vacancies are distributed at random over the available positions. However, ordering can lead to changes in cation coordination polyhedra and more complex structures, as is found in $Ba_5Ru_3Na_2O_{14}$. At temperatures above approximately 823 K, the structure can be imagined as an idealised 10H-$(cccch)_2$ structure, described as (c′chcc′), where the c′ layers have an oxygen deficiency, written as $Ba(O_{5/6}V_{1/6})_3$, with ⅙ of the oxygen sites unoccupied (Figure 3.12a). The cation distribution in the apex-sharing segment of structure is such that the

Table 3.6 *Some phases with the (c_ph) stacking sequence*

Phase	Space group	*a* (nm)	*c* (nm)
(cch)$_2$, 6H			
$BaTiO_3$ hex	H, $P6_3/mmc$ (194)	0.57238	1.39649
$BaRuO_3$	H, $P6_3/mmc$ (194)	0.57127	1.40499
$BaFeO_{3-\delta}$	H, $P6_3/mmc$ (194)	0.5673	1.3921
$Ba_3FeRu_2O_9$	H, $P6_3/mmc$ (194)	0.57248	1.40652
$Ba_3Ti_2RuO_9$	H, $P6_3mc$ (186)	0.56680	1.39223
$Ba_3Ti_2MnO_9$	H, $P6_3mc$ (186)	0.56880	1.39223
$Ba_3YRu_2O_9$	H, $P6_3/mmc$ (194)	0.58816	1.4501
$Ba_3LaRu_2O_9$	H, $P6_3/mmc$ (194)	0.59579	1.50058
$Ba_3NdRu_2O_9$	H, $P6_3/mmc$ (194)	0.59319	1.47589
$Ba_3SmRu_2O_9$	H, $P6_3/mmc$ (194)	0.59192	1.46788
$Ba_3YbRu_2O_9$	H, $P6_3/mmc$ (194)	0.5858	1.4432
$Ba_3CeRu_2O_9$	H, $P6_3mc$ (186)	0.58894	1.46476
$Ba_3PrRu_2O_9$	H, $P6_3/mmc$ (194)	0.58855	1.46071
$Ba_3TbRu_2O_9$	H, $P6_3/mmc$ (194)	0.58365	1.44217
$BaRu_{0.5}Mn_{0.5}O_3$	H, $P6_3/mmc$ (194)	0.56895	1.39309
$Ba_{0.7}Sr_{0.3}RuO_3$	H, $P6_3/mmc$ (194)	0.56749	1.39943
$Ba_3BiIr_2O_9$ (700 K)	H, $P6_3/mmc$ (194)	0.59974	1.51495
$Ba_3BaSb_2O_9$	H, $P6_3/mmc$ (194)	0.61812	1.59900
$BaFeO_2F$	H, $P6_3/mmc$ (194)	0.57635	1.42119
(ccch)$_2$,8H			
$Ba_4LiTa_3O_{12}$	H, $P6_3mc$ (186)	0.5802	1.90850
$Ba_4LiNb_3O_{12}$	H, $P6_3mc$ (186)	0.5803	1.90760
$Ba_8Ga_{0.8}Ta_{5.92}O_{24}$	H, $P6_3cm$ (185)	1.00836	1.89947
$Ba_8Ga_2Ta_{5.12}O_{24}$	H, $P6_3cm$ (185)	1.00769	1.89914
$Ba_8Ti_3Nb_4O_{24}$	H, $P6_3/mcm$ (193)	1.00677	1.89166
$Ba_8Ti_3Ta_4O_{24}$	H, $P6_3/mcm$ (193)	1.00314	1.88694
$Ba_8Ti_{2.2}Ta_{4.64}O_{24}$	H, $P6_3/mcm$ (193)	0.58038	1.8912
$Ba_8ZnTa_6O_{24}$	H, $P6_3cm$ (185)	1.00830	1.90659
$Ba_8NiTa_6O_{24}$	H, $P6_3cm$ (185)	1.00745	1.90122
$Ba_8CoNb_6O_{24}$	Tr, $P\bar{3}m1$ (164)	0.57898	1.88936
(cccch)$_2$, 10H			
$Sr_5Mn_{3.6}Fe_{1.4}O_{14.35}$	H, $P6_3/mmc$ (194)	0.545035	2.23753
$SrMn_{0.72}Fe_{0.28}O_{2.87}$	H, $P6_3/mmc$ (194)	0.54503	2.23735
$Ba_5Ru_3Na_2O_{14}$	H, $P\bar{6}2c$ (190)	0.59261	2.4400
$Ba_{10}Ti_{1.2}Ta_{7.04}O_{30}$	H, $P6_3/mmc$ (194)	0.57981	2.37547
$Ba_{10}Ti_{0.6}Ta_{7.52}O_{30}$	H, $P6_3/mmc$ (194)	0.58056	2.3860
$Ba_5W_3Li_2O_{15}$	H, $P6_3/mmc$ (194)	0.57559	2.3719
$Ba_{10}Mg_{0.25}Ta_{7.9}O_{30}$	H, $P6_3mc$ (186)	0.58105	2.38846

Note: Space group or **a**-axis variation is due to ordering of atoms or vacancies.

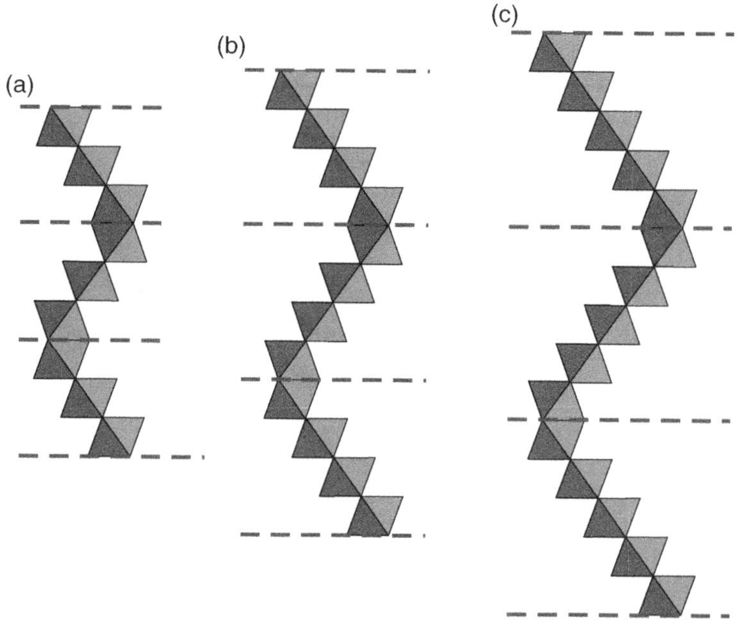

Figure 3.11 *The twinned motif of the structures in Figure 3.10: (a) (cch); (b) (ccch); (c) (cccch)*

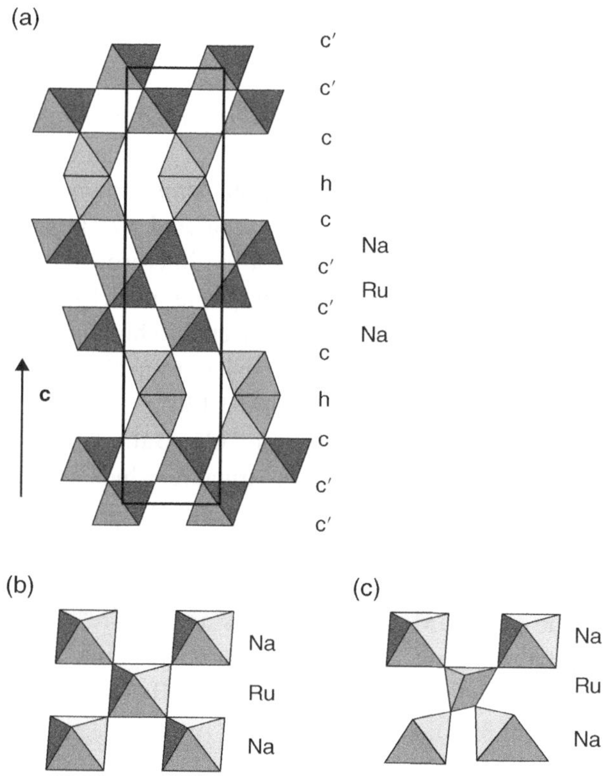

Figure 3.12 *10H—$Ba_5Ru_3Na_2O_{14}$. (a) The high-temperature structure; (b) octahedral coordination polyhedra of Na and Ru above 823 K; (c) distorted polyhedra at 298 K*

Table 3.7 *Some* $c_p h_q$ *intergrowth phases*

Phase	Space group	*a* (nm)	*c* (nm)
$(chcch)_3$, 15R			
$SrMn_{0.915}Fe_{0.085}O_{2.979}$	Tr, $R\bar{3}m$ (166)	0.54489	3.38036
$BaFeO_2F$	Tr, $R\bar{3}m$ (166)	0.57659	3.57149
$BaFeO_{2.25}F_{0.5}$	Tr, $R\bar{3}m$ (166)	0.57489	3.5894
$(chchh)_2$ 10H			
$BaIr_{0.3}Co_{0.7}O_{2.84}$	H, $P6_3/mmc$ (194)	0.57075	2.38462
$Ba_5Fe_5NiO_{13.5}$	H, $P6_3/mmc$ (194)	0.57713	2.45812
$BaMnO_{2.91}$	H, $P6_3/mmc$ (194)	0.56547	2.31936
(chchhh) 6H′			
$BaMnO_{2.92}$	H, $P\bar{6}m2$ (187)	0.56623	1.39993

central set of octahedra is preferentially occupied by Ru and its neighbours by Na. Surrounding polyhedra are fairly symmetrical (Figure 3.12b). As the temperature is lowered below 823 K, vacancy ordering causes considerable distortion of the Ru and Na coordination polyhedra so that, at 298 K the resemblance to octahedra is lost for the Ru and half the Na ions (Figure 3.12c). At lower temperatures still the distortions become greater, and the resemblance to the ideal structure is lost.

3.5.3 $c_p h_q$ Intergrowth Structures

The phases described here (Table 3.7) can be regarded as intergrowths between members of the $c_p h$ and ch_q series described earlier. Typical of these is the15R—$SrMn_{0.915}Fe_{0.085}O_{2.979}$ phase (Figure 3.13a). The stacking sequence $(chcch)_3$ contains pairs of apex-linked face-sharing octahedra, separated by single layers of apex-shared octahedra, and can be considered to be a 1 : 1 intergrowth between the phases 4H—$BaMnO_{2.65}$ (Figure 3.8d) and 6H—$BaTiO_3$ (Figure 3.10a). The 5H—$BaIrO_3$ structure which forms when 9R—$BaIrO_3$ is subjected to high pressure is a distorted variant of this structure, with a monoclinic cell: space group *C*2/*m* (12), $a = 0.9955$ nm, $b = 5.743$ nm, $c = 13.805$ nm, $\beta = 119.23°$ (Figure 3.13b).

A phase with a closely similar composition, 10H—$BaIr_{0.3}Co_{0.7}O_{2.84}$, with a packing sequence $(chchh)_2$ (Figure 3.14a) has a structure which is a 1 : 1 intergrowth between the 9R—$BaRuO_3$ structure (Figure 3.9a) and 4H—$BaMnO_{2.65}$ (Figure 3.8d) structures. These materials are oxygen deficient compared to the ideal and contain oxygen vacancies which appear to be distributed at random over the available sites in the structure. The 6H′-$BaMnO_{2.92}$ phase has a structure composed of blocks of two and four face-sharing octahedra, with a stacking sequence (chchhh) (Figure 3.14b). It is a 1 : 1 intergrowth between 4H—$BaMnO_{2.65}$ (Figure 3.8d) and 8H—$BaMnO_{2.875}$ (Figure 3.9b).

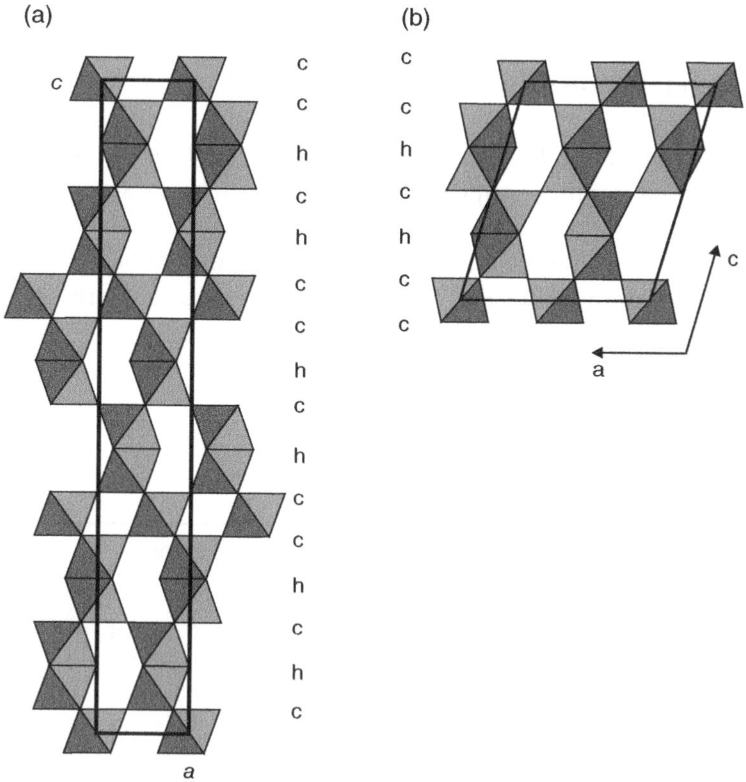

Figure 3.13 *Idealised (chcch) structures of (a) 15R—*$SrMn_{0.915}Fe_{0.085}O_{2.979}$*; (b) 5H—*$BaIrO_3$

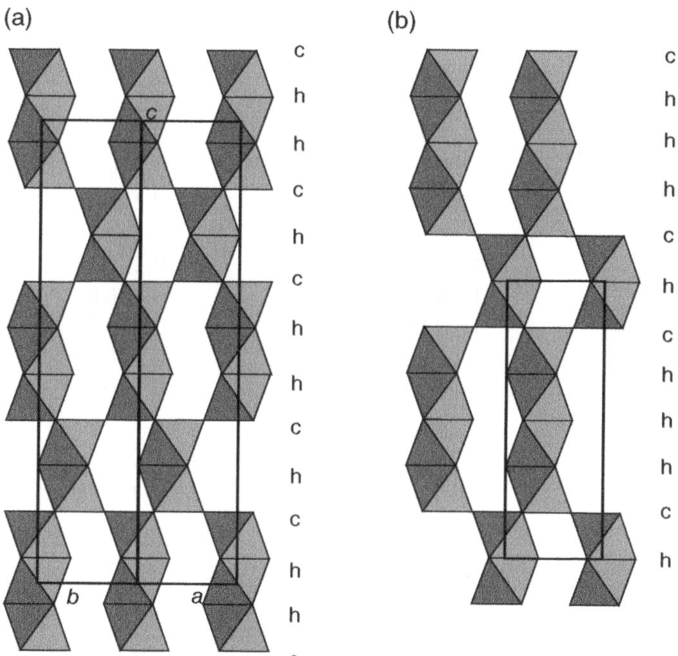

Figure 3.14 *Idealised structures of (a) (chchh) 10H—*$BaIr_{0.3}Co_{0.7}O_{2.84}$*; (b) (chchhh) 6H′—*$BaMnO_{2.92}$

3.6 Hexagonal Perovskites with $c_p$hh Stacking

The stacking sequence of the AO_3 layers in the compounds in this section is (cc…chh). The hh motif signifies a triplet of face-sharing octahedra, separated by $(p-1)$ sheets of apex-sharing octahedra. There are two major groups, firstly those with a full complement of cations, and an approximate composition of $A_nB_nO_{3n}$ and secondly a series of cation-deficient phases that more or less conform to the series formula $A_nB_{n-1}O_{3n}$ (Table 3.8).

Table 3.8 *Some phases with (cc…chh) structures*

Phase	Space group	a (nm)	c (nm)
(cchh) 12R			
$Ba_4Ti_2Mn_2O_{12}$	Tr, $R\bar{3}m$ (166)	0.56914	2.79186
$Ba_4CeMn_3O_{12}$	Tr, $R\bar{3}m$ (166)	0.57980	2.86070
$Ba_4PrMn_3O_{12}$	Tr, $R\bar{3}m$ (166)	0.57943	2.85716
$Ba_4ErRu_3O_{12}$	Tr, $R\bar{3}m$ (166)	0.58559	2.9014
$Ba_4TbRu_3O_{12}$	Tr, $R\bar{3}m$ (166)	0.58306	2.90132
$Ba_{12}Ca_3Mo_3Mn_6O_{36}$	Tr, $R\bar{3}m$ (166)	0.58364	2.8870
$Ba_{12}In_3Mn_9O_{34.5}$	Tr, $R\bar{3}m$ (166)	0.57382	2.8240
$Ba_{12}Ti_6Mn_6O_{36}$	Tr, $R\bar{3}m$ (166)	0.56914	2.79186
$Ba_{12}Ir_6Co_6O_{36.24}$	Tr, $R\bar{3}m$ (166)	0.57206	2.84073
$La_4Ti_3O_{12}$	Tr, $R\bar{3}$ (148)	0.55509	2.6178
(ccchh), 5H, 10H			
$Ba_5IrCo_4O_{14.5}$	Tr, $P\bar{3}m1$ (164)	0.57179	1.19865
$La_5Mg_{0.5}Ti_{3.5}O_{15}$	Tr, $P\bar{3}m1$ (164)	0.55639	1.09928
$Ba_5Ta_4O_{15}$	Tr, $P\bar{3}m1$ (164)	0.5776	1.182
$Ba_5Nb_4O_{15}$	Tr, $P\bar{3}m1$ (164)	0.57883	1.17782
$BaLa_4Ti_4O_{15}$	Tr, $P\bar{3}c1$ (165)	0.55671	2.24603
(cccchh), 18R			
$Ba_2La_4Ti_5O_{18}$	Tr, $R\bar{3}$ (148)	0.55809	4.10558
$Sr_6Nb_4SnO_{18}$	Tr, $R\bar{3}m$ (166)	0.5661	4.186
$Ba_5KNb_5O_{18}$	Tr, $R\bar{3}m$ (166)	0.57840	4.2532
$Ba_6TiNb_4O_{18}$	Tr, $R\bar{3}m$ (166)	0.57852	4.24886
$La_6Mg_{0.913}Ti_{4.06}O_{18}$	Tr, $R\bar{3}m$ (166)	0.55665	3.97354
(cccccchh), 8H			
$Ba_8Ti_3Nb_4O_{24}$	H, $P6_3/mcm$ (193)	1.0068	1.8917
$Ba_8Yb_{0.5}Ti_2Nb_{4.5}O_{24}$	—	0.5797	1.8905
(ccchhccccchh) $(5_1 6_1)$			
$Ba_{11}TiNb_8O_{33}$	Tr, $R\bar{3}m$ (166)	0.57863	7.78011

3.6.1 (cc…chh) $A_nB_nO_{3n}$ Structures

The simplest stacking sequence (cchh), can generate two alternative structures: (cchh) and (chch). This latter stacking is the (ch) sequence found in 4H–$BaMnO_{2.65}$ (Figure 3.8d). The 12R $(cchh)_3$ stacking is analogous to the $(chh)_3$ 9R–$BaRuO_3$ structure (Figure 3.9a) except that the blocks of three face-sharing octahedra involved are now separated by a sheet of apex-sharing octahedra. The phases recorded all have Ba^{2+} as the A cation and are represented by the 12R–$Ba_4LnRu_3O_{12}$ phases, where Ln represents a lanthanoid ion (Figure 3.15a).

In many of these compounds there is an ordering of the cations, often clarified by bond valence methods (Appendix A). For example, the 12R–$Ba_4LnRu_3O_{12}$ phases have Ru occupying the face-sharing octahedra, while the Ln ions occupy the apex-sharing octahedra. In 12R–$Ba_{12}In_3Mn_9O_{34.5}$, the Mn^{4+} cations occupy the columns of face-sharing octahedra, the In^{3+} cations are in the apex-shared octahedra, and the oxygen vacancies are distributed at random over the available sites. In the closely related 12R–$Ba_{12}Ca_3Mo_3Mn_6O_{36}$, the central octahedra of the face-sharing triplets are occupied by Mn^{4+} cations and its two neighbours by 1 : 1 Mn^{4+}/Mo^{6+} mixture, while the Ca^{2+} cations occupy the apex-sharing octahedra.

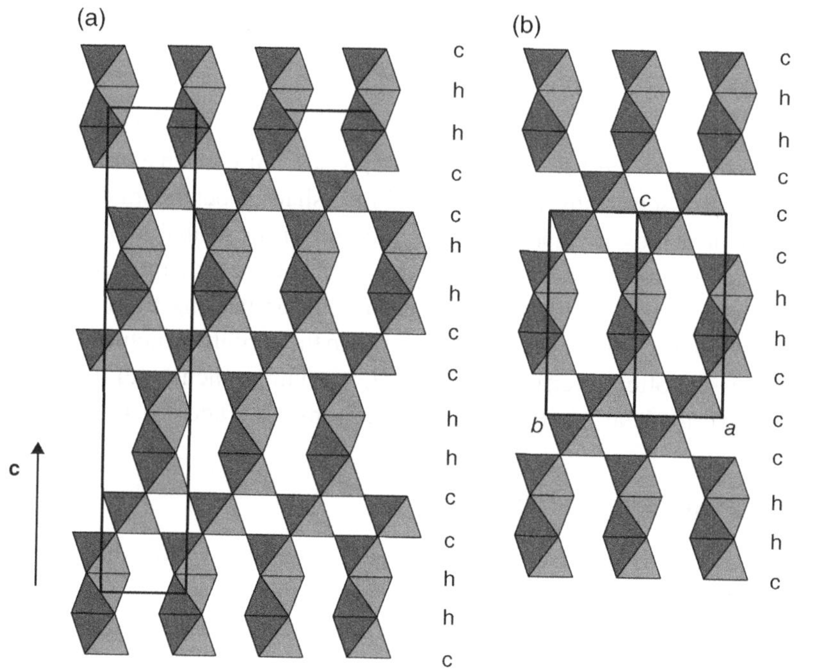

Figure 3.15 *The idealised structures of (a) the 12R–$Ba_4LnRu_3O_{12}$ phases; (b) 5H–$Ba_5IrCo_4O_{14.5}$*

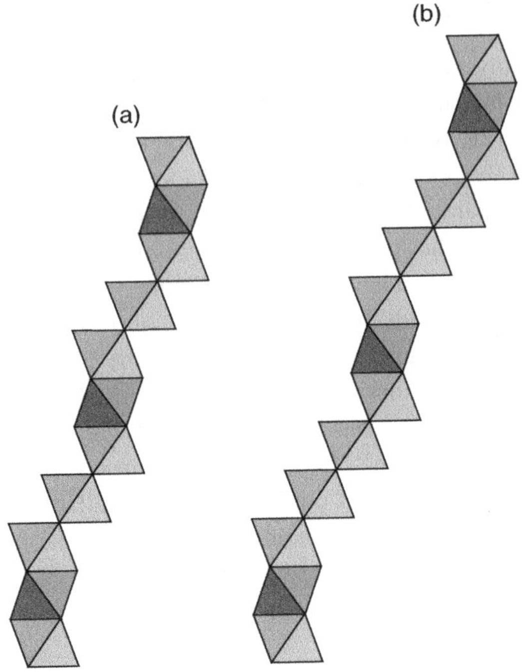

Figure 3.16 *The shifted motif of the structures shown in Figure 3.15: (a) (cchh); (b) (ccchh)*

The off-stoichiometric phase 5H—$Ba_5IrCo_4O_{14.15}$, with a packing sequence $(ccchh)_2$, is similar to the 12R—$Ba_4LnRu_3O_{12}$ phases, but with an extra layer of apex-shared octahedra separating the triplets of face-sharing octahedra (Figure 3.15b). The oxygen vacancies are distributed at random over the available oxygen sites.

Notice that in each of these structures, the sheets of apex-shared octahedra appear to be displaced laterally with respect to each other on passing from one to another parallel to the **c**-axis. This aspect is emphasised in the (ccchh) structure (Figure 3.16). These phases are often referred to as *shifted*, to contrast them with the c_ph *twinned* structures described earlier (Figure 3.11) (Sections 3.5.2 and 3.6.4).

3.6.2 (cc...chh) $A_nB_{n-1}O_{3n}$ Structures

The same (cc…chh) stacking sequence described earlier is found in an extensive series of cation-deficient hexagonal perovskite phases with a general formula $A_nB_{n-1}O_{3n}$ with *n* generally taking values of 4 and more. These are formed by many large A cations, typically Ca^{2+}, Sr^{2+}, Ba^{3+}, La^{3+}, Pr^{3+} and Nd^{3+}, and medium-sized B cations, such as Mg^{2+}, Zn^{2+}, Cd^{2+}, B^{3+}, Al^{3+}, Ga^{3+}, Cr^{3+}, Fe^{3+}, Ti^{3+}, Ti^{4+}, Zr^{4+}, Sn^{4+}, Ru^{4+}, Nb^{5+}, Ta^{5+}, W^{6+} and Re^{6+}. They are typified by the phases found in the $La_4Ti_3O_{12}$–$LaTiO_3$ and $La_4Ti_3O_{12}$–$BaTiO_3$ systems.

Figure 3.17 *Schematic cation-deficient structures: (a) (cchh)$_3$ 12R—La$_4$Ti$_3$O$_{12}$; (b) (ccchh)$_2$ 10H—La$_5$Ti$_4$O$_{15}$; (c) 18R—Ba$_2$La$_4$Ti$_5$O$_{18}$; (d) 8H—Ba$_8$Ti$_3$Nb$_4$O$_{24}$. Only the BX$_6$ octahedral skeleton of the structures is drawn, with A ions omitted. The unshaded octahedron does not contain Ti^{4+} cations and is to show the relationship between layers*

Although the stacking sequence of the AO_3 layers in these compounds is the same as in the previous materials, in these cation-deficient phases the central octahedron of the triplet groups does not contain any B cations. Because of this the $A_nB_{n-1}O_{3n}$ structures can be conveniently considered to consist of layers of corner-sharing cubic perovskite structure-type $(n-1)$ octahedra in thickness (Figure 3.17a–c). It is seen that on moving from one cubic perovskite layer to the next, the rows in the sheets of apex-linked octahedra are displaced or shifted laterally one relative to the other as in the stoichiometric $A_nB_nO_{3n}$ phases in the previous section.

The overall symmetry of the structure is related to the value of n in the series formula. In cases where n is a [(multiple of 3) − 1], the structure has a Bravais hexagonal lattice, while if n is given by a [(multiple of 3)] or a [(multiple of 3) + 1], the structure conforms to a Bravais rhombohedral lattice.

The smallest repeat structure is that of the $n=4$ phase $La_4Ti_3O_{12}$ (Figure 3.17a). The stacking sequence is (cchh), and as n is given by (3 + 1) the unit cell is rhombohedral with a packing sequence $(cchh)_3$ to form a 12R- phase. As $LaTiO_3$ or $BaTiO_3$ is added to the $n=4$ phase, the slabs of apex-sharing octahedra widen to form firstly $La_5Ti_4O_{15}$ and $BaLa_4Ti_4O_{15}$ ($n=5$) (Figure 3.17b). These have a stacking sequence of (ccchh), and as n is given by $(2\times3-1)$ the unit cell is hexagonal with a 5H- or 10H-repetition. The $n=6$ phases $La_6Ti_5O_{18}$ and $Ba_2La_4Ti_5O_{18}$ (Figure 3.17c) have a stacking sequence within each block of (cccchh), and as n is given by (2×3) the unit cell is rhombohedral with a packing sequence $(cccchh)_3$ and a 18R- repetition. The $n=7$ $(hhccccc)_3$ phases include 21R–$Ba_7Ta_4Zr_2O_{21}$ and the $(cccccchh)_2$ $n=8$ phases include 8H–$Ba_8Ti_3Nb_4O_{24}$ (Figure 3.17d).

Note that in some of these phases, the octahedra in the layers may be tilted in opposite directions in successive layers, similar to the octahedral tilts found in the ABX_3 perovskites (Chapter 1), a feature not emphasised in the figures.

3.6.3 (hhcc...chhcc...c) Intergrowth Phases

Many ordered and disordered intermediate phases lying between two adjacent members of the $A_nB_{n-1}O_{3n}$ series have been synthesised. The simplest example is the ordered intergrowth between the $n=4$ phase $La_4Ti_3O_{12}$ and the $n=5$ phase $BaLa_4Ti_4O_{15}$, with a composition $BaLa_8Ti_7O_{27}$, also described as (4_15_1). The stacking of the BaO_3 layers is the sum of the two separate sequences: (cchh) and (ccchh) (Figure 3.18). (This stacking sequence is one of the many arrangements of the (c_5h_4) arrangement (Table 3.3).) Similarly, the ordered intergrowth between the $n=5$ phase $Ba_5Nb_4O_{15}$ and the $n=6$ phase $Ba_6TiNb_4O_{18}$ to give $Ba_{11}TiNb_8O_{33}$, (5_16_1) structure has a packing of BaO_3 planes given by $(ccchhccccchh)_3$.

[Note that these ordered phase compositions do not correspond to the formula $A_nB_{n-1}O_{3n}$ because of the way in which this is derived, that is, the (4_15_1) intergrowth is

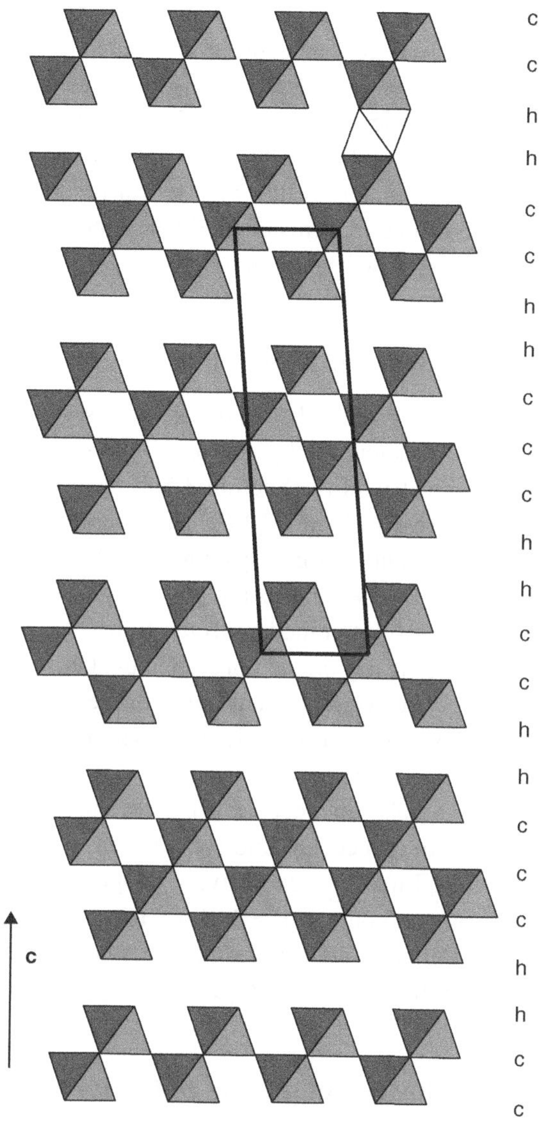

Figure 3.18 *Idealised structure of the n = 4.5 phase BaLa8Ti7O27*

made up of $La_4Ti_3O_{12}$+$BaLa_4Ti_4O_{15}$, giving the overall composition $BaLa_8Ti_7O_{27}$ (i.e. $A_9B_7O_{27}=A_nB_{n-2}O_{3n}$). However, for conformity with the series formula, the value of n should lie between 4 and 5 at 4.5, so dividing the composition by 2 yields $Ba_{0.5}La_4Ti_{3.5}O_{13.5}$, which agrees with the $A_nB_{n-1}O_{3n}$ expression. Similarly the composition formed by the (5_16_1) intergrowth has the composition $Ba_{11}TiNb_8O_{33}$ (i.e. $A_nB_{n-2}O_{3n}$), but reducing n to the value of 5.5 gives $Ba_{5.5}Ti_{0.5}Nb_4O_{16.5}$, which again is $A_nB_{n-1}O_{3n-1}$.]

Electron microscopy has shown that the nominally two-phase region between these intergrowth phases often contains multiple ordered sequences of two end members. For example, as the $n=5$ phase $BaLa_4Ti_4O_{15}$ is added to $n=4$ $La_4Ti_3O_{12}$, the following well-ordered sequences – (4_65_1) $BaLa_{28}Ti_{22}O_{87}$, (4_55_1) $BaLa_{24}Ti_{19}O_{75}$, (4_45_1) $BaLa_{20}Ti_{16}O_{63}$, (4_35_1) $BaLa_{16}Ti_{13}O_{51}$, (4_25_1) $BaLa_{12}Ti_{10}O_{39}$ and (4_15_1) $BaLa_8Ti_7O_{27}$ – have been encountered. It is also worth pointing out that many of these phases are better described in terms of modulated incommensurate structures rather than the idealised intergrowths depicted here.

3.6.4 (cc...ch) $A_nB_{n-1}O_{3n}$ Shift and Twinned Phases

The $A_nB_nO_{3n}$ and $A_nB_{n-1}O_{3n}$ phases described in the previous section, all with a stacking sequence cc...chh, are characterised by shifted blocks of cubic apex-sharing octahedra, while the cc...ch structures (Section 3.5.2) have these same blocks in a twinned relationship (Figures 3.11 and 3.17). The energetic difference between these two structures is slight, and in some systems both forms occur. For example, although the $n=4$ to $n=7$ $A_nB_{n-1}O_{3n}$ phases are generally of the cc...chh shift type, the $n=8$ phases are of the cc...ch twinned configuration.

The $n=7$ preparations are clearly on the borderline of stability between these two stacking types and often give disordered material with the (cc...chh) shift structure intergrown with (cc...ch) twinned stacking. It would seem that the balance between this pair of structures is very sensitive to the size and type of the B cations present. The 8H–$Ba_8Ti_2Nb_{4.5}Lu_{0.5}O_{24}$ takes the (cccccchh), $n=8$, shift structure; $Ba_8Ti_{2.5}Nb_{4.25}Lu_{0.25}O_{24}$ adopts a mixture of (cccccchh) and (ccch) $n=8$, mixed twinned and shift forms; while $Ba_8Ti_{2.75}Nb_{4.125}Lu_{0.125}O_{24}$, exclusively adopts the (ccch) $n=8$, twinned structure. Similarly the twinned phase 8H–$Ba_8Ga_{4-x}Ta_{4+0.6x}O_{24}$ exists with a composition ($1.8<x<3.2$). When the amount of Ta increases beyond this range, the $n=8$ phase is replaced by the shift phase $Ba_5Ta_4O_{15}$.

3.7 Anion-Deficient Phases Containing BaO_2 (c′) Layers

A large number of hexagonal perovskite oxide phases form complex structures which are anion deficient, especially those containing the variable valence ions Mn, Fe and Co (Table 3.9). In these phases a BaO_3 layer loses oxygen and is transformed typically into a BaO_2 layer. The packing of the ions in this layer remains cubic or nearly so with respect to the neighbouring layers and is often designated as c′ in stacking arrangements. The cations that are nominally in octahedral interstices on either side of the parent BaO_3 layer generally move into neighbouring (*isolated*) tetrahedral sites which are not linked to each other. Additionally, in some structures the coordination polyhedra are more varied and can include square pyramidal coordination.

Table 3.9 *Phases with BaO_2 (c′) layers*

Phase	Space group	*a* (nm)	*c* (nm)
(ccc′cch), 12H			
$Ba_6Ru_2Na_2V_2O_{17}$	H, $P6_3/mmc$ (194)	0.58506	2.96241
$Ba_6Ru_2Na_2Mn_2O_{17}$	H, $P6_3/mmc$ (194)	0.58323	2.95299
$Ba_6Nd_2Ti_4O_{17}$	H, $P6_3/mmc$ (194)	0.599283	2.99289
$Ba_6Y_2Ti_4O_{17}$	H, $P6_3/mmc$ (194)	0.593055	2.95239
(cc′chh), 5H			
$Ba_4Co_5O_{14}$	Tr, $P\bar{3}m1$(164)	0.5660	1.1970
$BaCoO_{2.74}$	Tr, $P\bar{3}m1$(164)	0.5657	1.19223
$BaMn_{0.2}Co_{0.8}O_{2.80}$	Tr, $P\bar{3}m1$(164)	0.56723	1.18813
(cc′chhh), 12H			
$Ba_{0.9}CoO_{2.60}$	H, $P6_3/mmc$ (194)	0.56612	2.84627
$BaCoO_{2.60}$	H, $P6_3/mmc$ (194)	0.5767	2.8545
$BaMn_{0.4}Co_{0.6}O_{2.83}$	H, $P6_3/mmc$ (194)	0.56935	2.85751
(ccc′cchh), 21R			
$Ba_7Ca_2Mn_5O_{20}$	Tr, $R\bar{3}m$(166)	0.58207	5.1359
(ccc′cchhh), 16H			
$Ba_4Ca_{0.9}Mn_{3.1}O_{11.3}$	H, $P\bar{6}m2$(187)	0.58005	3.8954

3.7.1 (c…c′…ch) Structures

These phases are the analogues of the cc…ch 'twinned' phases described in Section 3.5.2. As before, c and h layers have a stoichiometry AX_3, and the layers adjacent to tetrahedrally coordinated cations, labelled c′, have a stoichiometry AX_2. In these structures the tetrahedra are displaced across the interface and are not linked to each other. An example is given by 12H—$Ba_6Ru_2Na_2M_2O_{17}$, which has a stacking sequence $(ccc'cch)_2$ (Figure 3.19a) compared to an ideal $A_6B_6O_{18}$ phase with a stacking sequence $(ccccch)_2$ (Figure 3.19b). In these phases, M can be V, Cr, Mn, P or As, all of which are found in the tetrahedral positions in the structure. In addition, Ru can be replaced by Nb, Ta or Sb.

3.7.2 (c…c′…chh) Structures

These phases are the analogues of the cc…chh phases described in Section 3.6. A number were discovered, many using transmission electron microscopy, in the oxygen-deficit 5H—$BaCoO_{3-\delta}$ system. One of the first of these to be structurally characterised was $Ba_5Co_5O_{14}$ ($BaCoO_{2.74}$). This phase is related to the 5H (ccchh) $A_5B_5O_{15}$ structure (Figure 3.15b). To form idealised $BaCoO_{2.74}$, the central c layer of the ccc group is replaced by a BaO_2 (c′) layer, and the two layers of apex-sharing octahedra that lie to each side of it are converted to sheets of Co^{4+}-centred tetrahedra so that the stacking becomes $(cc'chh)_2$ (Figure 3.20a) which can be compared to the ideal (ccchh) structure

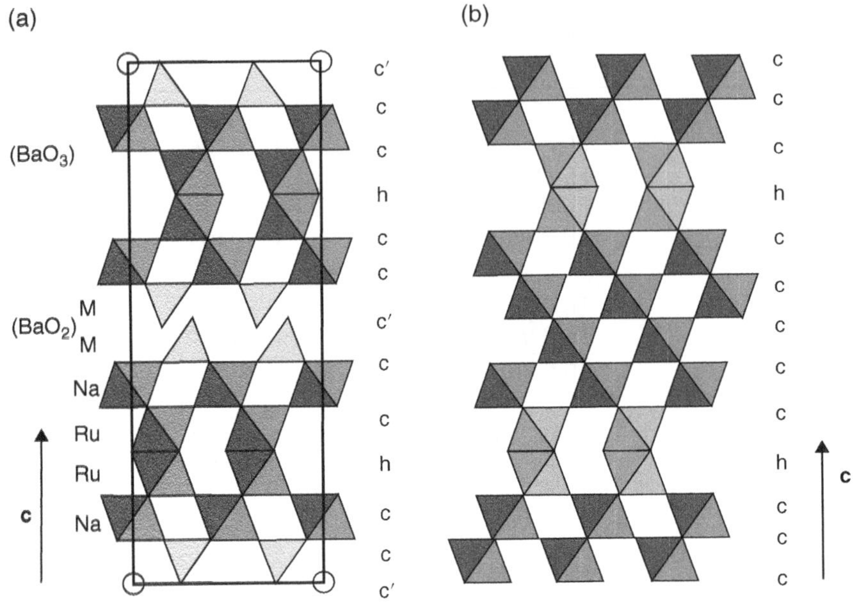

Figure 3.19 *(a) The (ccc′cch)*$_2$ *structure of 12H—*$Ba_6Ru_2Na_2M_2O_{17}$ *compared to (b) the ideal (ccccch)*$_2$ $A_6B_6O_{18}$ *phase. Only the octahedral and tetrahedral framework is shown; A cations are omitted*

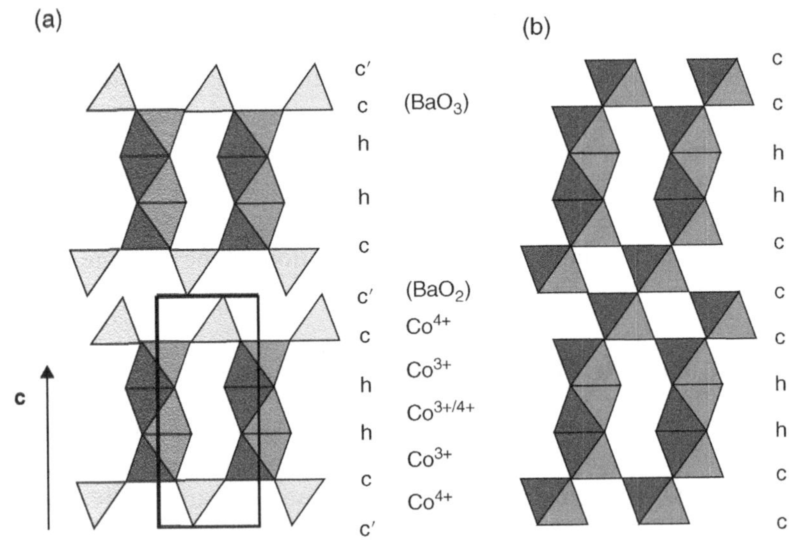

Figure 3.20 *(a) The (cc′chh)*$_2$ *structure of 5H—*$Ba_5Co_5CoO_{14}$ *compared to (b) the ideal (ccchh)*$_2$ $A_5B_5O_{15}$ *phase. Only the octahedral and tetrahedral framework is shown; A cations are omitted*

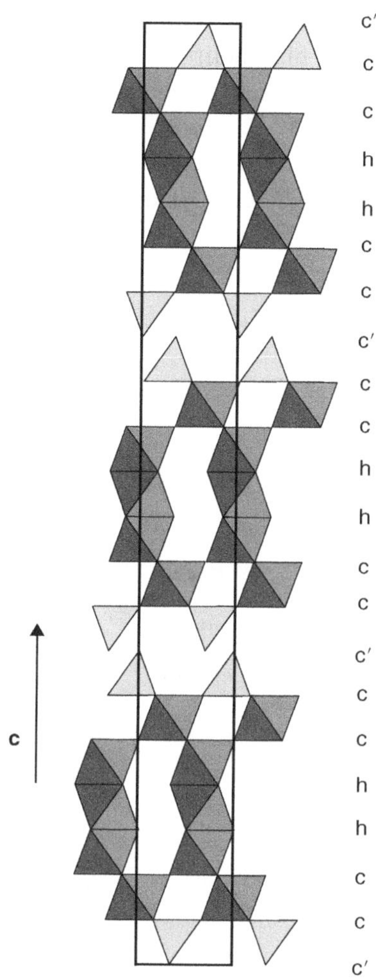

Figure 3.21 The (ccc′cchh)$_3$ structure of 21R—$Ba_7Ca_2Mn_5O_{20}$. Only the octahedral and tetrahedral framework is shown; A cations are omitted

(Figure 3.20b). It is found that Co^{4+} ions also share the middle of the face-shared triplet with Co^{3+}. The other two face-shared octahedra are occupied by Co^{3+} only.

The 21R—$Ba_7Ca_2Mn_5O_{20}$ phase, which is the $n=2$ member of the phase range shown by $Ba_{n+5}Ca_2Mn_{3+n}O_{3n+14}$, has a stacking sequence of (ccc′cchh)$_3$ (Figure 3.21).

3.7.3 (c…c′…chhh) Structures

The packing sequence cc′chhh is adopted by the oxygen-deficient phases 12H—$BaCoO_{2.60}$, 12H—$Ba_{0.9}CoO_{2.60}$ and 12H—$BaCo_{0.6}Mn_{0.4}O_{2.83}$, all approximately $Ba_6Co_6O_{18-\delta}$. The h and c planes have approximate compositions BaO_3 and the c′

planes a composition of BaO_2, and in addition oxygen and barium vacancies may also occur (Figure 3.22a). The hhh motif signifies that there are columns of four face-shared octahedra as the dominant structural feature, separated by cations in tetrahedral coordination.

The packing sequence with further cubic layers $(ccc'cchhh)_2$, is shown by 16H–$Ba_4Ca_{0.9}Mn_{3.1}O_{11.3}$ and the Cr-substituted phase 16H–$Ba_4Ca_{0.85}Mn_{2.65}Cr_{0.5}O_{11.5}$

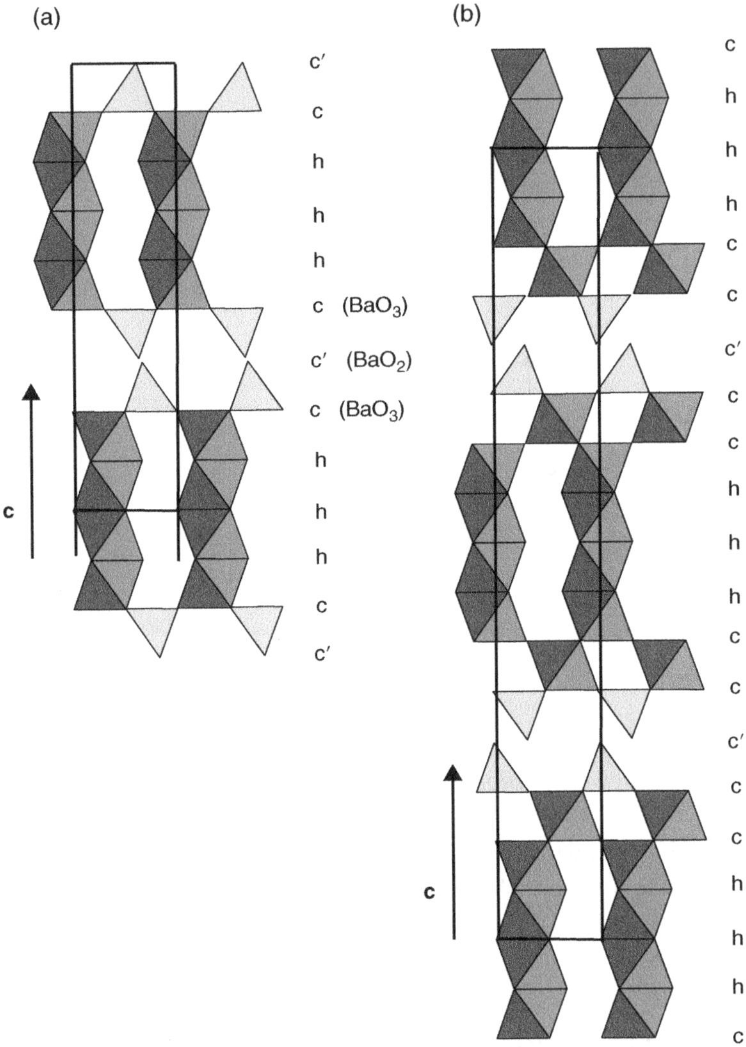

Figure 3.22 *(a) The $(cc'chhh)_2$ structure of 12H–$BaCoO_{2.60}$. (b) The $(ccc'cchhh)_2$ structure of 16H–$Ba_4Ca_{0.85}Mn_{2.65}Cr_{0.5}O_{11.5}$. Only the octahedral and tetrahedral framework is shown; Ba^{2+} cations are omitted*

(Figure 3.22b). It contains blocks of four face-sharing octahedra, and, as before h and c layers have the composition BaO_3, while the c′ layers have a nominal composition of BaO_2 and are bordered by ions in tetrahedral coordination. The positions of the additional vacancies required to make these compositions agree with the measured composition are not specified.

3.8 Anion-Deficient Phases with BaOX Layers

3.8.1 (h′) Layers

A number of hexagonal perovskite oxide phases form anion-deficient structures akin to those just described, in which the O^{2-} anions are partly replaced by halogen X^- anions. These typically arise with oxides of Mn, Fe and Co, which have variable valence. In these phases a BaO_3 layer loses oxygen and gains an X^- anion to form BaOX layers. The packing of the ions in this layer is no longer cubic with respect to the neighbouring layers, but takes on a hexagonal or approximately hexagonal (h′) position between two c layers, to form a ch′c motif between columns of face-sharing octahedra. The cations that are nominally in octahedral interstices on either side of a parent BaO_3 layer generally move into *apex-sharing* tetrahedral sites either side of the BaOX layer, but other coordination polyhedra, including square pyramidal, may also occur. In addition there is a variable amount of additional oxygen non-stoichiometry which is difficult to localise structurally. Moreover, the preparation conditions must be carefully controlled to obtain any of the desired forms, and slow cooling, rapid cooling (quenching) or the use of differing starting materials can all change the structure prepared, even within quite narrow composition ranges.

Only three structures have been characterised to date: the 6H-, 8H- and 10H- repetitions. The 10H- phases, typified by $Ba_5Co_5FO_{13-\delta}$, have a stacking sequence $(ch'chh)_2$ (Figure 3.23a). The nominal hexagonal unit cells of these phases are often distorted, and some phases are described in terms of lower symmetry unit cells. For example, hexagonal 10H—$Ba_5Fe_5O_{14}$ (Table 3.10), which forms in the absence of a halogen, is often described in terms of an orthorhombic cell (space group *Cmcm* (63), $a=0.57615$ nm, $b=0.99792$ nm, $c=2.4374$ nm) rather than a 10H- hexagonal cell. The correspondence between the two is $a(\text{ortho})=a(10\text{H})$, $b(\text{ortho})=\sqrt{3}a(10\text{H})$, $c(\text{ortho})=c(10\text{H})$.

The 6H structure (Figure 3.23b) which has columns of four face-sharing octahedra separated by BaOX layers, with a stacking sequence $(ch'chhh)_2$, is adopted by a number of compounds including $Ba_6Co_6FO_{16-\delta}$. In reality there is considerable difficulty in clarifying even the number of phases present in preparations, let alone the structures of the phases therein. This is so for the nominally single phase 6H—$Ba_6Co_{6-y}Mn_yClO_{16-\delta}$, which was originally described as orthorhombic: space group *Ammm* (65), $a=1.455$ nm, $b=0.569$ nm, $c=0.988$ nm. It is now considered to be a mixture of two phases with

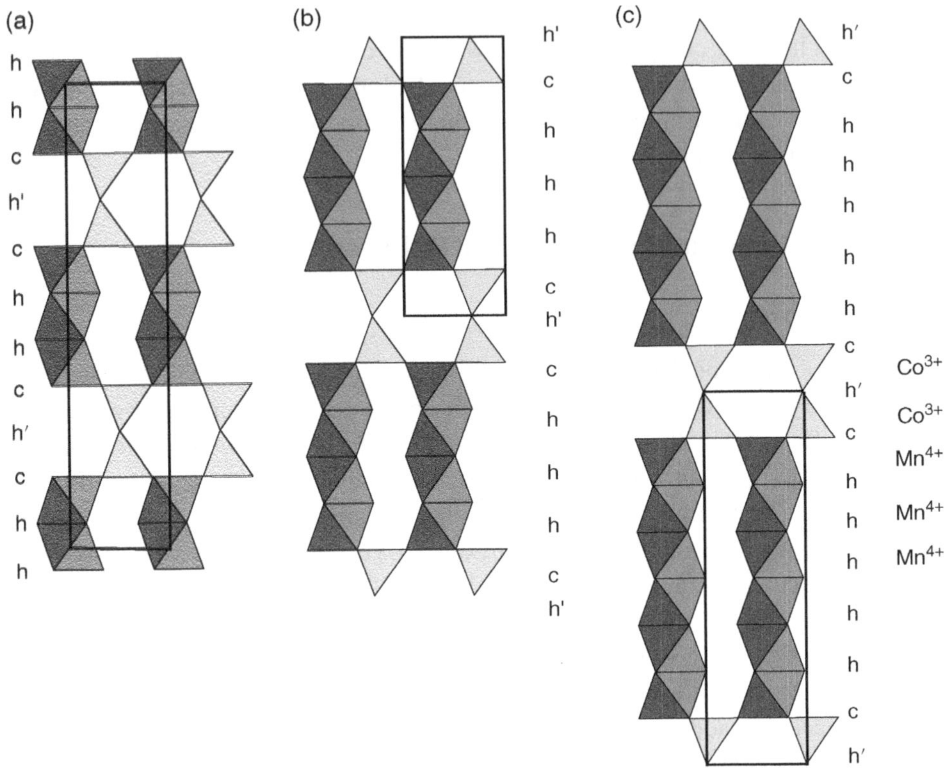

Figure 3.23 *Idealised structures of (a) 10H—$Ba_5Co_5FO_{13-\delta}$; (b) 6H—$Ba_6Co_6FO_{16-\delta}$; (c) 8H—$Ba_8Ca_2Mn_6ClO_{22}$. Only the octahedral and tetrahedral framework is shown; Ba^{2+} cations are omitted*

Table 3.10 *Some phases with BaOX layers*

Phase	Space group	*a* (nm)	*c* (nm)
(ch′chh), 10H			
$Ba_5Fe_5O_{14}$	H, $P6_3/mmc$ (194)	0.57794	2.46087
$Ba_5Co_5FO_{13-\delta}$	H, $P6_3/mmc$ (194)	0.56878	2.3701
$Ba_5Co_5ClO_{13}$	H, $P6_3/mmc$ (194)	0.56980	2.4469
$Ba_{10}Fe_8Pt_2Cl_2O_{25}$	H, $P6_3/mmc$ (194)	0.58034	2.4997
(ch′chhh), 6H			
$Ba_6Co_6FO_{16-\delta}$	H, $P\overline{6}m2$ (187)	0.56683	1.4277
$Ba_6Co_6ClO_{16}$	H, $P\overline{6}m2$ (187)	0.56698	1.44654
$Ba_6Co_5MnFO_{16-\delta}$	H, $P\overline{6}m2$ (187)	0.5680	1.4300
(ch′chhhhh), 8H			
$Ba_8Ca_2Mn_6ClO_{22}$	H, $P\overline{6}m2$ (187)	0.57207	1.94099
(cc′c′chhh), 14H			
$Ba_7Co_6BrO_{17}$	H, $P6_3/mmc$ (194)	0.56611	3.35672
(cc′c′chh), 18R			
$Ba_6Co_5BrO_{14}$	Tr, $R\overline{3}m$ (166)	0.56578	4.3166

a_1=0.569 nm, c_1=1.453 nm and a_2=0.567 nm, c_2=1.454 nm; these arise because of miscibility problems between the Mn and Co components.

The 8H structure (Figure 3.23c) has columns of six face-sharing octahedra separated by BaOX layers, with a packing sequence (ch′chhhhh). Only one example has so far been characterised: $Ba_8Ca_2Mn_6ClO_{22}$.

3.8.2 (c′c′) Layers

Two structures containing (c′c′) pairings are known, 14H—$Ba_7Co_6BrO_{17}$, with a stacking sequence $(cc'c'chhh)_2$, and 18R—$Ba_6Co_5BrO_{14}$, with a stacking sequence $(cc'c'chh)_3$. These contain two (BaOBr) sheets together in a (c′c′) motif. The phases are structurally similar to those described in Section 3.7, in that the tetrahedra are not linked, but now a layer of Br^- ions is interleaved between them (Figure 3.24a and b). The atoms close to the (BaOBr) layers are rather disordered in the preparations made to date.

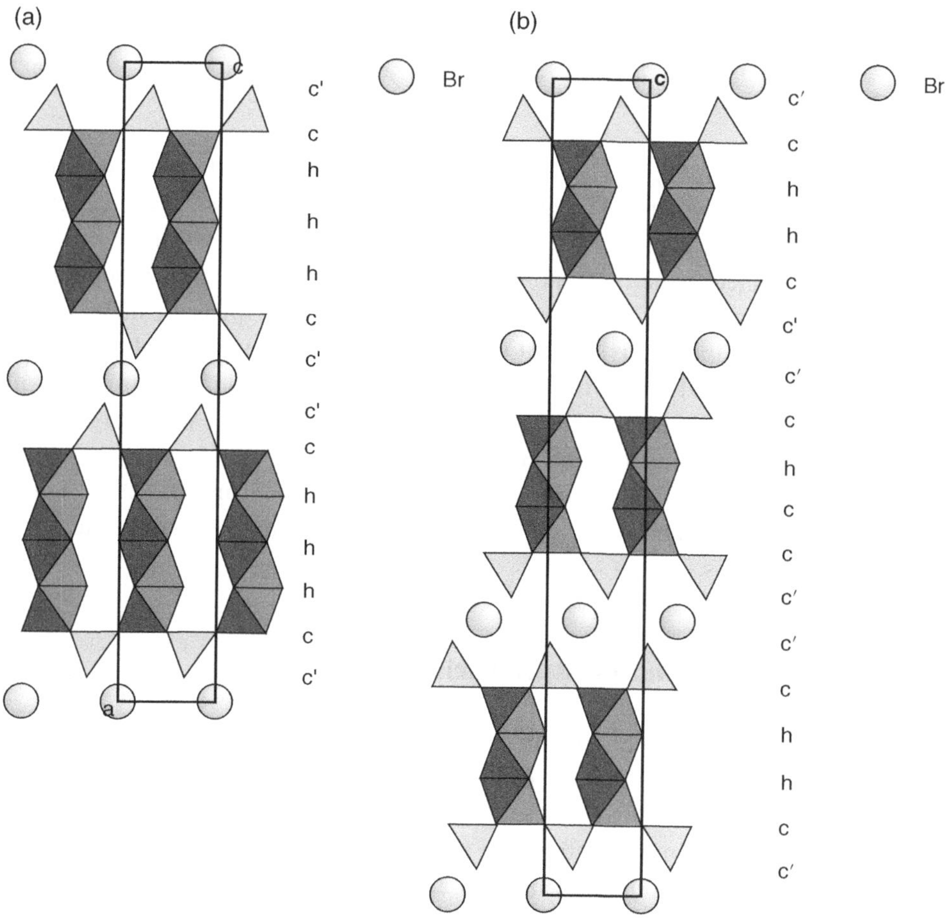

Figure 3.24 *Idealised structures of (a) 14H—$Ba_7Co_6BrO_{17}$; (b) 18R—$Ba_6Co_5BrO_{14}$. Only the octahedral and tetrahedral framework is shown; Ba^{2+} cations are omitted for clarity*

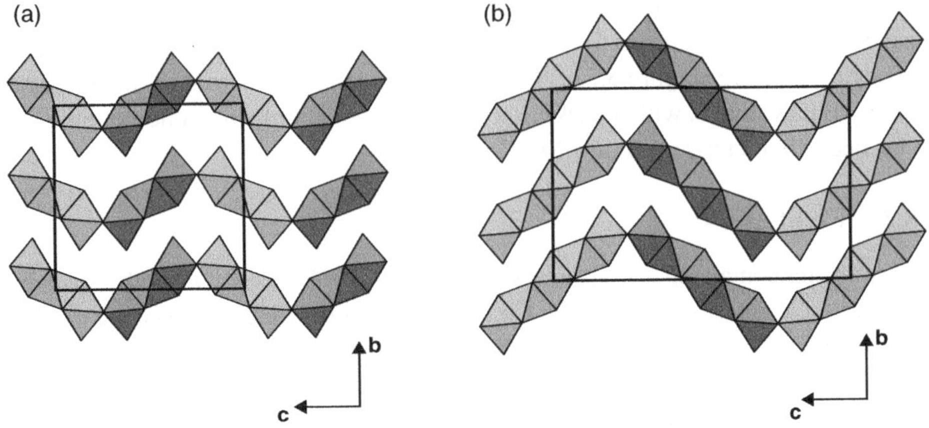

Figure 3.25 Idealised structures of (a) $Sr_4Mn_3O_{10}$; (b) $Ba_6Mn_5O_{16}$ projected down [100]. The chains with different shading lie at different heights along the **a**-axis. Only the octahedral framework is shown; A cations are omitted

3.9 $Sr_4Mn_3O_{10}$ and $Ba_6Mn_5O_{16}$

The two phases $Sr_4Mn_3O_{10}$ and $Ba_6Mn_5O_{16}$ have short columns of face-shared MnO_6 octahedra linked by two apical oxygen ions in a hinged arrangement. The structure of $Sr_4Mn_3O_{10}$ is analogous to 9R—$BaRuO_3$ (Figure 3.9a), with columns of three face-sharing octahedra and orthorhombic unit cell parameters: space group *Cmca* (64), $a = 0.54766$ nm, $b = 1.24659$ nm, $c = 1.25282$ nm. These triplets are offset in the unit cell parallel to the **a**-axis, so that the structures are not twinned, although the projection down [100] gives this impression (Figure 3.25a). The structure of $Ba_6Mn_5O_{16}$ has a similarity to 15R—$BaMnO_{2.90}$ (Figure 3.9c), but with hinged columns of five face-sharing octahedra (Figure 3.25b) and orthorhombic unit cell parameters: space group *Cmca* (64), $a = 0.57071$ nm, $b = 1.31856$ nm, $c = 1.99273$ nm. A number of fluorides such as $Cs_4Ni_3F_{10}$ and $Cs_6Ni_5F_{16}$ also adopt the same structures as these oxide phases.

3.10 Temperature and Pressure Variation

As with all perovskites, the structures described earlier are sensitive not only to changes in composition but also to changes in temperature and pressure. Just two examples are described.

In terms of temperature variation, the phase 6H—$Ba_3BiIr_2O_9$, with a hexagonal structure at high temperatures (Section 3.5.2, Table 3.6), adopts a low-temperature monoclinic unit cell: space group *C2/c* (15), $a=0.59531$ nm, $b=1.02792$ nm, $c=1.48892$ nm, $\beta=91.934°$. This decent in symmetry as the temperature falls is a common motif within all perovskites, and the example here is similar to many others cited earlier.

The simplified pressure–composition diagram for $BaIrO_3$–$BaMnO_3$ shows that changes between monoclinic, trigonal and hexagonal forms are pressure and composition dependent (Figure 3.26). At ambient pressure (1 atm = 0.000101 GPa), $BaIrO_3$ exists as a monoclinic (9M) form of the 9R $(chh)_3$ structure (Section 3.5.1). As $BaMnO_3$ is added, the monoclinic distortion decreases until an approximate composition of $BaIr_{0.6}Mn_{0.4}O_3$ is reached, at which point the structure adopts the more symmetrical 9R form. This phase persists until an approximate composition of $BaIr_{0.22}Mn_{0.78}O_3$ is reached. Thereafter a two-phase region is found. Here the 9R form, composition $BaIr_{0.22}Mn_{0.78}O_3$, coexists with the 2H—$BaMnO_3$ structure type, with an approximate composition $BaIr_{0.065}Mn_{0.935}O_3$. Further addition of $BaMnO_3$ does not change this structure, which persists up to pure 2H—$BaMnO_3$.

A similar two-phase region is encountered if pure 2H—$BaMnO_3$ is subjected to increasing pressure. At approximately 0.95 GPa the 2H structure loses stability. Over the approximate pressure range of 0.95–2.25 GPa the 2H structure coexists with the 9R structure, and at higher pressures a single-phase 9R—$BaMnO_3$ form is stable. Increasing pressure on monoclinic 9M—$BaIrO_3$ leads to the formation of a monoclinic 5M phase and then to a monoclinic 6M phase. Increase in $BaMnO_3$ content causes the high-pressure 6M phase to transform initially back to the 5M form and then to a new structure, the 4H—$BaMnO_{2.65}$ structure, with a (ch) stacking.

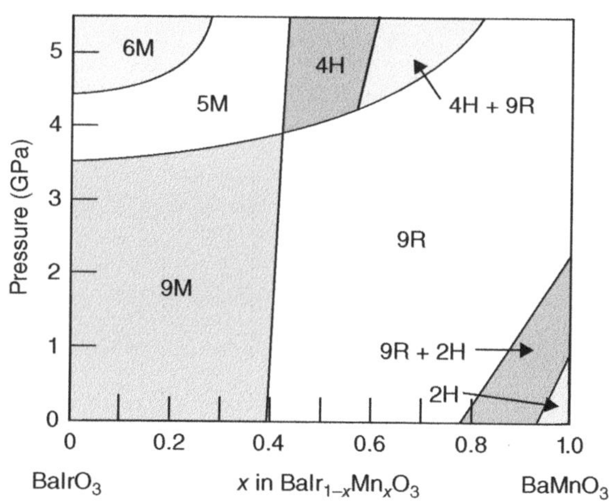

Figure 3.26 *Simplified pressure–composition phase diagram for the $BaIrO_3$–$BaMnO_3$ system (Original data in Zhou* et al. *(2010))*

Reference

J. G. Zhou *et al.*, *J. Solid State Chem.*, **183**, 720–726 (2010).

Further Reading

The structural chemistry of the $BaNiO_3$-derived hexagonal perovskites containing octahedra and trigonal prisms is surveyed in the following:

G. R. Blake, J. Sloan, J. F. Vente, P. D. Battle, *Chem. Mater.*, **10**, 3536–3547 (1998).

J. Darriet, M. A. Subramanian, *J. Mater. Chem.*, **5**, 543–552 (1995).

J. M. Perez-Mato, M. Zakhour-Nakhl, F. Weill, J. Darriet, *J. Mater. Chem.*, **9**, 2795–2808 (1999).

The close packing of spheres, the structures so generated and the terminology employed are described in the following:

H. Jagodzinsky, *Acta Crystallogr.*, **2**, 201 (1949).

P. Krishna, D. Pandy, Close-packed Structures, International Union of Crystallography, Pamphlet 5, at http://www.iucr.org/education/pamplets (Accessed November 13, 2015).

A. F. Wells, *Structural Inorganic Chemistry*, 5th edition, Oxford University Press, Oxford (1984), p578–589, Chapters 4 and 13.

A full list of Ru-containing perovskite phases of the 3C and 6H structures as well as other information and 83 references to previous studies is given in the following:

E. Quarez, F. Abraham, O. Mentré, *J. Solid State Chem.*, **176**, 137–150 (2003).

4

Modular Structures

Many phases related to the perovskite structure are modular in nature, being built from slabs of a parent perovskite interleaved with other structures. These phases are able to accommodate composition change by changing the thickness of the perovskite units or the inter-slab species. When the modular components are ordered, regular homologous series of phases occur. When modules are disordered, the crystals contain large numbers of planar faults and may be best viewed as non-stoichiometric. Table 4.1 summarises the structural relationships between four important series, excluding the superconducting cuprates (Tables 4.7 and 4.8).

4.1 K_2NiF_4 (A_2BX_4) and Ruddlesden–Popper Phases

4.1.1 The K_2NiF_4 (T or T/O) Structure

The K_2NiF_4 structure, which is also called the T or T/O structure in the superconductor literature, is built from slabs consisting of a single layer of apex-linked octahedra. The Ni ions occupy the octahedra, and the K ions are located at the centre of a square of oxygen ions formed by the vertices of the nearest octahedra and directly below the apical oxygen of the nearest octahedron in the layer above or below (Figure 4.1). The slabs can be considered to be slices of the ideal perovskite structure ($KNiF_3$) cut parallel to the (100) planes and one octahedron in thickness, each of

Perovskites: Structure–Property Relationships, First Edition. Richard J. D. Tilley.

Table 4.1 *Modular perovskites*

Name	Interlayer formula	Perovskite layer formula	Perovskite slab orientation	Perovskite slab displacement	Series formula	Examples
Ruddlesden–Popper	AO–AO	$A_{n-1}B_nO_{3n-1}$	$[100]_p$	$(\mathbf{a}_p+\mathbf{b}_p)/2$	$A_{n+1}B_nO_{3n+2}$	Sr_2TiO_4 ($n=1$) $Sr_3Ti_2O_7$ ($n=2$)
Dion–Jacobson	A′	$A_{n-1}B_nO_{3n+1}$	$[100]_p$	0 (Type I); $\mathbf{a}/2$ (Type II) $(\mathbf{a}_p+\mathbf{b}_p)/2$ (Type III)	$A'A_{n-1}B_nO_{3n+1}$	$LiLaNb_2O_7$ ($n=2$) $RbCa_2Ta_3O_{10}$ ($n=3$)
Aurivillius	Bi_2O_2	$A_{n-1}B_nO_{3n+1}$	$[100]_p$	$(\mathbf{a}_p+\mathbf{b}_p)/2$	(Bi_2O_2) $(A_{n-1}B_nO_{3n+1})$	Bi_2WO_6 ($n=1$) $Bi_4Ti_3O_{12}$ ($n=3$)
$Ca_2Nb_2O_7$	—	$A_nB_nO_{3n+2}$	$[110]_p$	$(\mathbf{a}_p+\mathbf{b}_p)/2$	$A_nB_nO_{3n+2}$	$Ca_4Ti_4O_{14}$ ($n=4$)

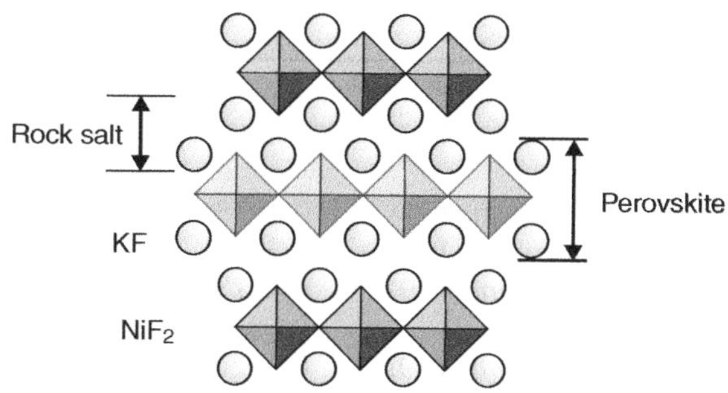

Figure 4.1 *The ideal K_2NiF_4 structure; the perovskite layer in lighter shading is displaced relative to its neighbours by $(a_p+b_p)/2$*

composition K_2NiF_4. Alternatively the structure is made up of a succession of rock-salt (RS) structure KF layers, separated by NiF_2 sheets in the sequence ...KF, KF, NiF_2.... The phases are ideally tetragonal, with $a=b\approx a_p$, $c\approx 3.5a_p$. Each slab is displaced relative to its neighbours by (in the ideal case) a vector $(\mathbf{a}_p+\mathbf{b}_p)/2$, where $\mathbf{a}_p$ and $\mathbf{b}_p$ are the ideal cubic unit cell vectors. The structure is adopted by a large number of both halide and oxide phases (Table 4.2). If the phases are treated as ionic for the purposes of classification, with the charge on the cations, q_A and q_B, being balanced by the charge on the anions, q_X, then

$$q_X=-1,(q_A,q_B)=(1,2);\quad \text{for example, } K_2NiF_4;$$
$$q_X=-2,(q_A,q_B)=(1,6);\quad \text{for example, } K_2UO_4;$$
$$(q_A,q_B)=(2,4);\quad \text{for example, } Sr_2TiO_4;$$
$$(q_A,q_B)=(3,2);\quad \text{for example, } La_2NiO_4.$$

Octahedral distortion and tilting frequently lower the symmetry of these phases. For example, La_2CuO_4 at room temperature is described as orthorhombic (sometimes monoclinic), mainly due to a Jahn–Teller elongation of the CuO_6 octahedra. These now contain two long (0.240 nm) bonds parallel to the unit cell **c**-axis and four short (0.189 nm) bonds parallel to the **a**- and **b**-axis at room temperature, resulting in the orthorhombic symmetry. As the temperature rises, this distortion lessens, and at temperatures above approximately 523 K, the structure adopts the ideal tetragonal form.

Doping and substitution on the A-, B- and X-sites by isovalent or aliovalent ions can be viewed in terms of charge balance, exactly as the ABX_3 phases. Thus, mixed oxyhalide phases can be described in which the charge imbalance generated by halide ion substitution for an oxide ion is balanced by the appropriate changes in the A and B cations, as in Sr_2FeO_3F, in which $2Sr^{2+}+Fe^{3+}$ balance the

Table 4.2 *K_2NiF_4 and related structures*

Phase	Space group	*a* (nm)	*b* (nm)	*c* (nm)
n = 1: K_2NiF_4				
K_2NiF_4	T, *I4/mmm* (139)	0.4013		1.3088
K_2MnF_4	T, *I4/mmm* (139)	0.4174		1.3272
K_2CoF_4	T, *I4/mmm* (139)	0.4073		1.3087
K_2CuF_4	T, *Bbcm* (64)	0.59043		1.2734
$HLaTiO_4$	T, *P4/nmm* (129)	0.37201		1.22914
$DLaTiO_4$	T, *P4/nmm* (129)	0.37232		1.23088
Sr_2TiO_4	T, *I4/mmm* (139)	0.38859		1.2597
$DLaTiO_4$	T, *P4/nmm* (129)	0.37232		1.23088
$DNdTiO_4$	T, *P4/nmm* (129)	0.37039		1.20883
$NaLaTiO_4$	T, *P4/nmm* (129)	0.37998		1.3273
$NaPrTiO_4$	T, *P4/nmm* (129)	0.37686		1.2940
$NaNdTiO_4$	T, *P4/nmm* (129)	0.37551		1.2848
$NaSmTiO_4$	T, *P4/nmm* (129)	0.37613		1.2634
$CaPrCoO_4$	T, *I4/mmm* (139)	0.37486		1.22437
$CaNdCoO_4$	T, *I4/mmm* (139)	0.373817		1.19168
La_2NiO_4	T, *I4/mmm* (139)	0.38616		1.26672
La_2CuO_4, 300 K	O, *Bmab* (64)	0.53357	0.54058	1.31432
La_2CuO_4, 1073 K	T, *I4/mmm* (139)	0.38329		1.3313
Ca_2SnO_4	O, *Pccn* (56)	0.57290	0.57352	1.25811
Ca_2RuO_4	O, *Pbca* (61)	0.53869	0.56334	1.17349
Sr_2RuO_4	T, *I4/mmm* (139)	0.38751		1.27278
K_2UO_4	T, *I4/mmm* (139)	0.43321		1.31382
Ca_2FeO_3Cl	T, *P4/nmm* (129)	0.38391		1.36732
Ca_2FeO_3Br	T, *P4/nmm* (129)	0.38440		1.41775
Sr_2FeO_3F	T, *P4/nmm* (129)	0.38664		1.31773
Sr_2FeO_3Cl	T, *P4/nmm* (129)	0.39204		1.42966
$Sr_2CuO_2F_2$	T, *I4/mmm* (139)	0.39670		1.28160
Sr_2TaO_3N	T, *I4/mmm* (139)	0.40390		1.26007
Ba_2TaO_3N	T, *I4/mmm* (139)	0.41151		1.33718
n = 2: $A_3B_2O_7$				
$La_2SrAl_2O_7$	T, *I4/mmm* (139)	0.37712		2.0197
$Sm_2SrAl_2O_7$	T, *I4/mmm* (139)	0.37159		1.9876
$Sr_3Ti_2O_7$	T, *I4/mmm* (139)	0.38988		2.03078
$Sr_3Fe_2O_7$	T, *I4/mmm* (139)	0.38637		2.01464
$Ca_3Mn_2O_7$	T, *I4/mmm* (139)	0.3683		1.9575
$Ca_{2.5}La_{0.5}Mn_2O_7$	O, $Cmc2_1$ (36)	1.93154	0.53683	0.53403
$Sr_3Mn_2O_6$	T, *P4/mbm* (127)	1.08682		2.02051
$Sr_{1.76}La_{1.24}Mn_2O_7$	T, *I4/mmm* (139)	0.3871		2.0191
$Sr_{1.6}La_{1.4}Mn_2O_7$	T, *I4/mmm* (139)	0.38682		2.0274
$La_3Ni_2O_7$	T, *I4/mmm* (139)	0.38742		2.0055
$La_3Ni_2O_{6.35}$	T, *I4/mmm* (139)	0.38741		2.0055
$Li_2LaTa_2O_7$	T, *I4/mmm* (139)	0.39250		1.9089

Table 4.2 *(continued)*

Phase	Space group	*a* (nm)	*b* (nm)	*c* (nm)
n = 3: $A_4B_3O_{10}$				
$Sr_3Ti_2O_{10}$	T, *I4/mmm* (139)	0.3903		2.814
$Sr_3PrFe_3O_{10}$	T, *I4/mmm* (139)	0.38582		2.79609
$(Sr_{0.775}La_{0.225})_4Fe_3O_{10}$	T, *I4/mmm* (139)	0.38878		2.81813
$Sr_3PrFe_3O_{10}$	T, *I4/mmm* (139)	0.38582		2.79609
$Sr_3PrFe_{1.5}Co_{1.5}O_{10}$	T, *I4/mmm* (139)	0.38263		2.77714
$La_4Ni_{2.7}Fe_{0.3}O_{10}$	O, *Cmca* (64)	0.54267	2.8013	0.54805
$Ca_4Mn_3O_{10}$	O, *Pbca* (61)	0.52650	0.52614	2.68171
$Li_2Ca_2Ta_3O_{10}$	T, *I4/mmm* (139)	0.38927		2.6591
$Na_2Ca_2Ta_3O_{10}$	T, *I4/mmm* (139)	0.38872		2.8655
n = 7				
$Li_4Sr_3Nb_6O_{20}$	T, *I4/mmm* (139)	0.3953		2.6041

X-site charge of −7, derived from the O^{2-} and F^- ions. In these sorts of cases, where several ionic species ions are distributed over identical sites, they are frequently distributed at random, as in $Sr_2Fe(O_3F)$, in which the O^{2-} and F^- ions occupy the same X-sites at random, $Ca_2(Ru, Cr)O_4$ with the octahedral B-sites (Ru, Cr), occupied at random $(Ln, Ca)_2(Mn, Ni)O_4$ and $(La, Sr)_2(Fe, Ru)O_4$ (in which both the interlayer A-sites and the octahedral B-sites are occupied at random). In a smaller number of phases the cations have been found to order in a fashion analogous to the double perovskites. An example is provided by a number of $NaLnTiO_4$ compounds. Here the Ln cations are ordered so that the structure has the layer sequence ···(-LnO-TiO_2-NaO-NaO-TiO_2-LnO-)···. The analogous phases in which the Na atoms are replaced by hydrogen or deuterium, $HLnTiO_4$ and $DLnTiO_4$, are similarly ordered.

4.1.2 Ruddlesden–Popper Phases

The Ruddlesden–Popper phases are a homologous series built from slabs of the perovskite-like $SrTiO_3$ with series formula $Sr_{n+1}Ti_nO_{3n+1}$. The simplest member of the series, Sr_2TiO_4 ($n = 1$), adopts the K_2NiF_4 structure and is described as a stacking of single slabs of ideal $SrTiO_3$ (Figure 4.2a). Increasing the perovskite slab thickness while keeping the inter-slab structure the same generates the phases $Sr_3Ti_2O_7$ ($n = 2$) and $Sr_4Ti_3O_{10}$ ($n = 3$), with $SrTiO_3$ itself corresponding to $n = \infty$ (Figure 4.2b–d). As with K_2NiF_4 and Sr_2TiO_4, adjacent perovskite sheets are displaced by a vector $(\mathbf{a}_p + \mathbf{b}_p)/2$, and the A ions sit at the centres of four

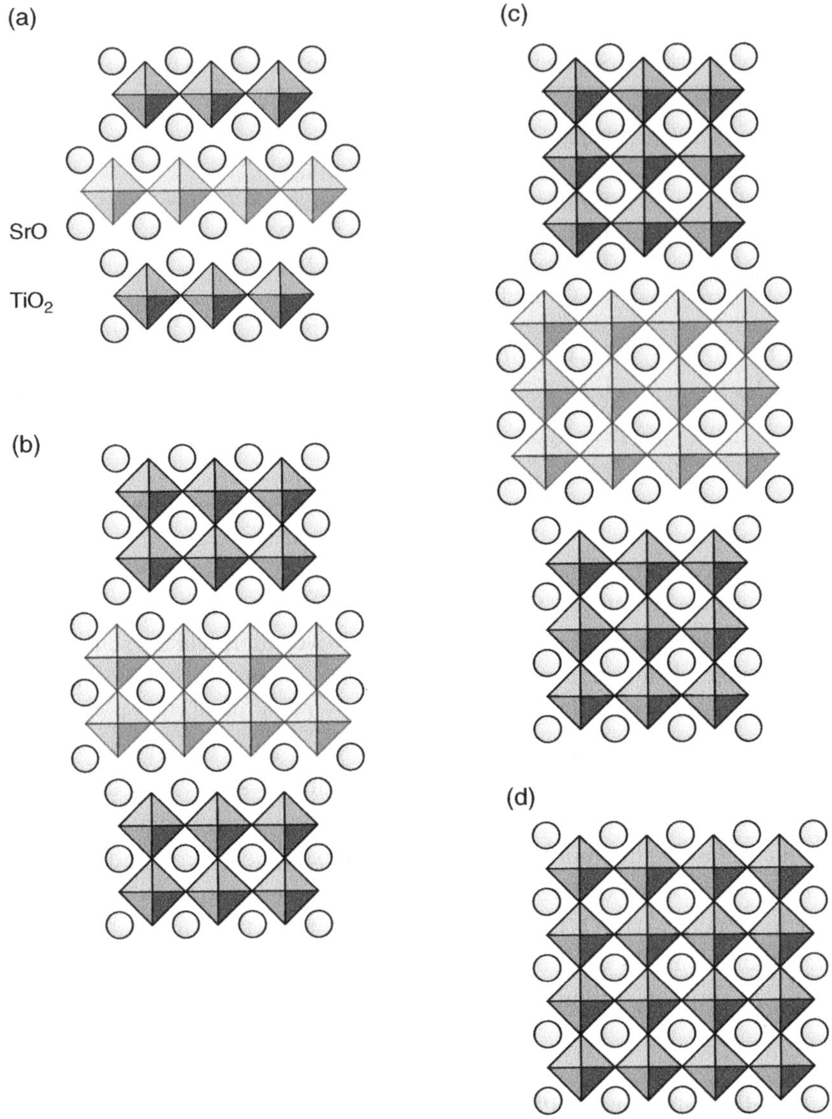

Figure 4.2 *The idealised Ruddlesden–Popper series; (a) n = 1, Sr_2TiO_4; (b) n = 2, $Sr_3Ti_2O_7$; (c) n = 3, $Sr_4Ti_3O_{10}$; (d) n = ∞, $SrTiO_3$*

apical oxygen ions of the four neighbouring octahedra and directly below the apical oxygen ion of the octahedron in the neighbouring layer. Phases with values of n higher than 3 do not occur in the Sr_2TiO_4–$SrTiO_3$ system, but crystals within this composition range are built of disordered regions of $SrTiO_3$ and the $n = 3$ phase.

Many examples of oxide-based $A_{n+1}B_nO_{3n+1}$ Ruddlesden–Popper analogues have been synthesised (Table 4.2). Assuming that ionic charges occur, the charges can be balanced according to

$$(n+1)q_A + (n)q_B = 2(3n+1)$$

This charge balance can be fulfilled in a large number of ways: for $n=2$, $q_A=(3, 3, 2)$ and $q_B=3$ is found in $La_2SrAl_2O_7$ and for $n=3$, $q_A=(3, 2, 2, 2)$ and $q_B=(3, 4, 4)$ occurs in $PrSr_3Fe_3O_{10}$.

The structures can be described in a number of ways, each of which is suited to an explanation of differing properties. Each entire perovskite-like slab, including the edges of the sheets, has a formula $A_{n+1}B_nO_{3n+1}$, but if the region between these slabs needs to be singled out, it is convenient to describe the structures as a succession of [AO–AO–($A_{n-1}B_nO_{3n-1}$)] planes and slabs, where the (AO) pairs adopt the RS structure to form $(AO)_2(A_{n-1}B_nO_{3n-1})$ compounds. For comparison with the Dion–Jacobson (Section 4.3) and Aurivillius phases (Section 4.4), it is convenient to separate the two outer sets of A atoms in each perovskite-like sheet from the rest of the slab to write the formula as $A_2(A_{n-1}B_nO_{3n+1})$.

In practice, many phases are disordered and contain random or partly ordered wider or narrower bands of perovskite-like structure corresponding to different n values. Non-stoichiometry is common for phases containing Co, Mn or Fe as the B-site cation. In addition, octahedral tilting and distortion are present in the perovskite slabs, and these both lower the symmetry of the structures and influence their physical properties.

4.2 The Nd_2CuO_4 (T′) and T* Structures

The structure of Nd_2CuO_4, which, in the superconductor literature, is called the T′ structure, is very similar to La_2CuO_4, the principal difference lying in the disposition of the oxygen atoms, as the cations in the two structures are in almost identical positions (Figure 4.3a). In Nd_2CuO_4 the octahedral sheets of the K_2NiF_4 structure are replaced with CuO_2 layers in which the Cu^{2+} ions are in square planar coordination (Figure 4.3b). The Nd^{3+} ions and adjacent O^{2-} ions form Nd_2O_2 layers that are similar in structure to fluorite (CaF_2). The structure may be described as an intergrowth of the perovskite and fluorite structures, and although the analogy is not as obvious as in the K_2NiF_4 phases, this description is convenient when the structures and properties of the superconducting cuprates are described.

A number of A_2BO_4 phases are isostructural to Nd_2CuO_4, including the lanthanoid cuprates Pr_2CuO_4 and Sm_2CuO_4 (Table 4.3). It has also been found that

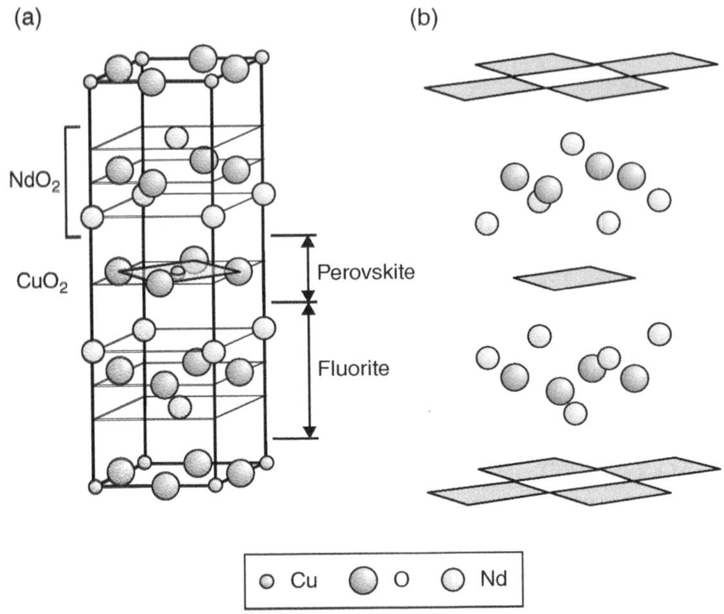

Figure 4.3 *The idealised Nd_2CuO_4 structure: (a) atomic positions; (b) idealised structure showing CuO_4 square planar groups*

Table 4.3 *Nd_2CuO_4 and related structures*

Phase	Space group	*a* (nm)	*c* (nm)
Nd_2CuO_4	T, *I4/mmm* (139)	0.39414	1.21626
Pr_2CuO_4	T, *I4/mmm* (139)	0.39609	1.22210
Sm_2CuO_4	T, *I4/mmm* (139)	0.39136	1.19708
La_2CuO_4 (T′)	T, *I4/mmm* (139)	0.40095	1.25408
$Sr_2CuO_2F_2$	T, *I4/mmm* (139)	0.3967	1.2860

La_2CuO_4 can adopt this structure as a metastable form that reverts to the K_2NiF_4 type on heating. Rather surprisingly, the phase $Sr_2CuO_2F_2$ also adopts the Nd_2CuO_4 structure, although the fluorine-rich compound $Sr_2Cu_2O_2F_{2+\delta}$ takes on the La_2CuO_4 structure type.

The phase (Ce, Nd, Sr)$_2$CuO$_4$, called the T* structure, is a 1 : 1 intergrowth of the two structure types, La_2CuO_4 and Nd_2CuO_4, with a stacking sequence of [(Nd, Sr) O–(Nd, Sr)O]$_{\text{rocksalt}}$ - CuO_2–[(Ce,Nd)-O_2-(Ce, Nd)]$_{\text{fluorite}}$ layers.

4.3 Dion–Jacobson and Related Phases

If the pair of A_2 atoms at the boundaries of the perovskite-like sheets in the Ruddlesden–Popper phases are replaced with just one A′ atom, the series takes the formula $A'(A_{n-1}B_nO_{3n+1})$, and the materials are called Dion–Jacobson phases (Table 4.4). The perovskite slabs are cut parallel to the [100] planes of an ideal perovskite parent and have a composition $(A_{n-1}B_nO_{3n+1})$. As with the

Table 4.4 *Dion–Jacobson phases*

Phase	Space group	*a* (nm)	*b* (nm)	*c* (nm)
n = 2: $A'AB_2O_7$				
$LiLaNb_2O_7$	T, *I4/mmm* (139)	0.38798		2.03597
$NaLaNb_2O_7$	T, *I4/mmm* (139)	0.39022		2.11826
$RbBiNb_2O_7$	O, *P2₁am* (26)	1.1232	0.5393	0.5463
$RbLaNb_2O_7$	O, *Imma* (74)	0.5501	2.2000	0.5489
$CsBiNb_2O_7$	O, *P2₁am* (26)	0.54953	0.54225	1.13766
$CsNdNb_2O_7$	O, *P2₁am* (26)	0.54722	0.54722	1.11695
$BaSrTa_2O_7$	O, *Immm* (71)	0.39937	0.78428	2.01609
$LiSrNb_2O_6F$	T, *I4/mmm* (139)	0.38304		2.0837
$NaSrNb_2O_6F$	—	0.3848		2.01796
$RbSrNb_2O_6F$	T, *P4/mmm* (123)	0.38503		1.12841
n = 3: $A'A_2B_3O_{10}$				
$CsPr_2Ti_2NbO_{10}$	T, *P4/mmm* (123)	0.38376		1.53015
$CsCaLaTiNb_2O_{10}$	T, *P4/mmm* (123)	0.3896		1.5235
$KCa_2Nb_3O_{10}$	O, *Cmcm* (63)	0.38802	2.9508	0.7714
β-$NaCa_2Nb_3O_{10}$	T, *P4₂/ncm* (138)	0.54731		2.90138
$RbCa_2Nb_3O_{10}$	T, *P4/mmm* (123)	0.38587		1.49108
$RbNa_xCa_2Nb_3O_{10}$	T, *P4₂/ncm* (138)	0.54790		2.88264
$HCa_2Nb_3O_{10}(H_2O)_{0.5}$	T, *P4/mbm* (127)	0.54521		1.4414
$RbLa_2Ti_2TaO_{10}$	T, *P4/mmm* (123)	0.38342		1.52776
$RbLa_2Ti_2TaO_{10}$	T, *P4/mmm* (123)	0.38342		1.52776
$CsPr_2Ti_2TaO_{10}$	T, *P4/mmm* (123)	0.38376		1.53015
$CsNd_2Ti_2TaO_{10}$	T, *P4/mmm* (123)	0.38319		1.52764
$CsSm_2Ti_2TaO_{10}$	T, *P4/mmm* (123)	0.38195		1.52371
$NaCa_2Ta_3O_{10}$	T, *I4/mmm* (139)	0.38607		2.9216
$KCa_2Ta_3O_{10}$	O, *C222* (21)	0.38657	2.9777	0.3852
$LiCa_2Ta_3O_{10}$	T, *I4/mmm* (139)	0.38515		2.8339
$RbCa_2Ta_3O_{10}$	T, *I4/mmm* (139)	0.38573		1.50440
$CsCa_2Ta_3O_{10}$	T, *I4/mmm* (139)	0.38659		1.52538
n = 4: $A'A_3B_4O_{13}$				
$NaCa_2NaNb_4O_{13}$	O, *Immm* (71)	0.3885	0.3884	3.615
$RbCa_2NaNb_4O_{13}$	T, *P4/mmm* (123)	0.38727		1.89115

Ruddlesden–Popper structures, large numbers of cation combinations can give rise to phases of this type. Frequently the A′ and A ions are in a (+1/+3) pairing, combined with B^{5+} in oxides with $n=2$, for example, $KLaNb_2O_7$, or a (+1/+2) pairing combined with B^{5+} in oxides with $n=3$, for example, $RbCa_2Ta_3O_{10}$.

There are three different structures that have been found to accommodate the $A'(A_{n-1}B_nO_{3n+1})$, stoichiometry to date, which are called types I, II and III. Type I structures are characterised by having no displacement between the perovskite layers, which are aligned one above the other (Figure 4.4a and b). This structure forms for the largest A′ ions, typically Rb^+ and Cs^+, and are represented by the $n=3$ phases $RbCa_2Ta_3O_{10}$ and $CsCa_2Ta_3O_{10}$. Ideally the structures are tetragonal with $a=b\approx a_p$, with the **c**-axis taken as perpendicular to the perovskite slabs. The oxygen coordination polyhedron of the A′ cations is cubic.

Type II structures have a displacement of $\mathbf{a}_p/2$ between the successive perovskite layers (Figure 4.4c and d), that is, ½ a unit cell displacement, either along the **a**- or

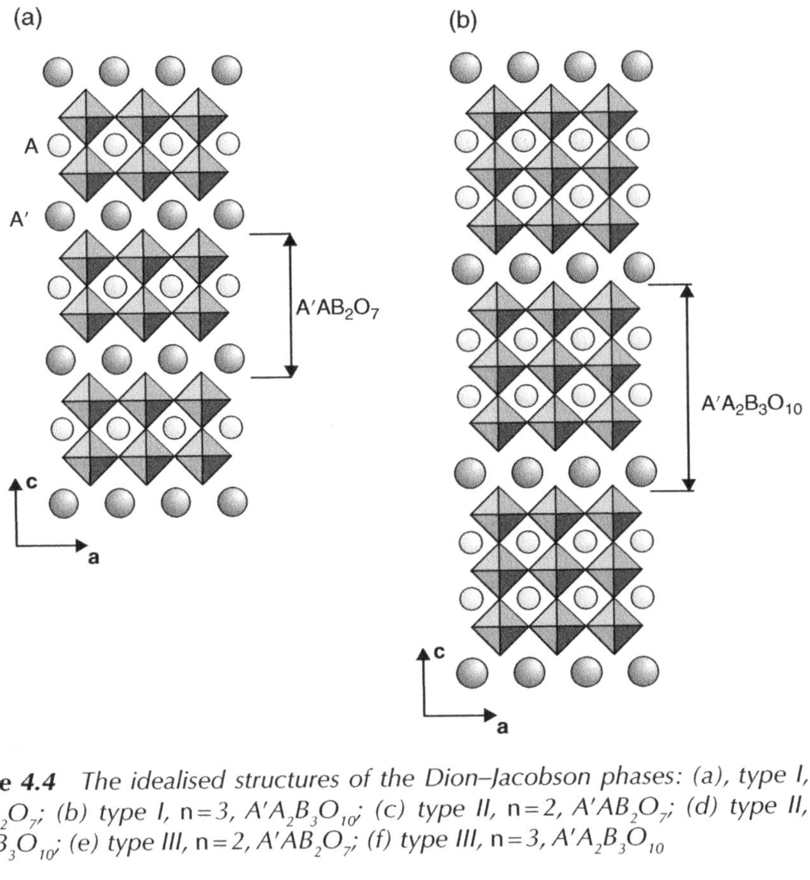

Figure 4.4 *The idealised structures of the Dion–Jacobson phases: (a), type I,* n=2*,* $A'AB_2O_7$*; (b) type I,* n=3*,* $A'A_2B_3O_{10}$*; (c) type II,* n=2*,* $A'AB_2O_7$*; (d) type II,* n=3*,* $A'A_2B_3O_{10}$*; (e) type III,* n=2*,* $A'AB_2O_7$*; (f) type III,* n=3*,* $A'A_2B_3O_{10}$

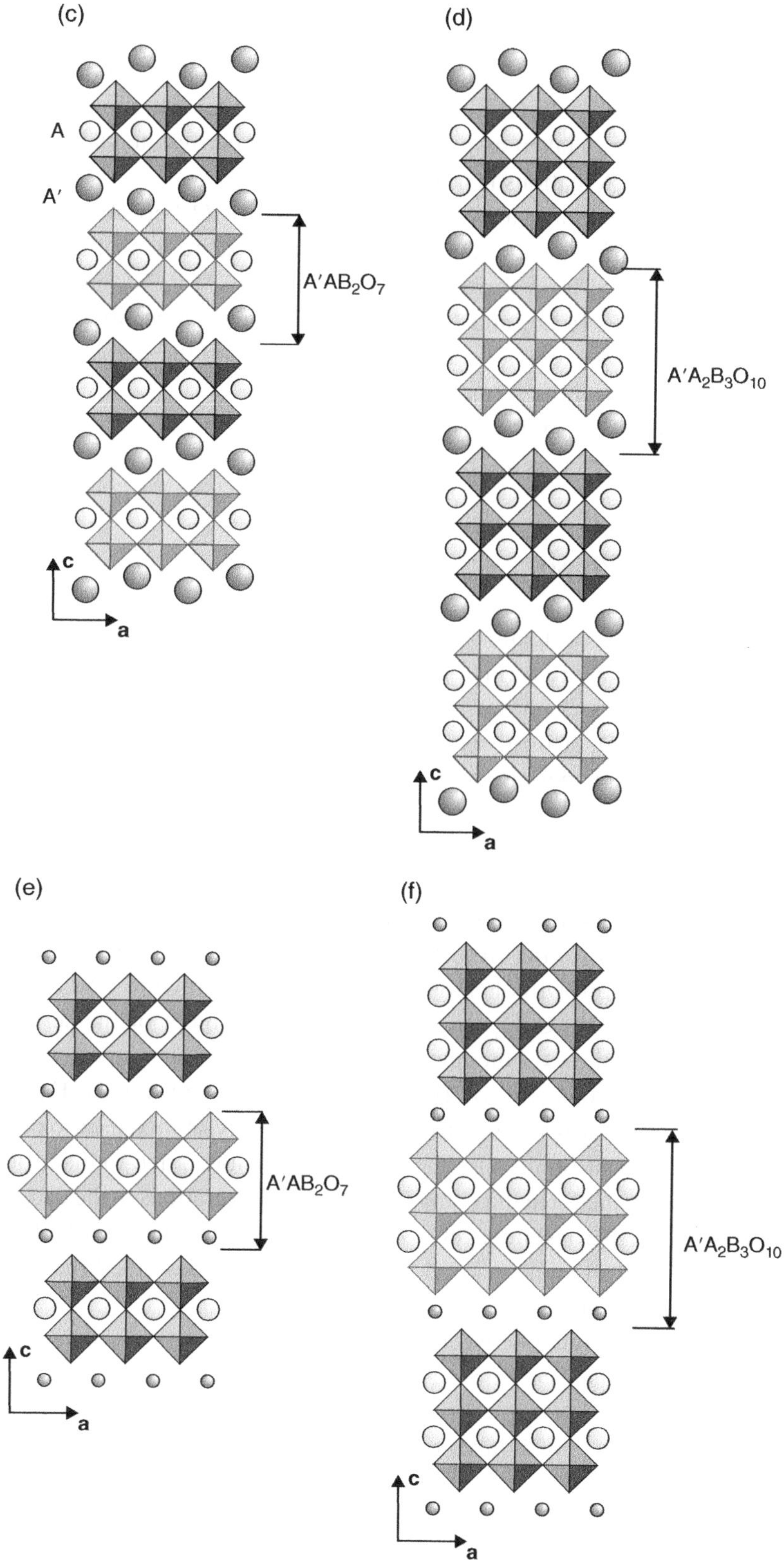

Figure 4.4 *(Continued)*

b-axis. This structure forms for large A′ cations, K^+ and Ba^{2+}, and are represented by the $n=3$ phase $KCa_2Ta_3O_{10}$. The ideal structures are orthorhombic with $a \approx a_p$, $b \approx 2a_p$, with the **c**-axis taken as perpendicular to the perovskite slabs. The oxygen coordination of the A′ cations is trigonal prismatic.

Type III structures have a displacement of $(\mathbf{a}_p + \mathbf{b}_p)/2$ between the successive perovskite layers (Figure 4.4e and f), which is the same as in the Ruddlesden–Popper phases. These form for the smallest A′ cations, Li^+, Na^+ and Ag^+, and are represented by $LiCa_2Ta_3O_{10}$ and $NaCa_2Ta_2O_{10}$. The ideal structures are tetragonal with $a = b \approx a_p$, with the **c**-axis taken as perpendicular to the perovskite slabs. The oxygen coordination of the A′ cations is tetrahedral.

As with the Ruddlesden–Popper phases, non-stoichiometry is common for phases containing Co, Mn or Fe as the B-site cation.

4.4 Aurivillius Phases

The Aurivillius phases again contain slabs of perovskite sliced along the ideal [100] direction. They are formed by replacement of the interlayer A_2 structures in the Ruddlesden–Popper phases and A′ in the Dion–Jacobson phases with a layer of composition Bi_2O_2. This gives the series a general formula $(Bi_2O_2)(A_{n-1}B_nO_{3n+1})$, sometimes written in an ionic form $(Bi_2O_2)^{2+}(A_{n-1}B_nO_{3n+1})^{2-}$ (Table 4.5). As before, the perovskite slabs have the formula $(A_{n-1}B_nO_{3n+1})$, where A is a large cation nominally occupying the perovskite A-site cages and B a medium-sized cation occupying the octahedra. The displacement between adjacent slabs is $(\mathbf{a}_p + \mathbf{b}_p)/2$ as in the Ruddlesden–Popper series. The index n generally runs from 1 to 5. The structure of the Bi_2O_2 layer is similar to that of fluorite, CaF_2, so that the series is also represented by an intergrowth of fluorite and perovskite structure elements and has a resemblance to the Nd_2CuO_4 structure. The $n=1$ member of the series is represented by Bi_2WO_6, the $n=2$ member, by Bi_3TiNbO_9, but the best known member of this series is the ferroelectric $Bi_4Ti_3O_{12}$, in which $n=3$ and A is Bi (Figure 4.5a).

As with the other structures described, the symmetry of mixed cation materials is often lower than that implied by the idealised structures, due to distortions of the metal–oxygen octahedra in the perovskite slabs and alterations in the cation arrangements in the inter-perovskite layers. Structures are often given similar but different space groups, and in some cases it is deemed best to described the structures as incommensurate modulated forms.

For compositions lying between two members of the series, ordered intergrowth of slabs of more than one thickness, especially n and $(n+1)$, is often found. For example, an ordered intergrowth of $n=1$ Bi_2WO_6 and $n=2$ Bi_3TiNbO_9 has a composition $Bi_5TiNbWO_{15}$, with alternating perovskite slabs 1 and 2 octahedra in

Table 4.5 Aurivillius phases

Phase	Space group	*a* (nm)	*b* (nm)	*c* (nm)
n = 1				
Bi_2WO_6	O, *P*2$_1$*ab* (29)	0.54559	0.54360	1.64297
n = 2				
Bi_3TiNbO_9	O, *A*2$_1$*am* (36)	0.54398	0.53941	2.25099
$PbBi_2Nb_2O_9$	O, *A*2$_1$*am* (36)	0.54909	0.54998	2.55313
$SrBi_2Nb_2O_9$	O, *A*2$_1$*am* (36)	0.55164	0.55141	2.50829
$Bi_{2.5}Na_{0.5}Ta_2O_9$	O, *A*2$_1$*am* (36)	0.54763	0.54478	2.49710
n = 1, *n* = 2				
$Bi_5TiNbWO_{15}$	O, *I*2*cm* (46)	0.54231	0.54027	4.1744
n = 3				
$Bi_4Ti_3O_{12}$	O, *Aba*2 (41)	0.54373	3.30610	0.54621
$Bi_3LaTi_3O_{12}$	O, *B*2*cb* (41)	0.54164	0.54157	3.29133
$Bi_2La_2Ti_3O_{12}$	T, *I*4/*mmm* (139)	0.38233		3.29366
$Bi_{2.5}Na_{1.5}Nb_2O_{12}$	O, *B*2*cb* (41)	0.55024	0.54622	3.2735
$Bi_2CaNaNb_3O_{12}$	O, *B*2*cb* (41)	0.54836	0.54585	3.2731
$Bi_2SrNaNb_3O_{12}$	T, *I*4/*mmm* (139)	0.39007		3.2926
n = 4				
$Bi_{2.5}Na_{2.5}Nb_4O_{15}$	O, *A*2$_1$*am* (36)	0.55095	0.54783	4.0553

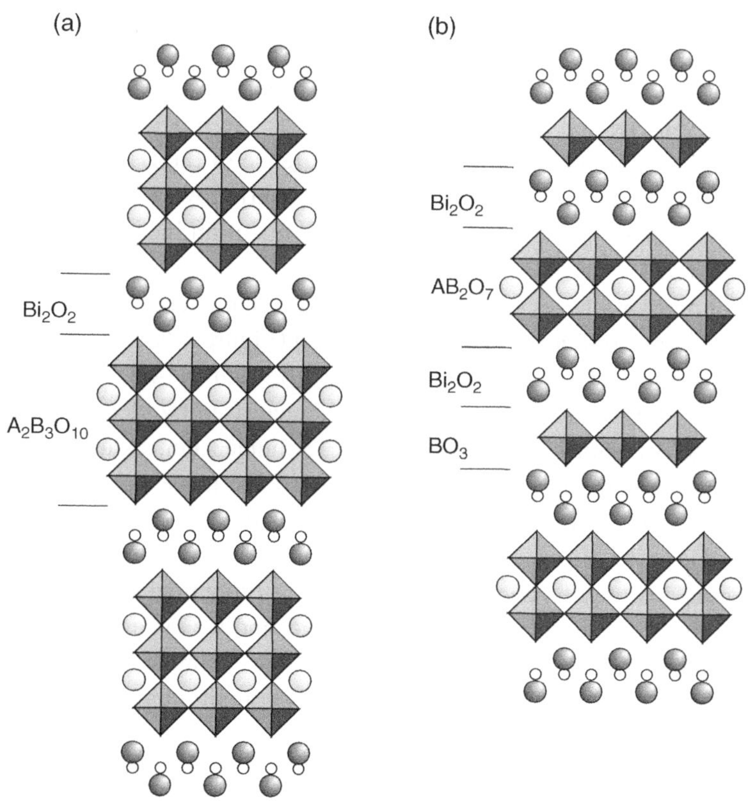

Figure 4.5 *Idealised structures of the Aurivillius phases: (a) n = 3, $(Bi_2O_2)(A_2B_3O_{10})$; (b) ordered (n = 1) : (n = 2) intergrowth Bi_5TiNbO_{15}*

thickness (Figure 4.5b). These intergrowth structures can become complex as narrower and narrower composition ranges are explored.

4.5 The $Ca_2Nb_2O_7$-Related Phases

The phases related to $Ca_2Nb_2O_7$, as in the phases described previously, are built from slabs of the perovskite structure, this time cut into slabs parallel to ideal perovskite $[110]_p$ planes. The first materials of this type were found in the system bounded by the end members $Ca_2Nb_2O_7$ and perovskite $NaNbO_3$. The structure of $Ca_2Nb_2O_7$ is, in reality, composed of slabs four octahedra in thickness and for this reason is better written $Ca_4Nb_4O_{14}$ (Figure 4.6a). The perovskite slabs are stacked along the **c**-axis of the idealised unit cell, and each slab is displaced from its neighbours by $(\mathbf{a}_p+\mathbf{b}_p)/2$, as in the Ruddlesden–Popper phases. The BO_6 octahedra have a crenellated appearance viewed down $\mathbf{a}_p$ and corner-linked rows when viewed along $\mathbf{b}_p$ (Figure 4.6b).

In these phases the formula of each perovskite slab is $A_nB_nO_{3n+2}$, where n is the number of complete BO_6 octahedra in the slice, and as there are no additional cations between the perovskite layers, this is also the formula of the phases. As the slices becomes thicker, the formula of the solid approaches that of perovskite, ABO_3, for which $n=\infty$. Ideally the structure has $a \approx a_p$, $b \approx \sqrt{2}a_p$, $c \approx (n+1)\sqrt{2}a_p$. As with the other phases described, octahedral tilting and distortion reduce this symmetry. In the main the BO_6 octahedra are most regular at the centre of the blocks and most deformed towards the block borders. Similarly, bond valence computations suggest that the valence of the B cations tends to be classical at the perovskite slab borders and frequently lower at the block centres. In a number of these phases a description in terms of an incommensurate structure is preferred.

A large number of phases of this type have been characterised (Table 4.6). Apart from phases corresponding to a single value of n (Figure 4.6a and c), innumerable repeating sequences involving two values (n_1, n_2), such as (4,4,4,5), (4,4,5), (4,5) (Figure 4.6d), (4,5,5) and so on have been characterised by electron microscopy. This can lead to some ambiguity in the formulae. For example, the phase $Pr_3Ti_2TaO_{11}$, nominally an $n=3$ member of the series, has a stacking sequence of (2, 4), not (3, 3) (Figure 4.6e). Structures such as this are sometimes labelled as $n=3$ (type II).

Disordered [110] faults, consisting of slabs of varying thickness, arise in a number of systems as a result of composition change, and for materials corresponding to $n=5$ or more, disorder tends to be the rule. For example, the reaction of the perovskite $SrTiO_3$ with Nb_2O_5 results in crystals containing randomly distributed

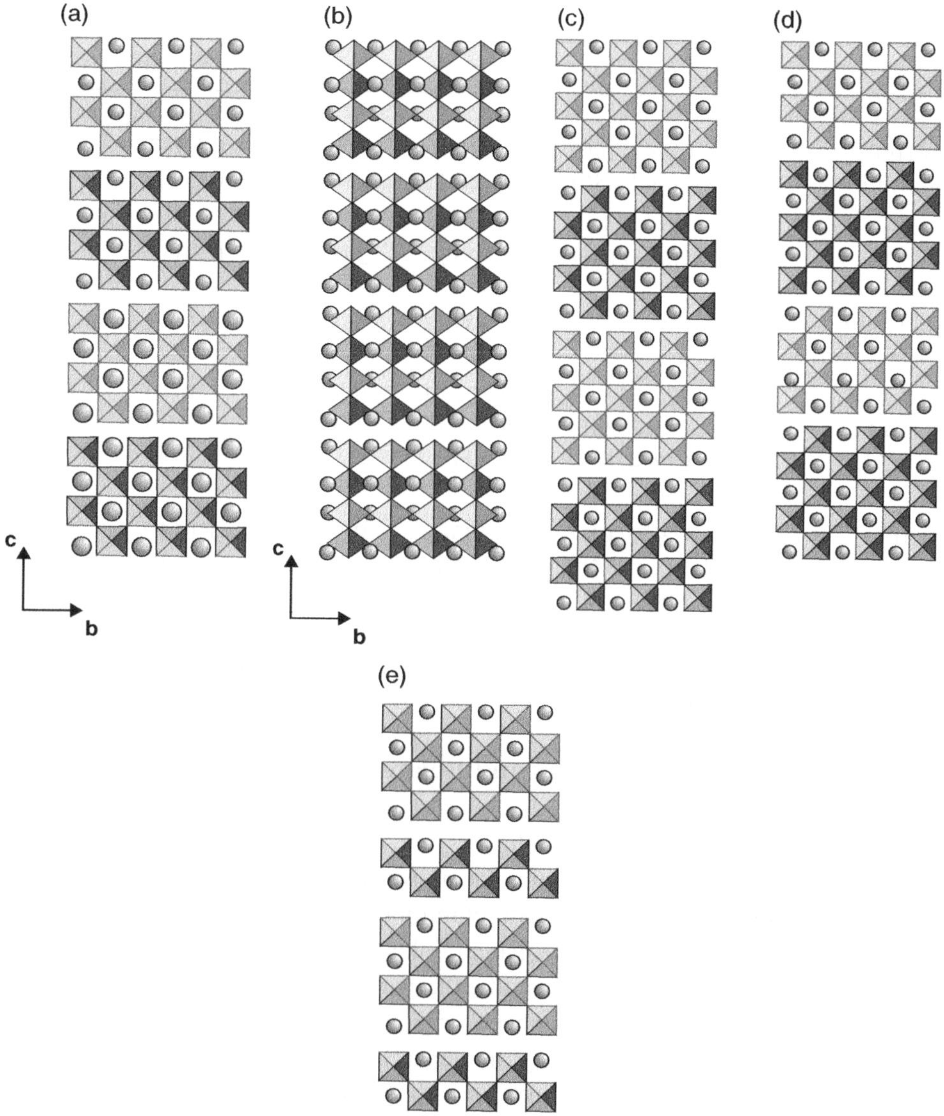

Figure 4.6 *$Ca_2Nb_2O_7$-related phases: (a) $Ca_4Ti_4O_{14}$ (n = 4) viewed along [100]; (b) $Ca_4Ti_4O_{14}$ (n = 4) viewed along [010]; (c) n = 5, $A_5B_5O_{17}$; (d) n = (4, 5), $La_{4.5}Ti_{4.5}O_{15.5}$; (e) n = (2, 4), $Pr_3Ti_2TaO_{11}$*

[110] defects. The incorporation of Nb_2O_5 gives each crystal a composition $Sr(Nbi_xTi_{1-x})O_{3+x}$, and each Nb^{5+} ion that replaces a Ti^{4+} ion requires that extra oxygen is incorporated into the crystal. The planar defects open the structure and allow this extra oxygen to be accommodated without introducing oxygen interstitials.

Table 4.6 *$Ca_2Nb_2O_7$ ($Ca_4Nb_4O_{14}$) phases*

Phase	Space group	*a* (nm)	*b* (nm)	*c* (nm)	*β* (°)
n=2					
$Ba_2Fe_2F_8$	O, *Cmc*2_1 (36)	0.4234	1.4856	0.5837	
$Ba_2Mn_2F_8$	O, *Cmc*2_1 (36)	0.4209	1.4708	0.5874	
n=3					
$Sr_2LaTa_3O_{11}$	O, *Immm* (71)	0.3965	2.0874	0.5632	
n=2, *n*=4					
$Pr_3Ti_2TaO_{11}$	O, *Pmc*2_1 (26)	0.38689	2.0389	0.55046	
n=4					
$La_4Ti_4O_{14}$	M, *P*2_1 (4)	0.78122	0.55421	1.30102	98.66
$Sr_4Nb_4O_{14}$	O, *Cmc*2_1 (36)	0.39544	2.6767	0.56961	
$Sr_4Ta_4O_{14}$	O, *Cmc*2_1 (36)	0.3940	2.715	0.5692	
n=5					
$La_5Ti_5O_{17}$	M, *P*2_1/*c* (14)	0.78580	0.55281	3.1449	97.17
$Ca_5Nb_5O_{17}$	M, P2_1/c (14)	0.77494	0.54928	0.32241	90.81
$Sr_5Nb_5O_{17}$	O, Pnnm (58)	3.2456	0.56714	0.3995	
n=6					
$Ca_2La_4Ti_6O_{20}$	O, *Pbn*2_1 (33)	0.38810	3.6771	0.55031	

4.6 Cuprate Superconductors and Related Phases

High-temperature cuprate superconductors are copper oxides that maintain the superconducting state to temperatures far above that of the alloy-based 'conventional' superconductors (Table 4.7). Structurally the phases are all built up of slices of a nominal $ACuO_3$ perovskite-type parent linked by layers (in the main) of the RS or fluorite type. In all these compounds, the part of the structure which leads to superconductivity consists of CuO_2 sheets originating in the perovskite slabs. When more than one sheet is present, they are separated by cation layers, Q, to give a sequence CuO_2–$(Q–CuO_2)_{n-1}$. The index *n* is the total number of CuO_2 layers in the phase, which is equal to the formula number of Cu atoms present. As in the layered structures described earlier, these differing *n* values characterise members of homologous series of phases, each member of which differs from the preceding one by having the perovskite slab one extra octahedron in thickness.

The copper valence in most compounds lies between the formal values of Cu^{2+} and Cu^{3+}. The appearance of superconductivity and the superconducting transition temperature, T_c, at which the superconducting state appears on cooling, are closely connected with the composition of these solids, which all exhibit considerable degrees oxygen composition variation (Section 8.4).

Table 4.7 *High-temperature cuprate superconducting oxides*

Compound[a]	T_c (K)	Space group	*a* (nm)	*b* (nm)	*c* (nm)
Single CuO_2 layers					
$La_{1.84}Sr_{0.16}CuO_4$	38	T, *I4/mmm* (139)	0.37985		1.3312
$Nd_{1.97}Ce_{0.07}CuO_4$	20	T, *I4/mmm* (139)	0.39353		1.20874
Double CuO_2 layers					
$YBa_2Cu_3O_{6.35}$	10	T, *P4/mmm* (123)	0.3858		1.17913
$YBa_2Cu_3O_{6.5}$	60	O, *Pmmm* (47)	0.3834	0.3878	1.174
$YBa_2Cu_3O_{6.95}$	93	O, *Pmmm* (47)	0.38136	0.38845	1.16603
$YBa_2Cu_3O_7$	—	O, *Pmmm* (47)	0.38255	0.38871	1.16618
$NdBa_2Cu_3O_7$	—	O, *Pmmm* (47)	0.38649	0.39163	1.1767
Bi_2O_2/Tl_2O_2 layers					
$Bi_2Sr_2CuO_6$	10	T, *I4/mmm* (139)	0.37962		2.39333
$Bi_2Sr_2CaCu_2O_8$	92	T, *I4/mmm* (139)	0.38361		2.93709
$Bi_2Sr_2Ca_2Cu_3O_{10}$	110	O, *A2aa* (37)	0.5411	0.5409	3.7082
$Tl_2Ba_2CuO_6$	92	T, *I4/mmm* (139)	0.38594		2.31273
$Tl_2Ba_2CaCu_2O_8$	119	T, *I4/mmm* (139)	0.38565		2.93122
$Tl_2Ba_2\ Ca_2Cu_3O_{10}$	128	T, *I4/mmm* (139)	0.38487		3.5662
$Tl_2Ba_2\ Ca_3Cu_4O_{12}$	119	T, *I4/mmm* (139)	0.38503		4.226
TlO/HgO layers					
$TlBa_2CuO_5$	—	T, *P4/mmm* (123)	0.3850		0.9540
$TlBa_2CaCu_2O_7$	103	T, *P4/mmm* (123)	0.3847		1.273
$TlBa_2\ Ca_2Cu_3O_{8.5}$	110	T, *P4/mmm* (123)	0.38478		1.586
$HgBa_2CuO_4$	94	T, *P4/mmm* (123)	0.38851		0.9526
$HgBa_2CaCu_2O_6$	127	T, *P4/mmm* (123)	0.38528		1.27216
$HgBa_2\ Ca_2Cu_3O_8$	133	T, *P4/mmm* (123)	0.38453		1.57713
$Hg_{0.8}Tl_{0.2}Ba_2Ca_2Cu_3O_{8.33}$	138	T, *P4/mmm* (123)	0.3849		1.5827
$HgBa_2Ca_3Cu_4O_{10}$	126	T, *P4/mmm* (123)	0.3858		1.9011

[a] The formulae are representative and do not always show the exact oxygen stoichiometry for the optimum T_c values given.

4.6.1 La_2CuO_4, Nd_2CuO_4 and $YBa_2Cu_3O_7$

The structurally simplest cuprate superconductors are doped La_2CuO_4 (the T or T/O structure), doped Nd_2CuO_4 (the T′ structure) and oxygen-deficient $YBa_2Cu_3O_7$. At room temperature, La_2CuO_4 adopts a distorted version of the K_2NiF_4 structure, in which the CuO_6 octahedra are Jahn–Teller lengthened along the **c**-axis (Figure 4.7). At higher temperatures the Jahn–Teller distortion is lost and the material adopts the undistorted K_2NiF_4 structure. The room temperature insulating stoichiometric compound is transformed into a superconductor by replacement of some of the La^{3+} cations with the alkaline earth cations Ba^{2+}, Sr^{2+} or Ca^{2+}. The structure of Nd_2CuO_4

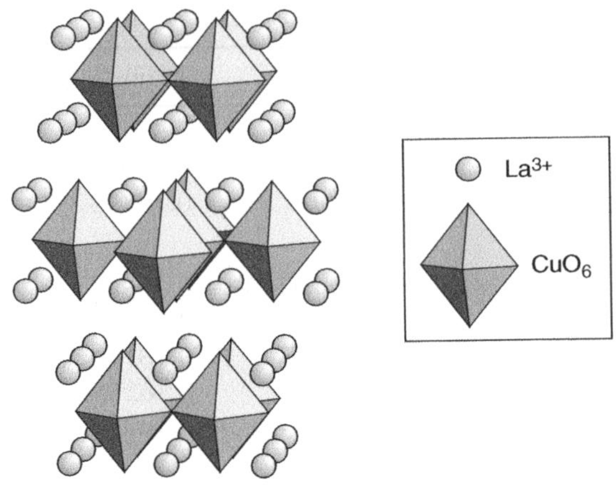

Figure 4.7 The K_2NiF_4-related structure of room temperature La_2CuO_4, containing sheets of Jahn–Teller elongated CuO_6 octahedra

(and the isostructural metastable La_2CuO_4) has a similar cation arrangement, but the CuO_2 sheets are linked by fluorite structure Nd_2O_2 layers so that overt octahedra do not occur (Figure 4.3).

The compound $YBa_2Cu_3O_7$ has been widely studied because it was the first superconductor discovered with a T_c above the boiling point of liquid nitrogen. The crystal structure of $YBa_2Cu_3O_7$ consists of three perovskite-like unit cells stacked one on top of the other (Figure 4.8a). The middle perovskite unit contains Y as the A cation and Cu as the B atom. The cells above and below this contain Ba as the A atom and Cu as the B atom to give a metal formula of YBa_2Cu_3 as one would expect for a tripled perovskite cell, $A_3B_3O_9$. However, instead of the nine O atoms expected, there are seven, arranged in such a way as to give the Cu atoms square pyramidal and square planar coordination, rather than octahedral (Figure 4.8b). If the ions are allocated the formal charges of Y^{3+}, Ba^{2+} and O^{2-}, the Cu must take an average charge of 2.33, which can be considered to arise from the nominal presence of two Cu^{2+} ions and one Cu^{3+} ion per unit cell. $YBa_2Cu_3O_7$ can lose oxygen down to a composition of $YBa_2Cu_3O_{6.0}$ (Figure 4.8c). The oxygen atoms are not lost at random during reduction but come solely from the CuO_4 square planar units and lie along the **b**-axis. This has the effect of converting the copper coordination from square planar to linear, resulting in CuO_2 chains running through the structure (Figure 4.8d).

There are a number of other structures similar to that of $YBa_2Cu_3O_7$, especially those in which the Y^{3+} cations are replaced with lanthanoid ions Ln^{3+}. In addition, phases with B-site substitution, for instance, $YSr_2Cu_2CoO_7$, are known, although many of these have structures that are not so obviously perovskite related. The compound $YSrCu_2CoO_7$, for example, is built from tetrahedra and square pyramids and does not show a simple structural relationship with perovskite.

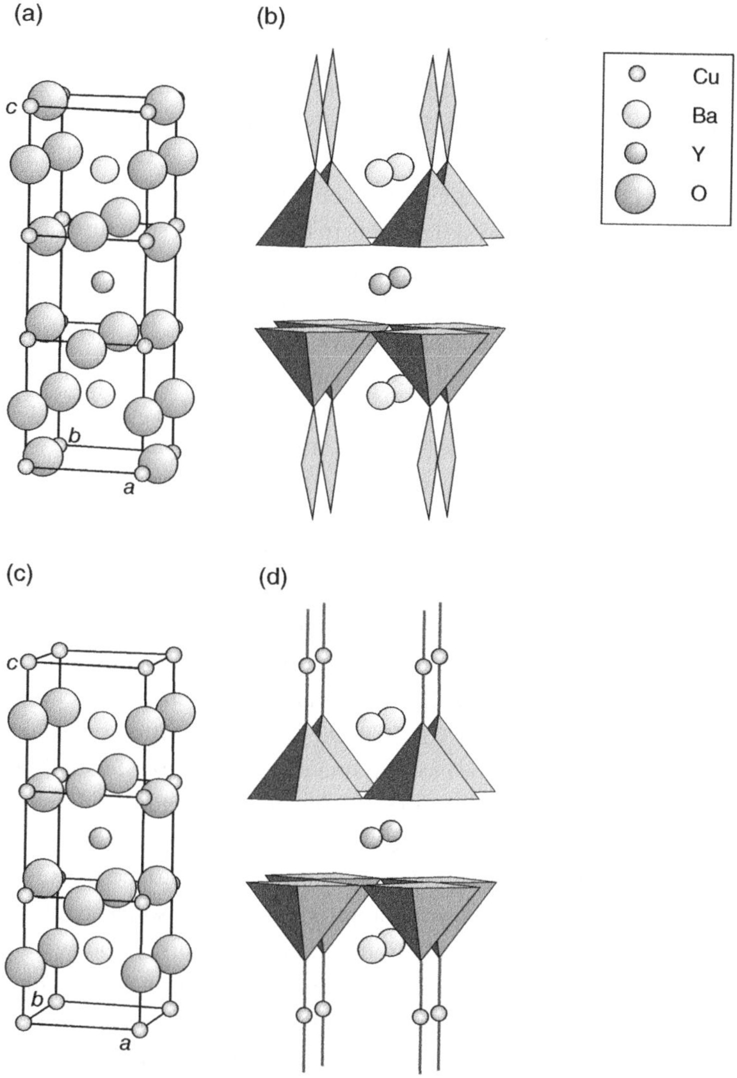

Figure 4.8 *(a) The structure of $YBa_2Cu_3O_7$; (b) idealised $YBa_2Cu_3O_7$ drawn as CuO_5 square pyramids and CuO_4 square planar polyhedra; (c) the structure of $YBa_2Cu_3O_6$; (d) idealised $YBa_2Cu_3O_6$ showing CuO_5 square pyramids and O–Cu–O linear groups*

4.6.2 Layered Perovskite Structures

The other superconducting cuprate structures are constructed from perovskite-like layers separated by a variety of different linking regions. The construction principle can be illustrated with reference to compounds in which the perovskite sheets are linked by Bi_2O_2 or Tl_2O_2 layers. The formula of the Bi_2O_2 series is $Bi_2Sr_2Ca_{n-1}Cu_nO_{2n+4}$

and that of the Tl_2O_2 series is $Tl_2Ba_2Ca_{n-1}Cu_nO_{2n+4}$, where it must be understood that the oxygen composition in the real compounds is slightly different from that given in the idealised formulae and that the oxygen non-stoichiometry is significant for the appearance of superconductivity. The idealised structures of the first three members of these series, $Bi_2Sr_2CuO_6$ (=$Tl_2Ba_2CuO_6$), $Bi_2Sr_2CaCu_2O_8$ (=$Tl_2Ba_2CaCu_2O_8$) and $Bi_2Sr_2Ca_2Cu_3O_{10}$ (=$Tl_2Ba_2Ca_2Cu_3O_{10}$), contain perovskite layers 1, 2 and 3 octahedra thick. The single perovskite sheets in the idealised structure of $Bi_2Sr_2CuO_6$ are complete and separated by Bi_2O_2 (or Tl_2O_2) layers (Figure 4.9a). In the other compounds, the oxygen structure needed to form the perovskite framework is incomplete. The nominal double layer of CuO_6 octahedra needed to form a sheet of idealised perovskite structure is replaced by square pyramids in $Bi_2Sr_2CaCu_2O_8$ and $Tl_2Sr_2CaCu_2O_8$ (Figure 4.9b). In the phases $Ba_2Ca_2Sr_2Cu_3O_{10}$ and $Tl_2Ca_2Sr_2Cu_3O_{10}$, the three CuO_6 octahedral perovskite layers have been replaced by two sheets of square pyramids and the middle layer by a sheet of CuO_4 square planar coordination polyhedra (Figure 4.9c).

The different homologous series of superconductors are described as built from different *charge reservoir layers* interleaved with $n = 1, 2, 3, 4, \ldots$ CuO_2–$(Q–CuO_2)_{n-1}$ *superconducting layers* (Table 4.8). Clearly, other members of these series can be envisaged, containing different n values, intergrowth of ordered or disordered n and $(n+1)$ slab thicknesses or series with different alternating charge reservoir slabs.

The real structures of the $Bi_2Sr_2Ca_{n-1}Cu_nO_{2n+4}$ series are often better described as containing a modulation along or close to one of the short axes, usually taken to be the **b**-axis of the structure. For example, the idealised structure of $Ba_2Sr_2Ca_2Cu_3O_{10}$ (Figure 4.10a and b) has an ideal b lattice parameter of approximately a_p. The modulation is associated with the incorporation of extra oxygen into the Bi—O layers of the structure and can be represented as an approximately sinusoidal buckling of these layers in order to incorporate the additional atoms. This distortion is transmitted to the perovskite-like blocks so that the whole structure takes on a wave-like structure. The complexity of the real structures is considerable. The modulation wavelength is rarely commensurate with the ideal b lattice parameter. The modulation waves in one set of Bi—O planes can correspond with those in the adjacent layers (as in Figure 4.10c) to give an approximately orthorhombic unit cell or be displaced by a small amount to give an approximately monoclinic cell (Figure 4.10d). Moreover, the modulation waves may also run at an angle to the **b**-axis in some or all of the layers.

4.6.3 Structures Related to the Layered Cuprate Phases

A number of other copper-containing and non-copper-containing phases have structures related to these cuprate phases. The oxide $Sr_4(Fe_{6-x}Co_x)O_{13\pm\delta}$ has a complex modulated structure in which the modulation vector is dependent on the

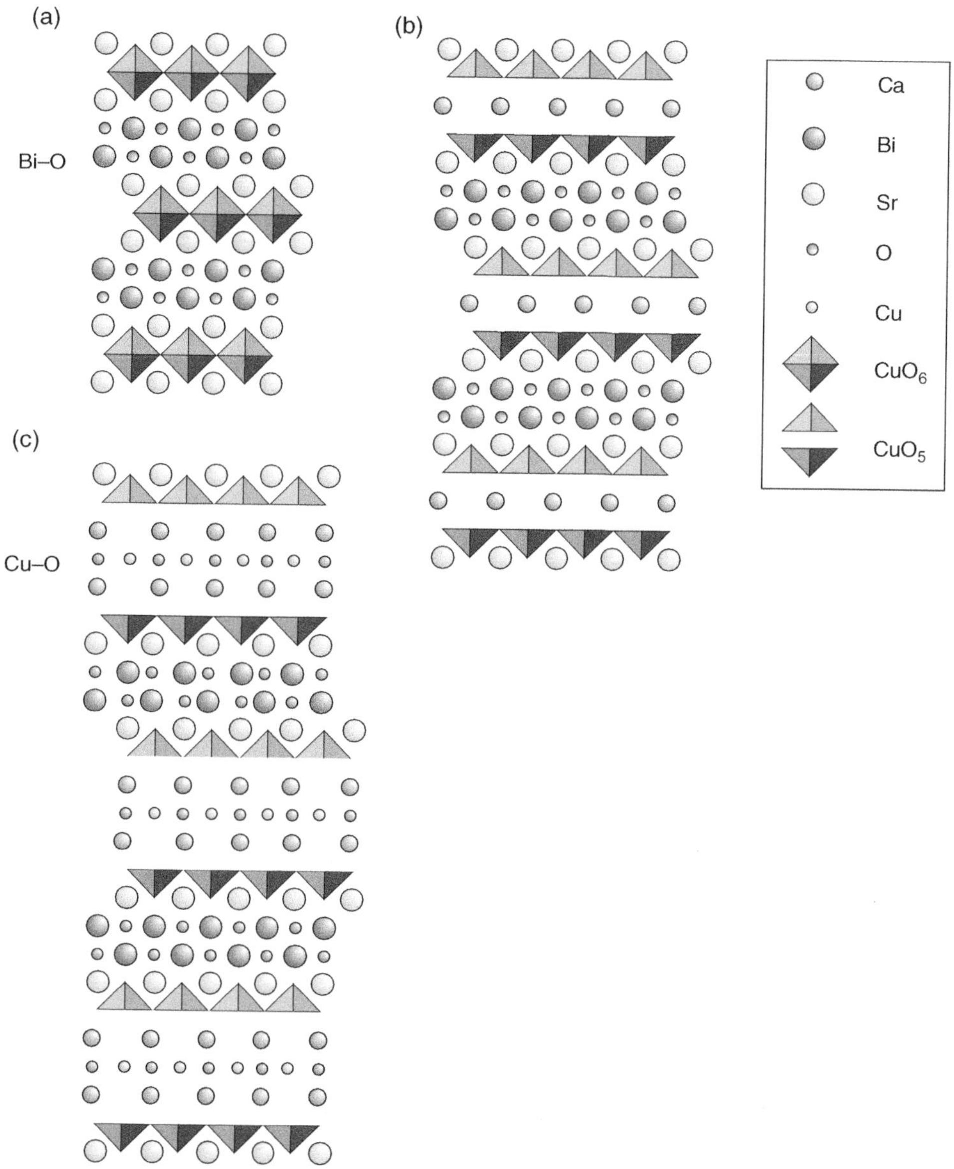

Figure 4.9 *Idealised structures of some cuprate superconductors: (a) $Bi_2Sr_2CuO_6$ and $Tl_2Ba_2CuO_6$; (b) $Bi_2Sr_2CaCu_2O_8$ and $Tl_2Ba_2CaCu_2O_8$; (c) $Bi_2Sr_2Ca_2Cu_3O_{10}$ and $Tl_2Ba_2Ca_2Cu_3O_{10}$*

Table 4.8 *Homologous series of cuprate superconductors*

Charge reservoir[a]	Charge reservoir formula	Superconducting slab formula	Idealised series formula[a]	Examples[b]
AX–AX	LaO	$CuO_2\ (CuO_2)_{n-1}$	$La_2Cu_nO_{2n+2}$	$La_2CuO_{4+\delta}$ ($n=1$) $La_{1-x}Sr_xCuO_4$ ($n=1$)
	SrF	$CuO_2\ (CuO_2)_{n-1}$	$Sr_2Cu_nO_{2n}F_2$	$Sr_2CuO_2F_{2+\delta}$ ($n=1$)
A–O_2–A	Nd_2O_2	$CuO_2\ (CuO_2)_{n-1}$	$Nd_2Cu_nO_{2n+2}$	Nd_2CuO_4 ($n=1$)
AO–M–AO	BaO–Cu–BaO	$CuO_2\ (Y\ CuO_2)_{n-1}$	$Ba_2Y_{n-1}Cu_{n+1}O_{2n+2}$	$YBa_2Cu_3O_6$ ($n=2$)
		$CuO_2\ (Ca\ CuO_2)_{n-1}$	$CuBa_2Ca_{n-1}Cu_nO_{2n+2}$	$Ba_2Ca_2Cu_4O_8$ ($n=3$) $Ba_2Ca_3Cu_5O_{10}$ ($n=4$)
	BaO–Hg–BaO	$CuO_2\ (Ca\ CuO_2)_{n-1}$	$HgBa_2Ca_{n-1}Cu_nO_{2n+2}$	$HgBa_2CuO_4$ ($n=1$) $HgBa_2CaCu_2O_6$ ($n=2$) $HgBa_2Ca_2Cu_3O_8$ ($n=3$)
AO–MO–AO	BaO–TlO–BaO	$CuO_2\ (Ca\ CuO_2)_{n-1}$	$TlBa_2Ca_{n-1}Cu_nO_{2n+3}$	$TlBa_2CuO_5$ ($n=1$) $TlBa_2CaCu_2O_{7+\delta}$ ($n=2$) $TlBa_2Ca_2Cu_3O_{9+\delta}$ ($n=3$)
	BaO–BiO–BaO	$CuO_2\ (Ca\ CuO_2)_{n-1}$	$BiBa_2Ca_{n-1}Cu_nO_{2n+3}$	$BiBa_2CuO_5$ ($n=1$) $BiBa_2CaCu_2O_7$ ($n=2$) $BiBa_2Ca_2Cu_3O_9$ ($n=3$)
	SrO–GaO–SrO	$CuO_2\ (Y,Ca\ CuO_2)_{n-1}$	$GaSr_2(Y,\ Ca)_{n-1}Cu_nO_{2n+3}$	$GaSr_2(Y,Ca)Cu_2O_7$ ($n=2$) $GaSr_2(Y,Ca)_2Cu_3O_9$ ($n=3$)
AO–MO–MO–AO	SrO–BiO–BiO–SrO	$CuO_2\ (Ca\ CuO_2)_{n-1}$	$Bi_2Sr_2Ca_{n-1}Cu_nO_{2n+4}$	$Bi_2Sr_2CuO_6$ ($n=1$) $Bi_2Sr_2CaCu_2O_8$ ($n=2$) $Bi_2Sr_2Ca_2Cu_3O_{10}$ ($n=3$)
	BaO–TlO–TlO–BaO	$CuO_2\ (Ca\ CuO_2)_{n-1}$	$Tl_2Ba_2Ca_{n-1}Cu_nO_{2n+4}$	$Tl_2Ba_2CuO_6$ ($n=1$) $Tl_2Ba_2CaCu_2O_8$ ($n=2$) $Tl_2Ba_2Ca_2Cu_3O_{10}$ ($n=3$)
AO–MO–M′–MO–AO	SrO–PbO–Cu–PbO–SrO	$CuO_2\ (Y,Ca\ CuO_2)_{n-1}$	$Pb_2Sr_2(Y,\ Ca)_{n-1}Cu_{n+1}O_8$	$Pb_2Sr_2Y_{0.5}Ca_{0.5}Cu_3O_8$ ($n=2$)

[a] The charge reservoir layer is non-stoichiometric. This is not indicated in the idealised formulae given.
[b] The composition is variable, due to the non-stoichiometric nature of the charge reservoir slabs; δ indicates this and may be either positive or negative.

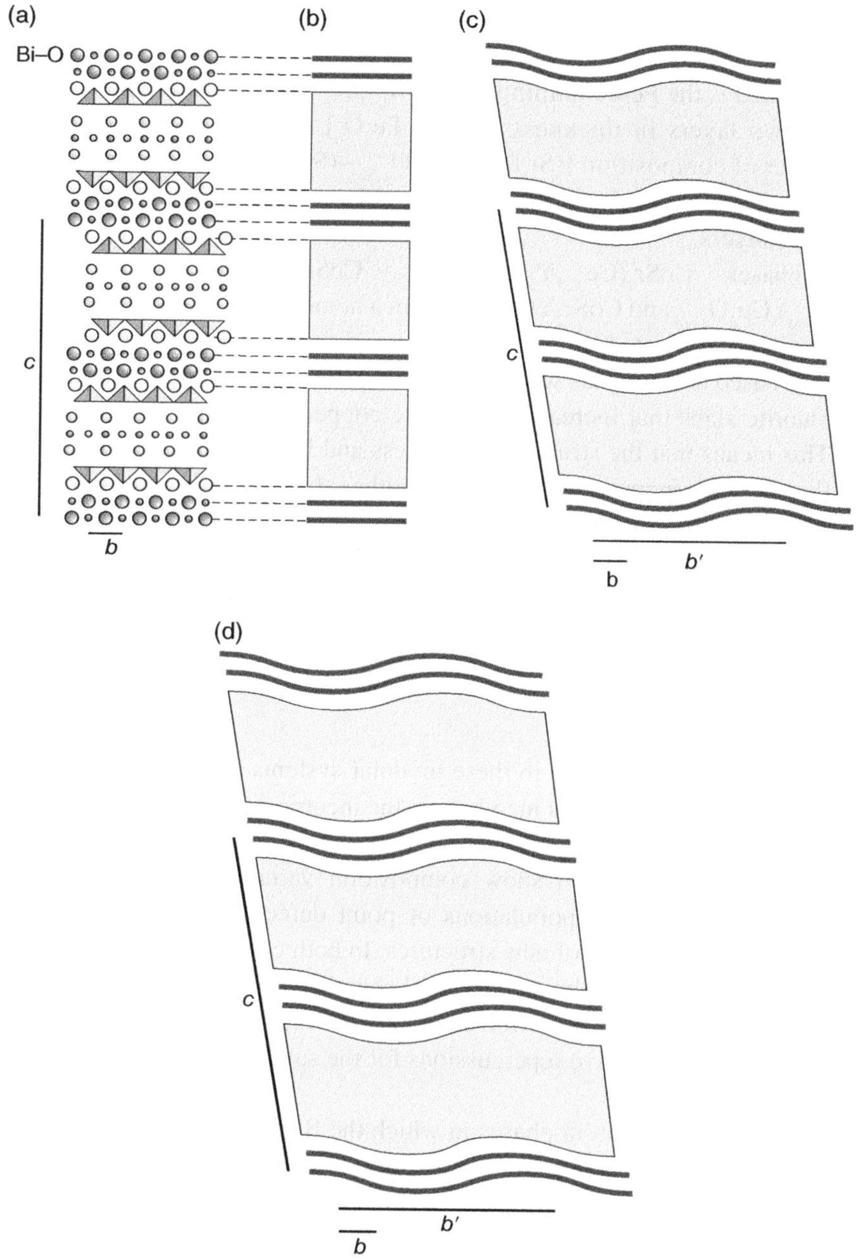

Figure 4.10 *Modulation in $Bi_2Ca_2Sr_2Cu_3O_{10+\delta}$: (a) idealised structure projected down [100]; (b) structure drawn as stacked perovskite and Bi–O units; (c) incommensurate modulation (exaggerated) along the b-axis with a period* $b' \approx 5.8\ b$ *producing orthorhombic symmetry; (d) the same modulation as in (c), but with the modulation waves displaced, producing monoclinic symmetry*

oxygen composition δ. It is built of single perovskite layers [$SrFeO_3$] separated by complex [SrO–$Fe_2O_{2.5\pm½\delta}$–SrO] layers which have a RS-associated form. Closely related is the Fe-containing phase $Bi_{1.33}Sr_{3.67}Fe_8O_{19\pm\delta}$. Here the perovskite slabs are two layers in thickness [$(Bi,Sr)_2Fe_2O_6$] separated by a complex RS-related layer of composition [(Sr,Bi)O–$Fe_2O_{1.5\pm\delta}$–(Sr,Bi)O]. A similar structure is adopted by $Bi_{2+x}Sr_{3-x}Fe_2O_{9+\delta}$, which has double perovskite layers separated by [BiO–BiO] layers.

The phases $CoSr_2(Ce_{0.25}Y_{0.75})_2Cu_2O_{9+\delta}$, $CoSr_2(Ce_{0.4}Nd_{0.6})_2Cu_2O_{9+\delta}$, $CoSr_2(Ce_{0.67}Y_{0.33})_3Cu_2O_{11+\delta}$ and $CoSr_2YCu_2O_{7+\delta}$ form a homologous series in which double CuO_2 planes are separated by 1, 2 and 3 layers of fluorite-related structure. Unlike the copper-based homologous series phases described earlier, here it is the thickness of the fluorite slabs that increases, while the copper-containing layers remain the same. This means that the structures show less and less resemblance to perovskite as the fluorite slabs increase in thickness. Another phase with fluorite slabs separating CuO_2 planes is $Cu(Ba_{0.67}Eu_{0.33})_2(Ce_{0.33}Eu_{0.67})_2Cu_2O_{9\pm\delta}$ containing fluorite (F)-related [(Ce,Eu)–O_2–(Ce,Eu)] and (Ba, Ln)O RS structure-related blocks in the sequence (RS)-($CuO_{1\pm\delta}$)-(RS)-(CuO_2)-(F)-(CuO_2).

4.7 Composition Variation

Gross stoichiometry variations in these modular systems are often accommodated by the formation of other series members or the incorporation of planar faults, often equivalent to isolated lamellae of other series members, into the structure. However, many individual phases also show composition variation which is sometimes accommodated by random populations of point defects and in other phases by ordering and the generation of new structures. In both cases composition flexibility is important in influencing the physical and chemical properties of these materials. In particular this effect has been well studied in the cuprate superconductors, where changes in composition have repercussions for the superconducting transition temperature, T_c, of the phase.

Oxygen loss is common in phases in which the B cations can take on a variable valence. In some phases these are distributed at random over the normal oxygen positions, thereby transforming some BO_6 octahedra into BO_5 square pyramids. This occurs in the $n = 2$ Ruddlesden–Popper phase $La_3Ni_2O_7$ when reduced in hydrogen to a composition of $La_3Ni_2O_{6.35}$. In the reduced structure, almost 2/3 of the nickel atoms are in square pyramidal coordination, which is well suited to the nominal charge state of Ni^+ (Ni'_{Ni}), thus generating

$$2Ni_{Ni} + O_O \rightarrow ½O_2 + V_O^{\bullet\bullet} + 2Ni'_{Ni}$$

The oxygen vacancies appear to be disordered in this phase. Similarly, in the $n=3$ Ruddlesden–Popper phase $Sr_3Fe_2O_{6+\delta}$ in which δ lies between 0 and 0.73, the oxygen vacancies appear to be disordered. The oxygen loss is balanced by the Fe^{3+}/Fe^{4+} ratio, as nominally $Sr_3Fe_2^{3+}O_6$ gradually transformations to $Sr_3Fe^{3+}_{0.74}Fe^{4+}_{1.46}O_{6.73}$. The reaction for the oxidation of $La_3Fe_2O_6$ is

$$\tfrac{1}{2}O_2 \rightarrow O_O + 2h^{\bullet} \rightarrow O_O + 2Fe^{\bullet}_{Fe}$$

where $Fe^{\bullet}_{Fe}$ is equivalent to Fe^{4+}. For charge balance to be maintained, two Fe^{4+} ions are created for every oxygen ion added.

The cuprate La_2CuO_4 is a typically insulating ceramic. It can be made in an oxygen-rich form by heating in oxygen at higher pressures to give $La_2CuO_{4+\delta}$. The extra oxygen atoms are introduced into the structure as interstitial point defects which occupy sites midway between the LaO planes at random. They are arranged so as to be as far away from the La^{3+} ions as possible and sit more or less at the centres of La^{3+} tetrahedra.

The incorporation of oxygen can be written as

$$\tfrac{1}{2}O_2 \rightarrow O_i^{2'} + 2h^{\bullet}$$

The holes so formed are believed to be located upon Cu^{2+} ions (Cu_{Cu}) to form Cu^{3+} ions ($Cu^{\bullet}_{Cu}$):

$$2h^{\bullet} + 2Cu_{Cu} \rightarrow 2Cu^{\bullet}_{Cu}$$

Thus two Cu^{3+} ions are created per oxygen interstitial. The electronic properties of these oxygen-rich solids change drastically from insulating to become firstly a metal and then a superconductor. There is evidence to suggest that these interstitial anions are ordered over small volumes to form microdomains and this ordering strongly changes the observed superconducting transition temperature T_c.

The compound $Sr_2CuO_2F_2$, a poor conductor, can take on extra F^- ions to form $Sr_2CuO_2F_{2+x}$. The additional F^- ions occupy similar interstitial sites to those occupied in La_2CuO_4. Once again the change in stoichiometry results in the generation of holes:

$$\tfrac{1}{2}F_2 \rightarrow F_i' + h^{\bullet} \rightarrow F_i' + Cu^{\bullet}_{Cu}$$

where the holes are supposed to be located on Cu^{2+} ions to form Cu^{3+} ($Cu^{\bullet}_{Cu}$). In this case, one Cu^{3+} ion is created for each F^- interstitial created, and the fluorine-rich phase becomes a superconducting metal.

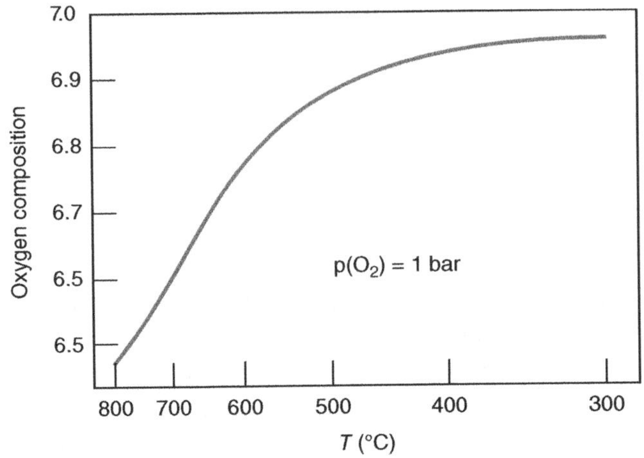

Figure 4.11 *The variation of the composition of $YBa_2Cu_3O_7$ when annealed in air*

The phase $YBa_2Cu_3O_{7-\delta}$ has a complete range of oxygen composition running from $YBa_2Cu_3O_6$ to $YBa_2Cu_3O_7$. The midpoint of this range, $YBa_2Cu_3O_{6.5}$, nominally corresponds to the Cu^{2+}-containing phase. For oxygen compositions richer than this, the phase contains nominally Cu^{2+} and Cu^{3+}, while for compositions below it the phase contains nominally Cu^{+} and Cu^{2+}. However, there is no indication that a stable Cu^{2+}-only phase exists at a composition $YBa_2Cu_3O_{6.5}$; this composition is simply part of a continuum over this composition range. The composition of any phases within this continuum depends closely upon the temperature of synthesis and the surrounding partial pressure of oxygen (Figure 4.11), and care during preparation must be taken if a particular composition is desired.

The phases described previously appear to prefer disordered point defects as a response to composition change, but other structures show ordering of the oxygen vacancies, particularly when the new coordination polyhedra formed are well suited to the cations present. For example, Sr_2MnO_4 with the K_2NiF_4 structure, which contains only Mn^{4+} cations, can be reduced to $Sr_2MnO_{3.5}$ with a monoclinic unit cell: space group $P2_1/c$ (14), $a = 0.68524$ nm, $b = 1.08131$ nm, $c = 1.08068$ nm, $\beta = 113.25°$ at 300 K. In this structure the oxygen vacancies are ordered in the MnO_2 layers to convert the octahedrally coordinated Mn^{4+} into square pyramidal Mn^{3+} (Mn'_{Mn}) units (Figure 4.12). The reduction reaction is

$$O_O \rightarrow ½O_2 + V_O^{\bullet\bullet} + 2Mn'_{Mn}$$

Figure 4.12 *Oxygen vacancy ordering (slightly idealised) in each perovskite sheet in* $Sr_2MnO_{3.5}$*; octahedra are converted into square pyramids. The* Sr^{2+} *ions are omitted for clarity*

A similar ordering is found within each of the perovskite layers in the reduced $n=3$ Ruddlesden–Popper phase $Sr_3Mn_2O_6$, derived from the fully oxidised $Sr_3Mn_2O_7$. Under normal preparation methods the ordering is confined to the perovskite layers and does not extend to three dimensions in the macroscopic crystal, although electron microscopy suggests that microdomains of such three-dimensionally ordered structures do exist. These structures are similar to those of Mn-containing brownmillerite-related compounds (Section 2.5.1).

Three-dimensional ordering of square pyramidal coordination does occur in a number of Fe-containing phases. An example is provided by the $n=3$ Ruddlesden–Popper phase $Sr_4Fe_3O_{10}$ in which the Sr is substituted by La. In the series so formed, replacement of Sr^{2+} by La^{3+} in the parent phase $Sr_4Fe_3^{4+}O_{10}$ generates an equal number of balancing Fe^{3+} ions to give La $Sr_{4-x}La_x\,(Fe^{4+}_{3-x}Fe^{3+}_x)\,O_{10}$ with counterbalanced defects $La^{\bullet}_{Sr}$ and Fe'_{Fe} (Fe^{3+}).

Loss of oxygen also generates Fe^{3+} from Fe^{4+} in order to maintain charge neutrality:

$$O_O \rightarrow \tfrac{1}{2}O_2 + 2e' + V_O^{\bullet\bullet} \rightarrow V_O^{\bullet\bullet} + 2Fe'_{Fe}$$

In the phase $Sr_{0.775}La_{0.225}Fe_3O_{10-\delta}$ the oxygen vacancy distribution changes at about 800°C from disordered to ordered. The ordered distribution of vacancies at

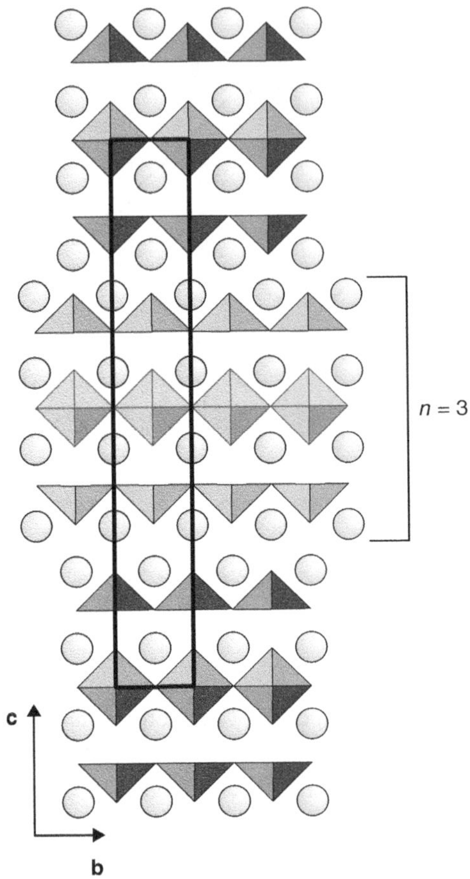

Figure 4.13 The idealised structure of $Sr_{0.775}La_{0.225}Fe_3O_{9.2}$ at 1000°C

approximately 1000°C in $Sr_{0.775}La_{0.225}Fe_3O_{9.2}$ creates layers of square pyramidal coordination polyhedra which are preferentially located in the outer octahedral sheets of the perovskite layers (Figure 4.13).

The similar group of $Ln_3Ba_2Mn_2Cu_2O_{12}$ materials (Ln is a lanthanoid) is derived from the Ruddlesden–Popper $n=4$ structure. In these phases the perovskite slab four octahedra thick is transformed into a slab made up from two outer MnO_6 octahedra and two inner CuO_5 square pyramids, each of which has lost one oxygen compared to the ideal structure, to give an ordered B-site perovskite slab (Figure 4.14). In addition, the Ln and Ba cations are ordered over the perovskite A-sites. Typical tetragonal lattice parameters are $Eu_3Ba_2Mn_2Cu_2O_{12}$: space group *I4/mmm* (139), $a=0.38850$ nm, $c=3.52949$ nm at 298 K; $Sm_3Ba_2Mn_2Cu_2O_{12}$: space group *I4/mmm* (139), $a=0.388362$ nm, $c=3.54928$ nm at 298 K.

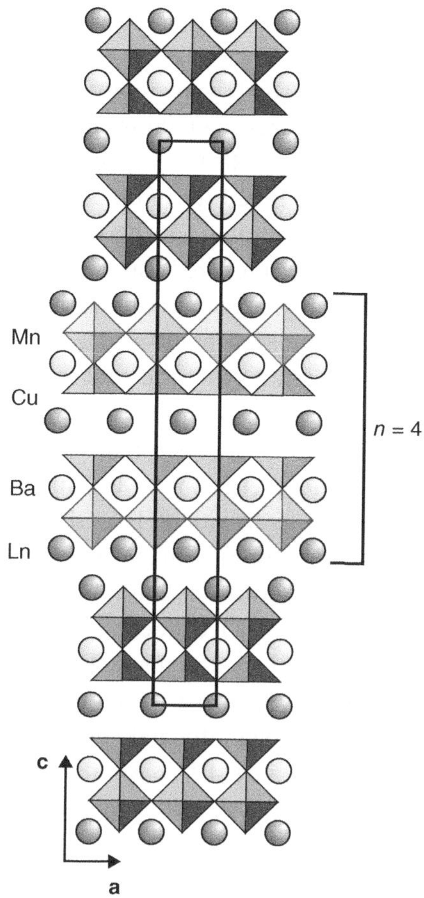

Figure 4.14 *The idealised structure of the* n = 4 *Ruddlesden–Popper-derived phases* $Ln_3Ba_2Mn_2Cu_2O_{12}$ *where Ln is a lanthanoid*

4.8 Intercalation and Exfoliation

Chemically the interlayers in the modular phases described previously are relatively reactive. This has been exploited to fabricate large numbers of phases that are not available via the normal solid-state chemistry preparative method of high-temperature synthesis from mixed oxides or oxide precursors. The technique generally employs low temperatures, typically room temperature to 300°C, and long reactions times, of the order of 1 week. This procedure, termed 'chimie douce', is often translated as 'soft chemistry' but is better called 'mild chemistry' or 'gentle chemistry'. Under such conditions, much of the structure remains intact, and the product phases have a strong topotactic relationship to the starting structures. For the modular perovskite phases, in the main this means that the perovskite blocks remain largely unchanged, while the interlayer layer regions are modified.

The first reactions of this type that were reported exploited Dion–Jacobson phases as the starting point. Structures containing the largest cations in the interlayer A′-sites, such as the $n=3$ phase $RbA_2B_3O_{10}$, can be synthesised by conventional solid-state reaction or a related technique at high temperatures. This technique, though, is too brutal to employ for the synthesis of many phases with smaller A′-site cations. These can be prepared using ion exchange at low or moderate temperatures, which replaces the large A′ cation with smaller A″ cations such as K, Na and Li, often using low melting point nitrates as the exchange medium. Immersing these Dion–Jacobson phases in nitric acid solution for several days will exchange the small A′ and A″ cations for H^+ to yield protonated acidic Dion–Jacobson structures. These protonated phases can be exfoliated, that is, the perovskite slabs can be separated in an intact form, by treatment with a reagent such as aqueous tetra(*n*-butyl)ammonium hydroxide (TBA^+OH^-). This initially penetrates the interlayers and eventually (sometimes with the aid of mechanical means such as ultrasonic agitation) produces a colloidal solution of exfoliated perovskite slabs. These are frequently called nanosheets. Treatment with aqueous HCl can cause the nanosheets to precipitate into restacked structures. These structures do not show a perfect crystallographic order, but have surprisingly regular stacking, with the various sheets being displaced both laterally and vertically, often containing interlayer water or other molecules (Figure 4.15). Similar reaction sequences can be described for Ruddlesden–Popper and Aurivillius phases. The materials so formed, especially exfoliated and restacked solids, have potential as photocatalysts and as components in fuel cells.

There are large numbers of these reactions described in the literature. The transformation of Dion–Jacobson phases into the corresponding Ruddlesden–Popper phases is often the first step in making new compounds. Ruddlesden–Popper phases can, in most cases, be made by heating the corresponding Dion–Jacobson solid in the vapour of an alkali metal, lithium being the exception. These Ruddlesden–Popper structures can then be transformed into modular structures separated by A′ and OH- or A′ and Cl- (Figure 4.16).

In addition to these inorganic perovskite-related phases and reactions, a number of layered organic–inorganic perovskites can be mentioned. Simple organic–inorganic perovskites such as $(CH_3NH_3)PbI_3$ possessing the perovskite structure can be imagined to be expanded to form layered materials similar to the Ruddlesden–Popper and other phases listed earlier. Here the inorganic perovskite sheets are separated by more complex molecular species. Two such materials have recently been studied as they offer the possibility of interesting and valuable multiferroic properties that can be exploited. These are the K_2NiF_4 analogues with methylammonium

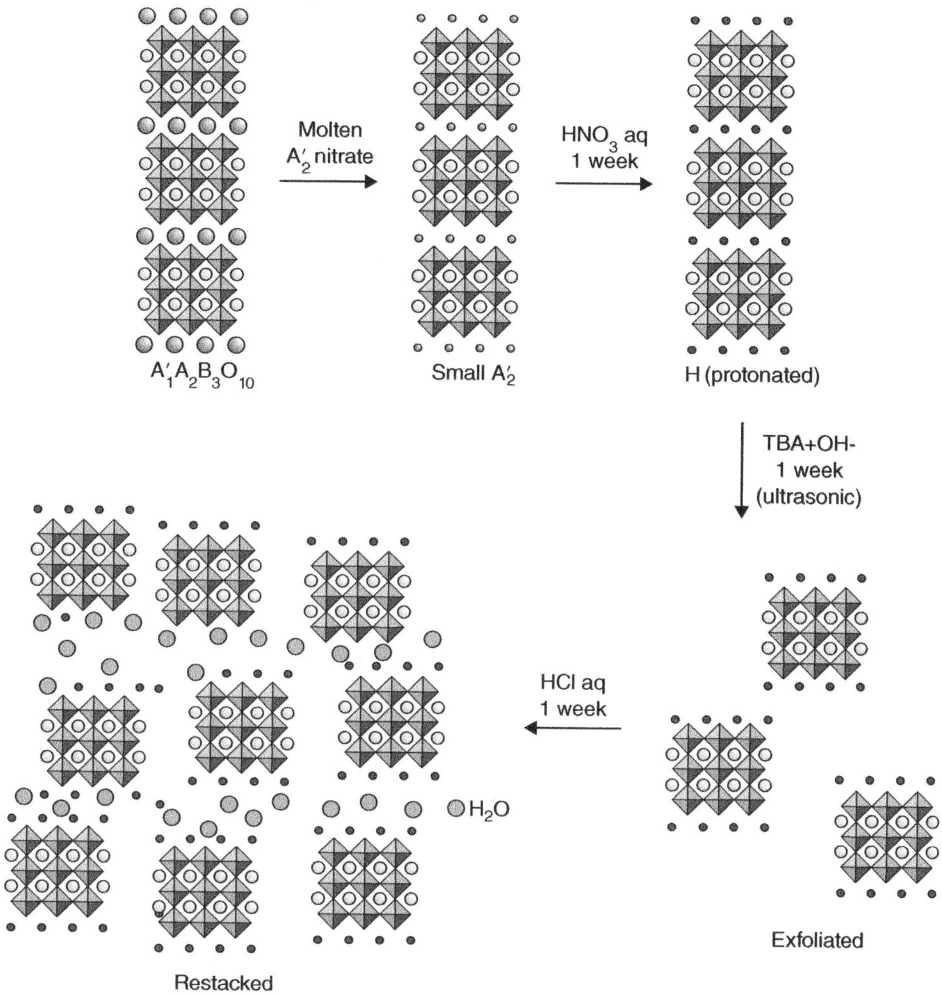

Figure 4.15 *Schematic flow chart for the formation of Dion–Jacobson phases with small cations and restacked structures starting from a large cation precursor such as* $RbCa_2Nb_3O_{10}$*. The exact chemical steps are more complex than those indicated in this outline*

[$(CH_3NH_3)^+$, or $(MA)^+$] to form $(MA)_2FeCl_4$ and ethylammonium [$(C_2H_5NH_3)^+$ or $(EA)^+$] to form $(EA)_2FeCl_4$.

Many other reactions could be illustrated, including the formation of ordered phases with n_1, n_2 stacking of perovskite slabs, ordered stacking of perovskite slabs and brucite-type double hydroxide layers or perovskite blocks with no interlayer cations present. For further information on the structural chemistry of these complex reactions, see Further Reading.

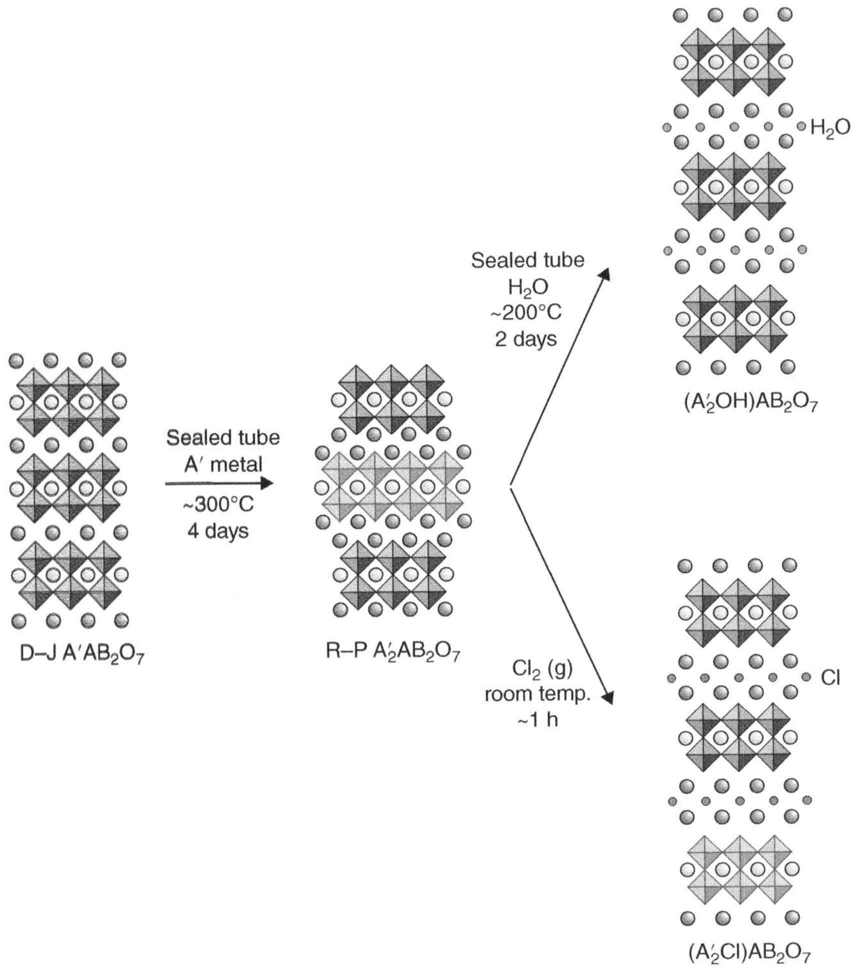

Figure 4.16 *Schematic flow chart for the formation of interlayer hydroxide and chloride phases starting from a Dion–Jacobson (D–J) large cation precursor such as* $RbLaNb_2O_7$ *via Ruddlesden–Popper (R–P) intermediates. The exact chemical steps are more complex than those indicated in outline*

Further Reading

Two lengthy reviews of layered perovskite structures are:

F. Lichtenberg, A. Herrnberger, K. Wiedenmann, *Prog. Solid State Chem.*, **36**, 253–387 (2008).

F. Lichtenberg, A. Herrnberger, K. Wiedenmann, J. Mannhart, *Prog. Solid State Chem.*, **29**, 1–70 (2001).

A review of the structures and properties of cuprates, cobaltites and manganites, with much information on modular phases, is given by:
B. Raveau, *Angew. Chem. Int. Ed.*, **52**, 167–175 (2013).

The symmetry consequences of octahedral tilting in the $Ca_2Nb_2O_7$ ($A_nB_nO_{3n+2}$) phases is described by:
I. Levin, L. A. Bendersky, *Acta Crystallogr.*, **B55**, 853–866 (1999).

The symmetry consequences of octahedral tilting, rotation, distortion and cation ordering in K_2NiF_4 phases is dealt with by:
P. V. Balachandran, P. Puggioni, M. Rondinelli, *Inorg. Chem.*, **53**, 336–348 (2014).

A review of the structures of superconductors is given in:
R. J. Cava, *J. Am. Ceram. Soc.*, **83**, 5–28 (2000).

A review of preparative methods involving modular perovskites is given by:
R. E. Schaak, T. E. Mallouk, *Chem. Mater.*, **14**, 1455–1471 (2002).

Two publications will provide a starting point for exploration of the chimie douce reactions of modular perovskites:
D. Montasserasadi *et al.*, *Inorg. Chem.*, **53**, 1773–1778 (2014).
T. Wang *et al.*, *Chem. Mater.*, **26**, 898–906 (2014).

5

Diffusion and Ionic Conductivity

This chapter is essentially concerned with the defect crystal chemistry of perovskites and the way in which it allows certain perovskite properties to be manipulated, typically diffusion and ionic conductivity.

5.1 Diffusion

Although native point defect populations in perovskites are sufficient for low diffusivity, the fast diffusion of cations and anions requires either the presence of a significant population of vacancies on the appropriate sublattices or fairly open regions such as occur in the layered phases. To increase diffusivity, perovskites with approximately cubic structures are doped in such a way as to generate vacancies on the appropriate sublattice. Changes in the partial pressure of the surrounding gases, especially oxygen and water vapour, can also accomplish this (see following text).

The diffusion of cations is generally specified in terms of the Arrhenius equation:

$$D = D_0 \exp\left(\frac{-E}{RT}\right)$$

where D is the diffusivity, D_0 the pre-exponential factor, E the activation energy for diffusion and T the temperature (K). The activation energy can be thought of as the

Perovskites: Structure–Property Relationships, First Edition. Richard J. D. Tilley.

energy required to move an ion from one stable position, usually through a bottleneck of surrounding ions, to an adjacent stable position.

To illustrate diffusion, B-site cation movement into lithium niobate, $LiNbO_3$, is described. This material has important optical properties (Chapter 9) and is employed in optical amplifiers, lasers and waveguides. However, the lithium niobate crystals suffer from optical damage when transmitting visible and near-infrared wavelengths. It has been found that B-site doping of surface layers can effectively reduce such damage, and there have been numerous studies on the diffusion of suitable cations such as Ti^{4+}, Zr^{4+}, Er^{3+} and Tm^{3+}, either in isolation or as co-diffusants, into crystals of $LiNbO_3$.

The optical behaviour of a $LiNbO_3$ component is sensitive to the direction of the light beam with respect to crystallographic orientation. The trigonal lithium niobate crystals are sliced in a variety of different orientations for these applications; these 'cuts' are described by a nomenclature that is identical to that used for trigonal low quartz using an X, Y and Z Cartesian coordinate system and hexagonal crystallographic axes (Figure 5.1a and b). An X-cut slice is taken as perpendicular to the hexagonal **a**-axis and has faces that are $(2\bar{1}\bar{1}0)$ planes. Y-cut slices are made in a direction perpendicular to the Y-axis and have faces that are $(1\bar{1}00)$ planes. Z-cut crystal slices are cut perpendicular to the **c**-axis and have (0001) faces.

For temperatures of the order of 1000–1100°C, typical values for diffusion of Er^{3+}, Tm^{3+} and Zr^{4+} into oriented slices of $LiNbO_3$ are

$$\begin{aligned}
&Er^{3+} \text{ into X-cut crystal}: D_0 = 3.20\times10^{9}\,\mu m^2 h^{-1}, E = 289 \text{ kJ mol}^{-1}\\
&Er^{3+} \text{ into Z-cut crystal}: D_0 = 3.89\times10^{9}\,\mu m^2 h^{-1}, E = 210 \text{ kJ mol}^{-1}\\
&Tm^{3+} \text{ into X-cut crystal}: D_0 = 4.37\times10^{8}\,\mu m^2 h^{-1}, E = 266 \text{ kJ mol}^{-1}\\
&Tm^{3+} \text{ into Z-cut crystal}: D_0 = 3.09\times10^{7}\,\mu m^2 h^{-1}, E = 231 \text{ kJ mol}^{-1}\\
&Zr^{4+} \text{ into Z-cut crystal}: D_0 = 2.14\times10^{14}\,\mu m^2 h^{-1}, E = 378 \text{ kJ mol}^{-1}
\end{aligned}$$

The movement of B-site cations involves passage across a triangular O^{2-}-O^{2-}-O^{2-} octahedral face into the A-site and then traversing another octahedral face into an adjoining B-site (Figure 5.1c). In the absence of a vacancy population on both sites, such movement is clearly prohibited. In the present case these vacant sites are the result of the normal point defect populations present in the crystal, which is reflected in the value of the pre-exponential factor D_0. The activation energy, however, represents the difficulty of moving a cation across the energy barrier mainly associated with the triangular bottleneck of oxygen anions. The activation energy for diffusion of the two lanthanoid cations is slightly higher when the cations penetrate an X-cut crystal slice than for the Z-cut orientation. Although the crystal structure at room temperature is trigonal, at the diffusion temperature it is cubic. Diffusion into a Z-cut face will be along a cubic [111] direction (as in Figure 5.1c), while diffusion

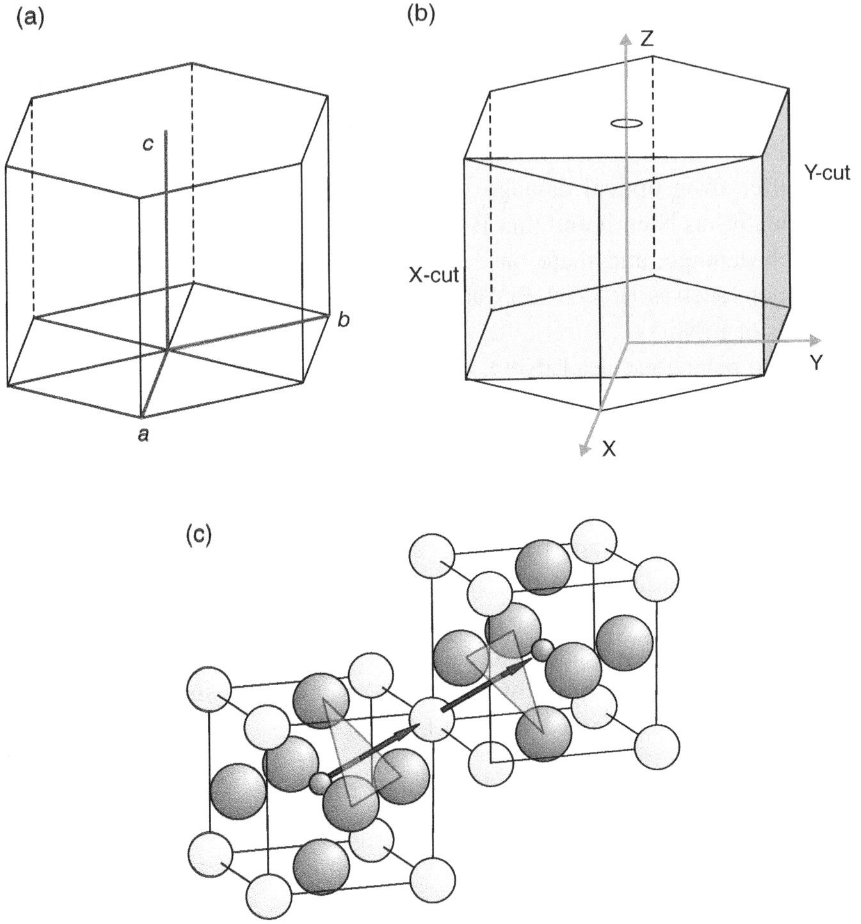

Figure 5.1 *(a) Single crystal form of $LiNbO_3$ crystal; (b) X-cut and Y-cut slices; (c) B-site cation diffusion path between adjacent octahedra*

perpendicular to an X-cut face will be perpendicular to this, which is approximately 20° away from [111]. The activation energy differences reflect this change in the overall diffusion pathway. In the case of Zr^{4+} the diffusion perpendicular to the Z-cut face will be along a [111] direction, as with the lanthanoid cations. The fact that the activation energy for the Ln^{3+} diffusion is lower than for Zr^{4+} diffusion reflects the change in charge state on the cation as in both cases the diffusion pathways will be identical. It appears that the high charge on the Zr^{4+} cations results in stronger bonding to the surrounding oxygen anions than is experienced by the Ln^{3+} cations, thus increasing the energy barrier that the Zr^{4+} cations have to jump. Clearly successful surface doping will be achieved after shorter heating times for lower valence species than higher valence ones for this system.

5.2 Ionic Conductivity

Ionic conductivity is the transport of cations and/or anions across the perovskite under the influence of an electric field. As with diffusion, for ionic conductivity of cations and anions in perovskites to occur the structure must either contain open regions or a significant population of vacancies on the appropriate sublattice to allow ionic movement. Substitution is again widely used to create vacancies in perovskites with approximately cubic structures so as to increase conductivity. A further requirement, for strictly ionic conductivity, is the absence of cations with a variable valence. In cases where variable valence cations are present, electronic conductivity may also occur and in such cases will invariably dominate, in magnitude, the ionic conductivity (Sections 5.4 and 5.5).

In cases where several ions contribute to the conduction process, the fraction of the conductivity that can be apportioned to each ion is called its transport number, t_c for cations and t_a for anions:

$$t_c + t_a = 1$$

Ionic conductivity can be described by equations similar to those for diffusion, leading to an Arrhenius-type equation often written as

$$\sigma T = \sigma_0 \exp\left(\frac{-E}{RT}\right) \qquad (5.1)$$

where σ is the ionic conductivity, T the temperature (K), σ_0 the pre-exponential factor and E the activation energy for conduction.

Enhanced O^{2-} anion conduction due to induced oxygen vacancy formation is found in derivatives of the substituted perovskite $LaInO_3$. The parent phase contains La^{3+} and In^{3+} cations. The movement of cations and anions in this structure involves the displacement of the diffusant into a suitable neighbouring site which is vacant. Anions must pass through the triangular face composed of the cations La^{3+}-La^{3+}-In^{3+}, while the most direct path for A-site cations requires them to squeeze between two O^{2-} anions (Figure 5.2a). The movement of B-site cations involves passage across a triangular O^{2-}-O^{2-}-O^{2-} octahedral face described earlier.

Because of this restriction, the oxide ion conductivity can be greatly increased by A-site substitution with Ca^{2+}, Sr^{2+} or Ba^{2+} (acceptor doping). The resulting charge imbalance cannot be corrected by valence change of the cations, and charge balance is achieved by the introduction of X-site vacancies to give a formula $La_{1-x}A_xInO_{3-x/2}$, where A represents Ca^{2+}, Sr^{2+} or Ba^{2+}. The resulting high concentration of vacancies leads to a considerable increase in oxide ion conductivity. The conductivity is found to obey Equation (5.1), and the slope of a plot of ln (σT) versus ($1/T$) will yield a value for the activation energy. These slopes are very similar for the three systems $La_{0.9}Ca_{0.1}InO_{2.95}$, $La_{0.9}Sr_{0.1}InO_{2.95}$ and $La_{0.9}Ba_{0.1}InO_{2.95}$ (Figure 5.2b), showing that there is only a small difference in the activation energy for the migration of O^{2-} anions in these phases, even though the bottleneck geometry can have three configurations, La-La-In, La-A-In or A-A-In, where A represents Ca^{2+}, Sr^{2+} or Ba^{2+}. Molecular dynamics simulations indicate, however, that smaller A cations have more strongly ionic bonds to oxygen than the larger A cations, which may be a factor in decreasing the conductivity of Ca^{2+}-doped phases compared to Sr^{2+}- and Ba^{2+}-doped materials (Figure 5.2b).

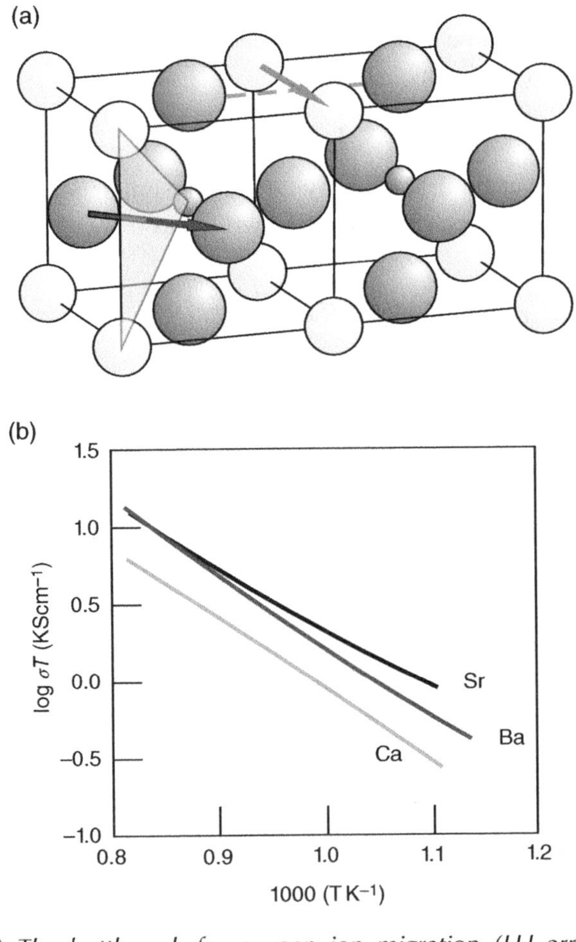

Figure 5.2 *(a) The bottleneck for oxygen ion migration (LH arrow) and for the migration of A-site cations, upper arrow; (b) plots of logσT versus 1000/T for the perovskites $La_{0.9}Ca_{0.1}InO_{2.95}$, $La_{0.9}Sr_{0.1}InO_{2.95}$ and $La_{0.9}Ba_{0.1}InO_{2.95}$*

A-site ionic diffusion enhanced by an A-site vacancy population is found in derivatives of $La_{2/3}TiO_3$. This phase contains Ti^{4+} and has 1/3 of the A-sites vacant. Lithium ion conductivity is of interest in materials used in Li batteries, and the doped phases $Li_{3x}La_{2/3-x}TiO_3$ are favoured as electrolytes because of the pre-existing A-site vacancy population allows for easy Li^+ diffusion. The substituted phases exist along a line in the phase diagram running from the parent $La_{2/3}TiO_3$ to the limiting composition $Li_{0.5}La_{0.5}TiO_3$ in which all of the A-sites are occupied. In all cases the Ti is in the fully oxidised Ti^{4+} state. Clearly the Li^+ ion conductivity will be low for the end members of the phase range and will increase as the Li^+ ion numbers increase as long as the A-site vacancy population remains significant. A maximum in conductivity would be expected when about half of the vacancies are occupied by Li^+. In practice, the highest conductivity phase has an approximate composition of $Li_{0.33}La_{0.56}TiO_3$, with a conductivity of approximately $10^{-3}\,S\,cm^{-1}$.

In reality materials have a partial segregation of the vacancies, so that the solid contains alternating La-rich and La-poor layers. This means that Li^+ conductivity is more akin to a two-dimensional process. Moreover, the segregation of the vacancies leads to microdomain formation and the presence of antiphase boundaries in the material, which may interfere with ionic conductivity, although in thin films of $Li_{3x}La_{2/3-x}TiO_3$ it has been found that that these structural features have only a small influence on the magnitude of the Li^+ ion conductivity.

Ionic conductivity has been studied extensively in the lanthanoid oxygen-rich Ruddlesden–Popper phases $Ln_2NiO_{4+\delta}$ and $La_2CuO_{4+\delta}$, again as these have potential applications as possible cathode materials in solid oxide fuel cells (SOFCs) (Section 5.7). The additional oxygen in these structures is accommodated in interstitial sites in the LnO-containing layers between the perovskite sheets (Figure 5.3a). The oxide ion conductivity is, as expected, far higher in this plane between the perovskite layers than in a direction perpendicular to them. Additionally, this interlayer conductivity remains high over a considerable range of temperatures, a feature of importance in fuel cell applications.

The oxide ion conductivity (and diffusion) takes place via an interstitialcy mechanism, in which the interstitial ion moves into a normally occupied site while simultaneously displacing the resident ion into an adjoining interstitial position (Figure 5.3b). The migration is aided by the fact that the octahedra in the layers are considerably tilted, with the tilt angle dependent upon the lanthanoid cation present. Theoretical calculations of the migration barrier show a correlation with octahedral tilt. As the O–Ni–O angle, derived from the parent $LnNiO_3$ perovskite phase, deviates more from the untilted 180° configuration, the more the migration barrier falls (Figure 5.3c). Moreover, as octahedral tilt is sensitive to pressure, calculations have been able to show that tensile strain lowers the potential barrier for migration, while compressive strain increases it. These effects are particularly important when epitaxial thin films are employed in devices.

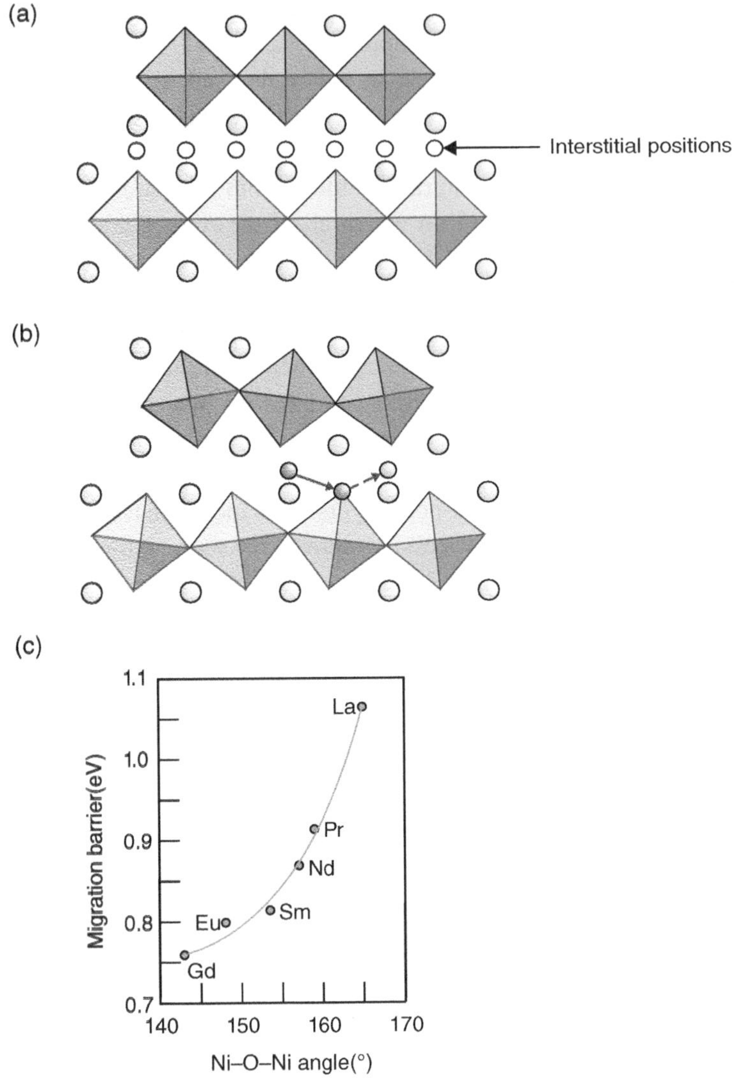

Figure 5.3 *Interstitialcy conduction in* $Ln_2NiO_{4+\delta}$*: (a) idealised structure showing interstitial sites; (b) interstitialcy migration mechanism between tilted octahedra; (c) calculated correlation between tilt angle in* $LnNiO_3$ *and migration barrier for oxide ion conductivity (For more extensive experimental data see Li and Benedek (2015))*

5.3 Proton Conductivity

Proton conducting perovskite oxides are the subject of numerous current studies, as these materials have a potential use in a number of electrochemical applications, including fuel cells, electrochromic displays and hydrogen sensors. Perovskite oxides can be manipulated using point defect chemistry to become efficient proton conductors. The most widely studied are the substituted (2, 4) $A^{2+}B^{4+}O_3$ perovskites $BaTiO_3$, $BaCeO_3$ and $BaZrO_3$ and the $A_2B_2O_5$ perovskite derivative $Ba_2In_2O_5$.

The oxide $Ba_2In_2O_5$ adopts the brownmillerite structure at lower temperatures (Section 2.4). The structure begins to disorder at approximately 900°C and forms a cubic perovskite phase with a high concentration of oxygen vacancies somewhere above this temperature. Proton conduction in the solid is brought about by the incorporation of water molecules. Water vapour can react directly with the oxide at higher temperatures, following disproportionation at the oxide surface:

$$H_2O \rightarrow OH^- + H^+$$

The OH^- ions occupy the oxygen vacancies in the disordered material as $OH_O^\bullet$ defects to complete or partially complete the perovskite structure, while the H^+ ions (i.e. protons) are normally assumed to be linked to existing oxygen anions to also form $OH_O^\bullet$ defects. The fully reacted materials have a composition $Ba_2In_2O_6H_2$ or $BaInO_3H$. The hydrogen ions in the structure are not bonded strongly to particular oxygen ions, but are linked by hydrogen bonds. This rather weaker bonding is taken to be the underlying reason why the material exhibits high proton conductivity.

The perovskites $BaCeO_3$ and $BaZrO_3$ are insulating solids when prepared in air. These are converted to oxygen-deficient phases by doping the B-sites with fixed valence trivalent M^{3+} ions, such as In^{3+}. Charge balance is not possible by valence induction and instead is achieved by the introduction of a population of oxygen vacancies. The presence of two M^{3+} substituents is then balanced by the formation of one oxygen vacancy. The formulae of the products are $BaCe_{1-x}M_xO_{3-x/2}$ or $BaZr_{1-x}M_xO_{3-x/2}$, for example, $BaZr_{1-x}In_xO_{3-x/2}$, with defect populations of $2In_{Zr}' : V_O^{\bullet\bullet}$. A typical dopant level of $x = 0.05$ is often used.

The proton conductivity is induced by reaction with water vapour, exactly as with $Ba_2In_2O_5$. Reaction with water vapour at higher temperatures results in disproportionation of water molecules at the oxide surface, with the resultant OH^- ions occupying the oxygen vacancies in the structure and the protons hydrogen bonded to existing lattice oxygen. The nominal formulae of the phases are $BaCe_{1-x}M_xO_3H_x$ and $BaZr_{1-x}M_xO_3H_x$, typified by $BaZr_{1-x}In_xO_3H_x$.

The hydration process in all proton conducting perovskites can be formally written as a defect equation:

$$H_2O(g) + O_O + V_O^{\bullet\bullet} \rightarrow 2OH_O^\bullet$$

The number of mobile protons is equal to the concentration of $OH_O^\bullet$ species. The equilibrium constant for the reaction, K, is

$$K = \left[OH_O^\bullet\right]^2 / \left[V_O^{\bullet\bullet}\right]\left[O_O\right] p_{H_2O}$$

The number of mobile protons available in the structure for conductivity rises in proportion to the partial pressure of water vapour according to the equation

$$\left[OH_O^{\bullet}\right]^2 = K\left[V_O^{\bullet\bullet}\right]\left[O_O\right] p_{H_2O}$$

which gives rise to a parabolic hydration isotherm (Figure 5.4), with a curve dependent upon the value of the dopant concentration. The values of $[V_O^{\bullet\bullet}]$ and the equilibrium constant K, which are determined by the thermodynamics of the reaction, can be deuced from the curve. This relationship between the water vapour pressure and conductivity can be exploited in humidity sensors. The reaction is also temperature sensitive, and higher water uptake and concomitantly higher proton conduction are achieved at lower temperatures.

The true stoichiometry of these compounds and the proton positions within them are difficult to determine using traditional diffraction techniques, which average over considerable volumes of structure. This fact makes it hard to be precise about the detailed mechanism of conduction. Under the influence of an electric field, the protons in the structure will tend to align themselves in a preferred direction with respect to the applied field (Figure 5.5a). However, the structure is not static and thermal vibrations are believed to be important in the proton transfer process. When lattice vibrations act in concert so as to tilt the octahedra and so shorten the distance between oxide ions or to weaken hydrogen bonds, a proton may then jump to an adjacent oxygen ion, either on the same coordination polyhedron or to a neighbour (Figure 5.5b). Theoretical approaches are now used to explore diffusion pathways and have been successful in describing low energy configurations of the solid, possible diffusion paths for proton migration and details of individual proton jumps.

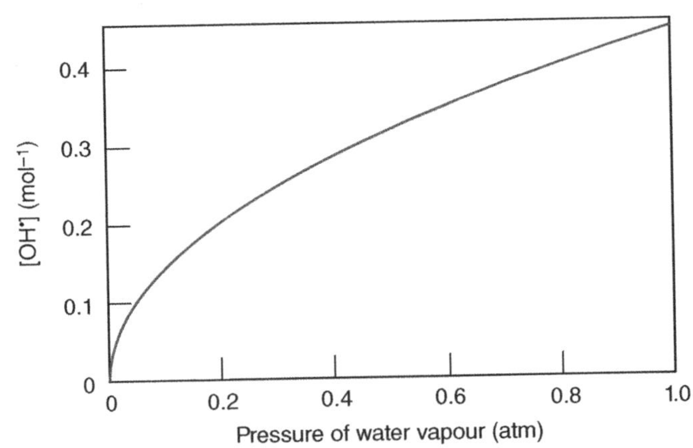

Figure 5.4 *Plot of $[OH_O^{\bullet}]$ (mobile proton concentration) versus water vapour pressure (hydration isotherm) for a material with 10 mol % vacancies and K = 2.0*

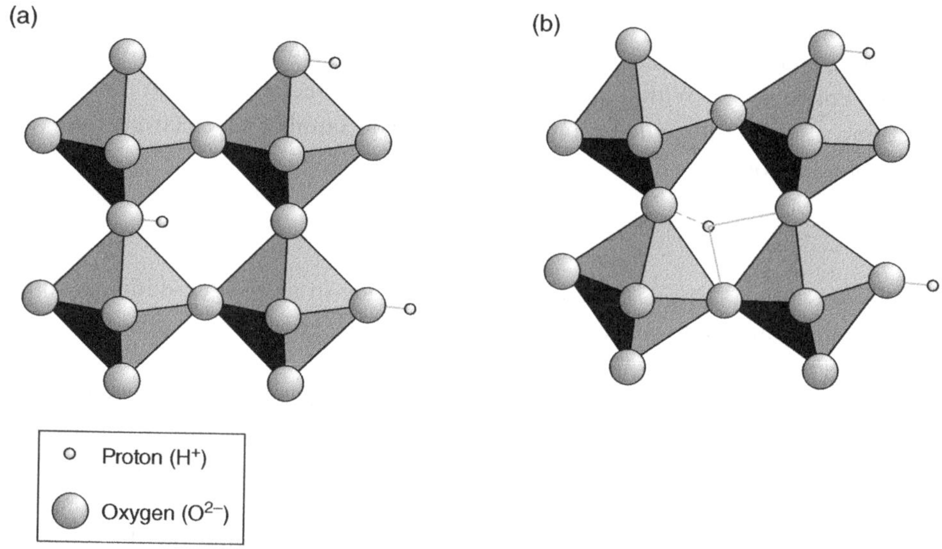

Figure 5.5 Proton diffusion: (a) an external electric field will tend to align protons; (b) suitable thermal vibrations may lengthen and weaken hydrogen bonds allowing migration to occur

5.4 Oxygen Pressure Dependence and Electronic Conductivity

The composition of many oxide perovskites is dependent upon the ambient oxygen pressure during preparation and in some circumstances can be made either oxygen rich or poor. In cases where no change of cation valence is possible, oxygen-poor compositions are accommodated by the formation of oxygen vacancies. Additional oxygen can be accommodated as interstitial ions if the structure is open enough to allow for this possibility. These options are often linked to changes in the ionic and electronic conductivity of the phase that can be of considerable importance.

As an example, the normally stoichiometric insulator $SrTiO_3$ can readily be made into an oxygen-deficient form $SrTiO_{3-\delta}$ by heating in a vacuum. The loss of oxygen can be written in defect terms as

$$O_O \leftrightarrow ½O_2 + V_O^{\bullet\bullet} + 2e' \tag{5.2}$$

The equilibrium constant for this reaction, K, is

$$K = \left[V_O^{\bullet\bullet}\right]\left[e'\right]^2 p_{O_2}^{½} \left[O_O\right]$$

The number of oxygen vacancies will be proportional to the partial pressure of oxygen as

$$\left[V_O^{\bullet\bullet}\right] \propto p_{O_2}^{-½}$$

A consequence of the presence of oxygen vacancies results in the appearance of a balancing population of electrons (Equation (5.2)), which will give rise to electronic (n-type) conductivity. Within the limits of the point defect equilibrium model, the electronic conductivity will be proportional to the number of electrons and the partial pressure of oxygen as

$$\sigma \propto [e'] \propto p_{O_2}^{-1/4}$$

[The closely related phase $BaTiO_3$ can also be made in an oxygen-deficient form and is transformed into a metallic conductor under greater reduction than that considered here (Section 8.2)].

In cases where a valence change of the cations (valence induction) is possible, this option is often preferred to the incorporation of nominally free electrons or holes and can lead to electronic conductivity by way of electron hopping from one cation to another. This is similar to the proton hopping described in the previous section. For example, the $n=3$ Ruddlesden–Popper phase $Sr_4Fe_3O_{10}$, containing only Fe^{4+}, is able to lose oxygen to give a population of oxygen vacancies. At higher temperatures these order to a new structure (Figure 4.13), but at lower temperatures they are disordered. In the disordered state the removal of oxygen from the crystal to the gas phase to create a vacancy must be accompanied by the creation of accompanying electrons (Equation 5.2), but these are located on two Fe^{4+} (Fe_{Fe}) cations to form Fe^{3+} (Fe'_{Fe}):

$$O_O + 2Fe_{Fe} \rightarrow \tfrac{1}{2}O_2 + V_O^{\bullet\bullet} + 2Fe'_{Fe}$$

The equilibrium constant for the reaction, K, is

$$K = [V_O^{\bullet\bullet}]\,[Fe'_{Fe}]^2\, p_{O_2}^{1/2} / [O_O]\,[Fe_{Fe}]^2$$

Under conditions where this point defect equilibrium accurately reflects the situation, electron n-type conductivity is possible due to the hopping of electrons from Fe^{3+} to Fe^{4+}. The electronic conductivity will depend upon the amount of Fe^{3+} present. This will depend upon the partial pressure of oxygen and can be derived from the equilibrium constant equation as

$$\sigma \propto [Fe'_{Fe}] \propto p_{O_2}^{-1/4}$$

However, the production of oxygen vacancies may also encourage oxygen ion conduction, and the total conductivity of a material may have several components, oxide ion, electron or hole contributions. These more complex situations are discussed in the following section.

5.5 Oxide Ion Mixed Conductors

In general, oxides that show both ionic and electronic conductivity are called mixed conductors. These have applications in electrochemical cells, particularly SOFCs (Section 5.7) where both ionic and electronic components are essential. The ionic conductivity is usually several orders of magnitude less than that of the electronic component and is often more difficult to quantify.

The Ruddlesden–Popper phase $Sr_3Fe_2O_{6+\delta}$ is similar to the $n = 3$ phase $Sr_4Fe_3O_{10}$, described earlier. The parent compound $Sr_3Fe_2O_7$ contains only Fe^{4+}. Normal preparation in air gives an oxide with a composition closer to $Sr_3Fe_2O_{6.73}$, corresponding to the mixed valence state $Sr_3Fe^{3+}_{0.54}Fe^{4+}_{1.46}O_{6.73}$. This persists unchanged upon heating in air up to approximately 400°C. Thereafter the phase loses oxygen down to a composition of approximately $Sr_2Fe_2O_6$ above 1000°C. This composition corresponds to Fe^{3+} only. The notional oxygen vacancies appear to be distributed at random over the equatorial sites surrounding the FeO_6 octahedra in the structure of these materials. At high oxygen partial pressure, the layer structure of the Ruddlesden–Popper phase is open enough to take in additional oxygen as interstitials in the interlayers between the perovskite blocks. The composition of the parent phase can thus traverse a considerable oxygen composition range. The total conductivity of the material will then contain possible contributions from oxygen ion vacancy migration, oxygen ion interstitialcy migration, electronic (n-type) conduction and hole (p-type) conduction. These possibilities can be explained in terms of the defect chemistry of the system under low and high oxygen partial pressures.

The oxygen ion conductivity, σ_i, is due to O^{2-} migration via the oxygen vacancy population at lower oxygen partial pressures and via O^{2-} interstitial migration at higher oxygen partial pressures. The change between the two mechanisms is found to depend only upon the temperature of the measurements. The ionic transport number varies with the composition and thus the surrounding oxygen partial pressure, showing a maximum of close to 0.4 at a partial pressure close to that giving a minimum in the electronic conductivity.

The oxygen vacancy conductivity is generally much smaller than the electron and hole conductivity. The n-type electronic conductivity depends both on the temperature and the surrounding partial pressure of oxygen in the following way. At low oxygen partial pressures, oxygen is lost to the surrounding atmosphere to create oxygen vacancies. The charged oxide ion is extracted as a neutral atom, leaving behind two electrons. The defect formation equation is

$$O_O \rightarrow \tfrac{1}{2}O_2 + V_O^{2\bullet} + 2e'$$

The equilibrium constant is given by

$$K = [e']^2 [V_O^{2\bullet}] p_{O_2}^{1/2} / [O_O]$$

The (n-type) conductivity is proportional to the electron concentration:

$$\sigma_n \propto [e'] \propto p_{O_2}^{-1/4}$$

Under low oxygen partial pressures the conductivity is n-type and decreases with oxygen partial pressure, in agreement with the experimental results (Figure 5.6).

At high oxygen pressures the incorporation of neutral oxygen to form oxygen interstitials requires the abstraction of two electrons to form oxide ions, thus generating an equal number of holes:

$$\tfrac{1}{2}O_2 \rightarrow O_i^{2'} + 2h^{\bullet}$$

The equilibrium constant is given by

$$K = [h^{\bullet}]^2 [O_i^{2'}] / p_{O_2}^{1/2}$$

The hole population results in p-type conductivity, the magnitude of which is proportional to the hole concentration:

$$\sigma_p \propto [h^{\bullet}] \propto p_{O_2}^{-1/4}$$

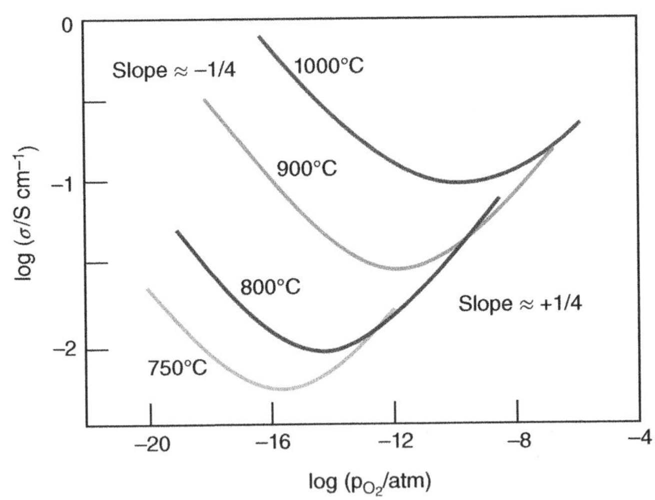

Figure 5.6 *Total conductivity of the Ruddlesden–Popper phase* $Sr_3Fe_2O_{6+\delta}$ *(For more extensive data see Shilova* et al. *(2002))*

Under higher oxygen partial pressure this equilibrium dominates and the conductivity increases with oxygen partial pressure to the power +¼, in agreement with the experimental results (Figure 5.6). The total conductivity is given by an equation of the form

$$\sigma(T, p_{O_2}) = \sigma_i(T) + \sigma_n(T) p_{O_2}^{-1/4} + \sigma_p(T) p_{O_2}^{1/4}$$

The form of the curve drops more or less symmetrically to a minimum which depends upon the temperature of the sample.

Many other perovskites containing mixed valence cations behave in a similar way, and these can be modified with A- or B-site substitutions in order to improve either aspects of stability or conductivity. For example, the perovskite $SrFe_{1-x}Sc_xO_{3-\delta}$ shows the same conductivity dependence as $Sr_3Fe_2O_{6+\delta}$:

$$\sigma(T, p_{O_2}) = \sigma_i(T) + \sigma_n(T) p_{O_2}^{-1/4} + \sigma_p(T) p_{O_2}^{1/4}$$

The slope of +¼ is found at high oxygen partial pressures and the slope of −¼ at low oxygen partial pressures, following the same defect chemistry as that described previously. However, the B-site substitution of Sc^{3+} prevents the transformation of the high-temperature disordered cubic phase into an ordered brownmillerite structure below a temperature of approximately 850°C. This improves the conductivity at lower temperatures compared to the un-substituted material.

5.6 Proton Mixed Conductors

In reality the conductivity of proton conducting phases (Section 5.3) is more complicated than described. Successfully doped materials are not electronic insulators, as the aforementioned equations imply, but are generally weak p-type semiconductors. This comes about because at high temperatures a small amount of oxygen can react with the defective perovskite to partially fill the vacancies and generate a population of holes:

$$V_O^{2\bullet} + ½O_2 \rightarrow O_O + 2h^\bullet$$

In addition, the presence of a population of oxygen vacancies allows for significant oxygen diffusion in the solid. This means that in moist atmospheres there is the possibility of mixed conductivity involving three charge carriers (H^+, O^{2-} and $h^\bullet$) or four, if electrons (e') are included.

The situation can be illustrated with respect to the B-site substituted perovskite $SrZr_{1-x}Y_xO_{3-0.5x}$, with dopant levels of $x = 0.05$–0.2. The B-site substitution of two lower valence Y^{3+} cations is balanced by the formation of one oxygen ion vacancy, allowing oxygen ion conductivity to occur. The doped phases, however, are generally p-type semiconductors. This arises from a combination of atmospheric oxygen with the vacancies to fill them and at the same time generate a hole population to maintain charge balance:

$$V_O^{2\bullet} + ½O_2 \rightarrow O_O + 2h^{\bullet} \quad (5.3)$$

On exposure to water vapour the oxygen vacancies can also react to form OH^-:

$$O_O + V_O^{2\bullet} + H_2O \rightarrow 2OH_O^{\bullet} \quad (5.4)$$

The resulting phases are good proton conductors. The total conductivity is due to contributions from oxygen ions, protons and holes:

$$\sigma(\text{total}) = \sigma(O^{2-}) + \sigma(H^+) + \sigma(h^{\bullet})$$

The contribution of each of these is strongly temperature dependent. A measurement of the transport numbers of these species as a function of temperature shows that at lower temperatures proton conductivity is of greatest importance, with $t(H^+)$ approaching 1.0; at middle temperatures oxygen ion conductivity becomes dominant, with $t(O^{2-})$ approaching 1.0; and at an high temperatures the material is predominantly a hole conductor, again with $t(H^+)$. approaching 1.0 (Figure 5.7). Between these temperatures, at approximately 350°C the solid is a mixed H^+ and O^{2-} conductor, while at approximately 650°C it is a mixed hole and O^{2-} conductor.

The B-site substituted Ba analogue, the proton conductor $BaZr_{0.8}Y_{0.2}O_{3-\delta}$, is similar. This material has been investigated for possible use in SOFCs, hydrogen sensors and pumps and as a catalyst. Doping the parent phase $Ba^{2+}Zr^{4+}O_3$ with trivalent Y^{3+} induces the formation of oxygen vacancies which can react with oxygen gas to generate holes or with water vapour to create hydroxyl as described previously. The defect population in the doped solid under moist conditions then consists of $V_O^{2\bullet}$, $OH_O^{\bullet}$, $h^{\bullet}$ and Y'_{Zr}. The domains over which each species is dominant for conductivity can be represented as phase fields in a diagram plotting oxygen partial pressure and hydrogen partial pressure in an analogous way to that used in Pourbaix diagrams for the compact description of corrosion reactions of metallic materials with water.

The conductivity fields can be delineated in several ways. Here they are bounded by lines tracing the locus where the transport number for a pair of defects is equal to 0.5. The boundary between the mainly oxygen ion conducting region and the mainly hole conducting region is defined by Equation (5.3). There is no involvement of water vapour in this equilibrium, so the boundary will be uninfluenced by hydrogen partial pressure and lie parallel to that axis, intersecting the oxygen partial pressure axis at a value appropriate to that at which the defect concentrations give rise to the conductivity relation $t(O^{2-}) = t(h^{\bullet}) = 0.5$. The boundary between the mainly

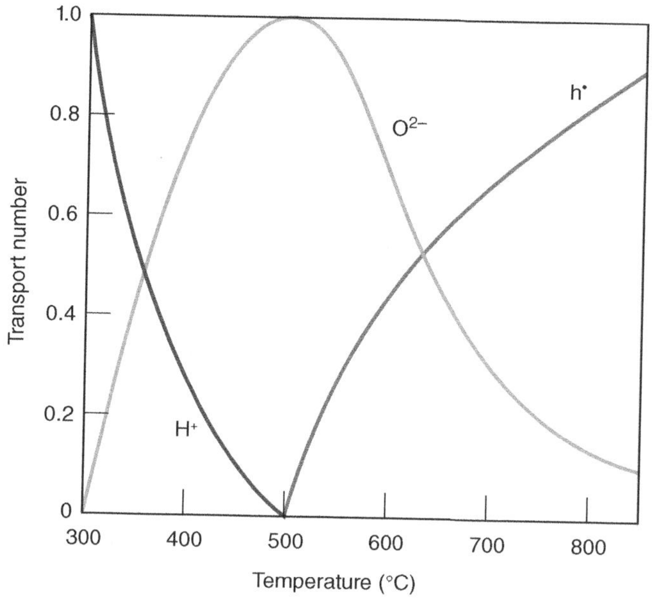

Figure 5.7 *The variation in transport number for proton, oxygen ion and hole conductivity for the mixed proton conductor* $SrZr_{1-x}Y_xO_{3-\delta}$ *(Data from Huang and Petric (1995))*

proton conducting domain and the mainly proton conducting domain will be defined by Equation (5.4). There is no involvement of oxygen gas in this equilibrium, so the boundary will be uninfluenced by oxygen partial pressure and lie parallel to that axis, intersecting the hydrogen partial pressure axis at a value appropriate to that at which the defect concentrations give rise to the conductivity relation $t(H^+)=t(h^\bullet)=0.5$. The boundary between the mainly oxygen conducting regime and the proton conducting regime depends upon both Equations (5.3) and (5.4) and is given by

$$\tfrac{1}{2}O_2 + 2V_O^{2\bullet} + H_2O \rightarrow 2OH_O^\bullet + 2h^\bullet$$

Because this equation involves both water vapour and oxygen, it will slope at an angle to the axes. The boundary is drawn where the conductivity relation $t(H^+)=t(O^{2-})=0.5$ holds (Figure 5.8).

The diagram shows the situation for a particular temperature and so can be used to determine the most important conductivity mechanism for a particular operating condition at the temperature of the diagram. Similar diagrams must be constructed for other temperatures, which, when stacked up, will give a three-dimensional representation of the way in which the predominant conduction mechanism changes with temperature. Another set of diagrams will also need to be constructed for other dopant concentrations.

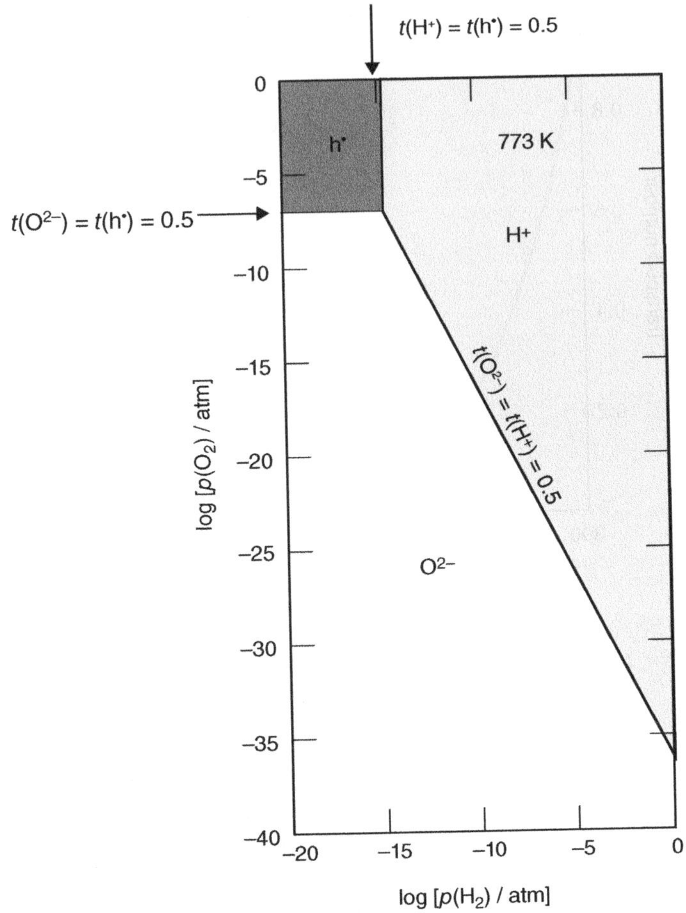

Figure 5.8 Phase-field map representation of the principal conductivity domains in $BaZr_{0.8}Y_{0.2}O_{3-\delta}$. Single-toned regions are areas within which conductivity of one species predominates (Original data in Nomura and Kageyama (2007))

5.7 Solid Oxide Fuel Cells

SOFCs generate electricity by the electrochemical oxidation of a continuously supplied 'fuel' within an electrochemical cell, generally using oxygen as the oxidising component. The most common forms of SOFCs use hydrogen or a number of hydrocarbons as fuel. Oxygen (in the form of air) is supplied to the cathode or *air electrode*, and the fuel is supplied to the anode or *fuel electrode*. For the reaction to proceed, either oxygen must be transported across the electrolyte to the fuel electrode, or fuel must be transported across the electrolyte to the air electrode. Moreover, the electrolyte must not show any significant degree of electronic conductivity.

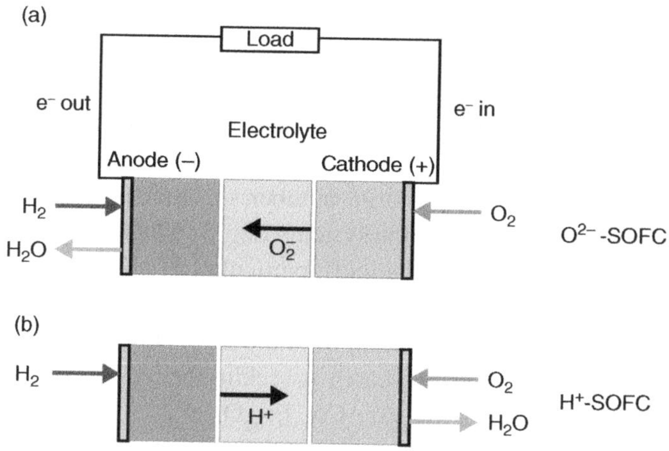

Figure 5.9 *Solid oxide fuel cells schematic: (a) oxygen ion conducting electrolyte; (b) proton conducting electrolyte, both with H_2 gas as fuel*

The principal types of cell in use have an electrolyte that conducts O^{2-} ions, so that the reaction products are generated at the anode (Figure 5.9a). Here the electrolyte is usually yttria-stabilised zirconia (YSZ) ($Y_xZr_{1-x}O_{2-x/2}$ containing $x/2$ oxygen vacancies). However, there is increasing interest in the use of electrolytes that are proton conducting, in which case the reaction products are produced at the cathode (Figure 5.9b). The electrolyte materials are identical to those described earlier (Section 5.3), typically $BaCeO_3$ and $BaZrO_3$, both B-site doped with a trivalent cation, usually Y, In or Ga. The most studied are $BaZr_{1-x}Y_xO_3$ and $BaCe_{1-x}Y_xO_3$, with x in the range of 0.1–0.2.

Taking an oxygen ion cell as an example, a typical unit contains a 20 μm thick cathode, a 3 μm thick barrier layer (often CeO_2 doped with Sm_2O_3), a 7 μm layer of YSZ electrolyte and a thick 550 μm layer of Ni-impregnated YSZ as the anode. The cell reaction can be illustrated using hydrogen as a fuel:

$$\begin{aligned}
&\text{Anode (fuel electrode) reaction:} && H_2(g) + O^{2-}(s) \rightarrow H_2O(l) + 2e^- \\
&\text{Cathode (air electrode) reaction:} && \tfrac{1}{2}O_2(g) + 2e^- \rightarrow O^{2-}(s) \\
&\text{Cell discharge reaction:} && H_2(g) + \tfrac{1}{2}O_2(g) \rightarrow H_2O(l)
\end{aligned}$$

The key cathode reaction is the oxygen reduction reaction, in which gaseous oxygen is converted into oxygen ions that can be transported across the electrolyte:

$$\tfrac{1}{2}O_2 + 2e' + V_O^{\bullet\bullet} \rightarrow O_O$$

Perovskites have two possible roles in these cells. The first of these, as a perovskite-derived proton conducting electrolytes, is described earlier (Section 5.3). In cells that use an oxide ion conducting electrolyte, perovskites are mainly employed as cathodes. The cathode can be an electron-only conducting perovskite, in which case the reaction must take place close to the electrolyte boundary. However, perovskites that are mixed conductors that can conduct both O^{2-} and electrons and show catalytic properties that assist in the chemical transformations are preferred as the reaction can then take place over a wider reaction region. The most widely used cathode material is $La_{0.8}Sr_{0.2}MnO_{3\pm\delta}$, but a large number of other perovskites have been explored in this capacity, including the systems $Ln_{1-x}Sr_xMnO_{3-\delta}$, $Ln_{1-x}Sr_xFeO_{3-\delta}$ and $Ln_{1-x}Sr_xCoO_{3-\delta}$ (where Ln is a lanthanoid), particularly $La_{0.6}Sr_{0.4}Co_{0.2}Fe_{0.8}O_{3\pm\delta}$ and $Sm_{0.6}Sr_{0.4}CoO_{3-\delta}$, due to their high electronic and ionic conductivity and relatively high stability under the operating conditions of the cells. However, none of these are completely perfect and continual research is seeking better combinations of these cations, in formulations such as $Ba_{0.5}Sr_{0.5}Co_{0.8}Fe_{0.2}O_{3-\delta}$.

A problem with the current generation of SOFCs is that diffusion of oxygen ions in the electrolyte is slow at room temperature and satisfactory cell operation is only accomplished when the electrolyte is held at high temperatures in the range 800–1000°C. This has many drawbacks. At these temperatures, the perovskite cathode material can become unstable, and surface segregation of some species can occur. This usually seriously hampers the catalytic activity of the cathode, which is important in facilitating the oxygen reduction reaction. In addition, the cathode may become corrupted by reaction with other components of the cell, such as Cr which can originate in alloy interconnects and result in surface formation of phases such as $SrCrO_4$, which reduces the effective performance of the cathode.

Because of this, and for obvious economic reasons, there is a continual search for means of lowering the operating temperatures of these cells, from the current temperature regime towards 650°C or lower. This has led to innovations in thin film, nanostructured and nanotubular cathode design.

References

P. Huang, A. Petric, *J. Mater. Chem.*, **5**, 53–56 (1995).
X. Li, N. A. Benedek, *Chem. Mater.*, **27**, 2647–2652 (2015).
K. Nomura, H. Kageyama, *Solid State Ion.*, **178**, 661–665 (2007).
Y. A. Shilova *et al.*, *J. Solid State Chem.*, **168**, 275–283 (2002).

Further Reading

For more information on defect chemistry of the material in this chapter and introduction to diffusion and conductivity, see:

R. J. D. Tilley, *Defects in Solids*, John Wiley & Sons, Inc., Hoboken (2008).
R. J. D. Tilley, *Understanding Solids*, 2nd edition, John Wiley & Sons, Inc., Chichester (2013), Chapter 7.

Ionic conductivity: for information on Li ion conductivity in thin films of $Li_{0.33}La_{0.56}TiO_3$ and an introduction to the literature of this material, see:
T. Ohnishi *et al.*, *Chem. Mater.*, **27**, 1233–1241 (2015).

Proton conductivity: for an introduction to the literature of this material, see:
J. Bielecki *et al.*, *J. Mater. Chem. A*, **2**, 16915–16924 (2014).
T. Norby, *Mater. Res. Soc. Bull.*, **34**, 923–928 (2009).

A comprehensive review of La, Sr and Mn perovskite cathode materials for solid oxide fuel cells is given in:
S. P. Jiang, *J. Mater. Sci.*, **43**, 6799–6837 (2008).

A series of review articles describing recent advances in solid oxide fuel cell design is to be found in:
E. Wachsman *et al.* (guest editors), *Mater. Res. Soc. Bull.*, **39**, 773–804, September (2014).

6
Dielectric Properties

6.1 Insulating Perovskites

Insulating perovskites are generally described as dielectrics (a term used to characterise a polarisable insulator), piezoelectrics, pyroelectrics and ferroelectrics. These names (which are general and not confined to perovskites) describe the *response* of the material, which is always an electric polarisation, to an applied external *stimulus*. In a dielectric, for example, the stimulus is an applied electric field, and the response is an electric polarisation of the material. Both the response and the stimulus must be described with respect to the crystal structure of the material, as vectors. Thus, for a dielectric, the response of the material, the electric polarisation, needs to be characterised by a vector **P**, and the stimulus, the applied electric field, needs to be specified as a vector **E**, both described with respect to the crystal structure of the perovskite. In a cubic crystal **P** is proportional and parallel to **E**, but for most crystals this is not true and the relationship between these two vectors needs to be described in tensor notation.

There is a well-established hierarchy of relationships between these insulating properties and the crystal structure of the phase. All insulating perovskites

Perovskites: Structure–Property Relationships, First Edition. Richard J. D. Tilley.

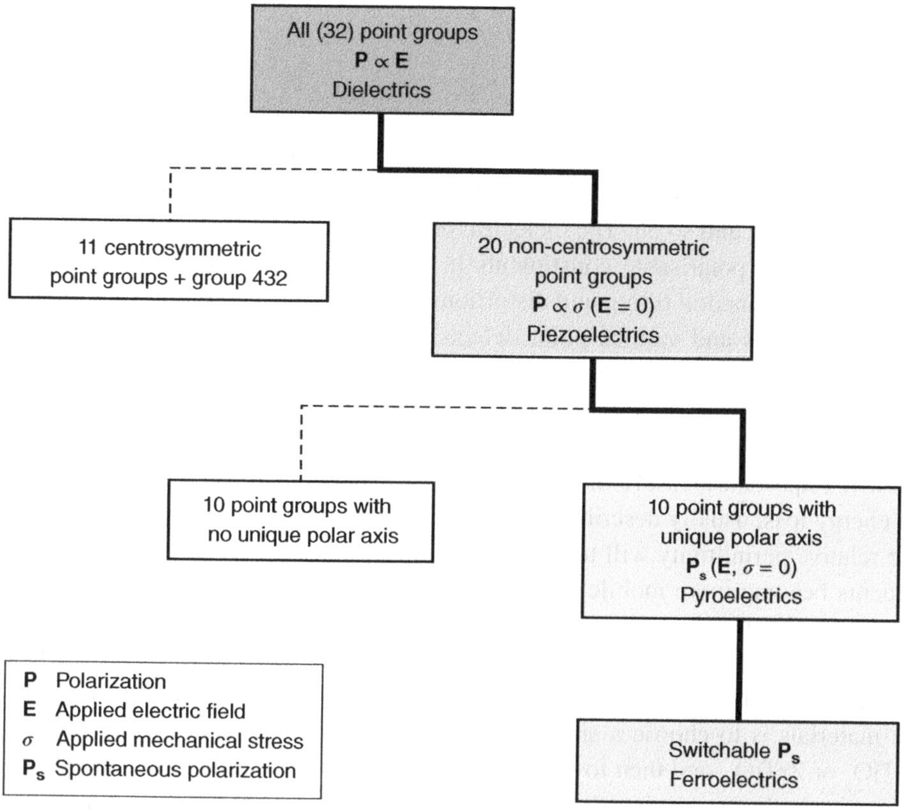

Figure 6.1 *The relationship between crystal point group symmetry and dielectric properties*

can be classified as dielectrics. Piezoelectric solids are a subset of dielectrics, in which polarisation arises as a result of an applied mechanical stress. Pyroelectric and ferroelectric crystals are a subset of piezoelectrics which show an electric polarisation, called the spontaneous polarisation, $\mathbf{P}_s$, even in the absence of an applied stimulus. Ferroelectrics are a subset of pyroelectrics. In these, the direction of the spontaneous polarisation vector, $\mathbf{P}_s$, can be switched in direction by the application of an electric field (Figure 6.1). This hierarchy mimics that of magnetic properties (Chapter 7).

In this chapter, the emphasis is upon dielectric and ferroelectric perovskites. As ferroelectrics are also piezoelectrics and pyroelectrics, it is found that most important applications for both piezoelectric and pyroelectric perovskites make use of ferroelectric materials.

6.2 Dielectric Perovskites

6.2.1 General Properties

Perovskites are vital circuit elements for many electronic purposes, from simple capacitors to dielectric resonators used in mobile phones, satellite communications, TV broadcasting and so on. The dielectric properties of bulk perovskites arise from the presence of polarisable constituents in the crystal. These include cation displacements, octahedral tilting and distortions as well as any defects present, such as grain boundaries and various point defects. The relative permittivity is the basic parameter describing a dielectric. In a static electric field this is written as ε_r (Table 6.1) but in varying electric fields is replaced by the complex relative permittivity, $\varepsilon_r' - i\varepsilon_r''$, which is a function of the frequency of the applied electric field. Another important measure of dielectric performance in a varying electric field is the energy loss, usually described in terms of the loss tangent, tan δ, equal to $\varepsilon_r''/\varepsilon_r'$. The relative permittivity will tend to increase with temperature as polarisable constituents become more mobile and fall off as the frequency of the applied electric field increases (Figure 6.2).

The dielectric properties of any simple perovskite are rarely ideal for any particular application. The general strategy for making improvements to perovskite dielectric materials is to choose a single material with desirable qualities, say, $CaTiO_3$, $BaTiO_3$ or $ZrTiO_3$, and then to substitute on the A- and B-sites so as to modify the properties in the appropriate direction. The resultant dielectric response depends on many factors and small quantities of dopant can lead to considerable changes in both relative permittivity and loss tangent. For example, A-site doping of the insulating phase $SrSnO_3$ with small amounts of Eu^{3+} is balanced by the formation of both Sn^{2+} cations on B-sites and oxygen vacancies. Substitution of 1% of the Sr^{2+} cations gives a phase of composition $Sr_{0.99}Eu_{0.01}Sn^{4+}_{0.92}Sn^{2+}_{0.08}O_{2.925}$, and substitution of 3% gives $Sr_{0.97}Eu_{0.03}Sn^{4+}_{0.76}Sn^{2+}_{0.24}O_{2.735}$. Although the dopant concentration difference is only 2%, there are significant differences between the dielectric performances of the two materials, especially at higher temperatures (Figure 6.3a and b).

The measured properties of polycrystalline ceramics are greatly influenced by the microstructure of the material. Of particular importance are the grain size of the crystallites, porosity, voids or gas bubbles within the ceramic and any impurity phases present. In addition, chemical defects such as point defects and mobile charge carriers within grains and the physics and chemistry of the grain boundaries will all have an influence on the measured properties of the solid. Because of this, ceramics are subjected to carefully controlled fabrication routes and the sintering temperature and time have a considerable effect upon the measured properties (Figure 6.3c).

Table 6.1 *The relative permittivity (ε_r) of some perovskites*

Compound	ε_r^a	Frequency (Hz)	Temperature (°C)
$BaZrO_3$	43	—	25
$BiFeO_3$	40	9.4×10^9	300
$Ba_4Ti_3O_{12}$	112	10^3	25
$CaCeO_3$	21	—	25
$CaTiO_3$	165	—	25
$CaCu_3Ti_4O_{12}$	$\sim10^5$	—	0 to ~450
$La_{2/3}Cu_3Ti_4O_{12}$	$\sim10^4$	10^2–10^6	25
Ba_2CoNbO_6	$\sim4\times10^3$	10^2–10^6	−173 to 40
$KNbO_3$	700	—	25
$KTaO_3$	242	2×10^5	298
$LiNbO_3^b$	$\varepsilon_{11}=\varepsilon_{22}=82$ $\varepsilon_{33}=30$	10^5	298
$LiTaO_3^b$	$\varepsilon_{11}=\varepsilon_{22}=53$ $\varepsilon_{33}=46$	10^5	25
$NaNbO_3^b$	$\varepsilon_{11}=\varepsilon_{22}=76$ $\varepsilon_{33}=670$	—	25
$NdAlO_3$	17.5	—	25
$NdScO_3$	27	—	25
$PbHfO_3$	390	10^5	300
$PbTiO_3$	~200	10^5	25
$Pb(Zn_{1/3}Nb_{2/3})O_3$	7	10^3	300
$PbZrO_3$	200	—	400
$SrTiO_3$	332	10^3	298
8H-$Ba_8Ga_{0.8}Ta_{5.92}O_{24}^a$	29	—	25
8H-$Ba_8Mg_{0.75}Ni_{0.28}Ta_6O_{24}$	27	—	25
H-$Ba_8Li_2Nb_6O_{24}^a$	37	—	25
8H-$Ba_8Li_2Ta_6O_{24}$	28	—	25
8H-$Ba_4LiSbTa_2O_{12}$	25	—	25
10H-$Ba_{10}Mg_{0.25}Ta_{0.79}O_{30}$	28	—	25
10H-$Ba_{10}Ti_{0.45}Sn_{0.75}Ta_{7.04}O_{30}$	17.6	—	25

[a] Where only one value is given, the sample is polycrystalline ceramic.
[b] $\varepsilon_{11}=\varepsilon_{22}$ are in the (001) plane, and ε_{33} is parallel to the hexagonal **c**-axis.

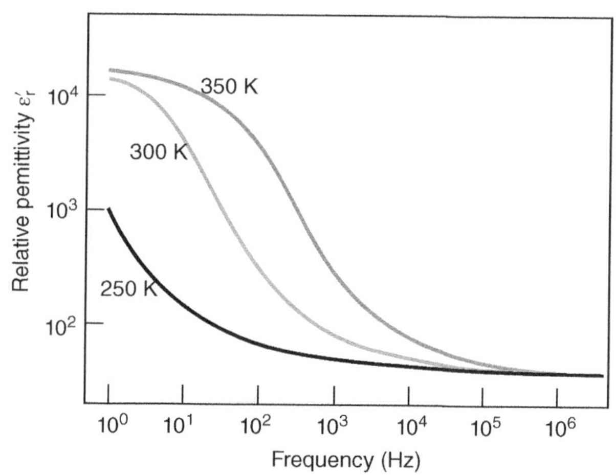

Figure 6.2 *The temperature and frequency dependence of the relative permittivity, ε_r', of $SrFe_{0.5}Ta_{0.5}O_3$ ceramics. Note the logarithmic scales on each axis*

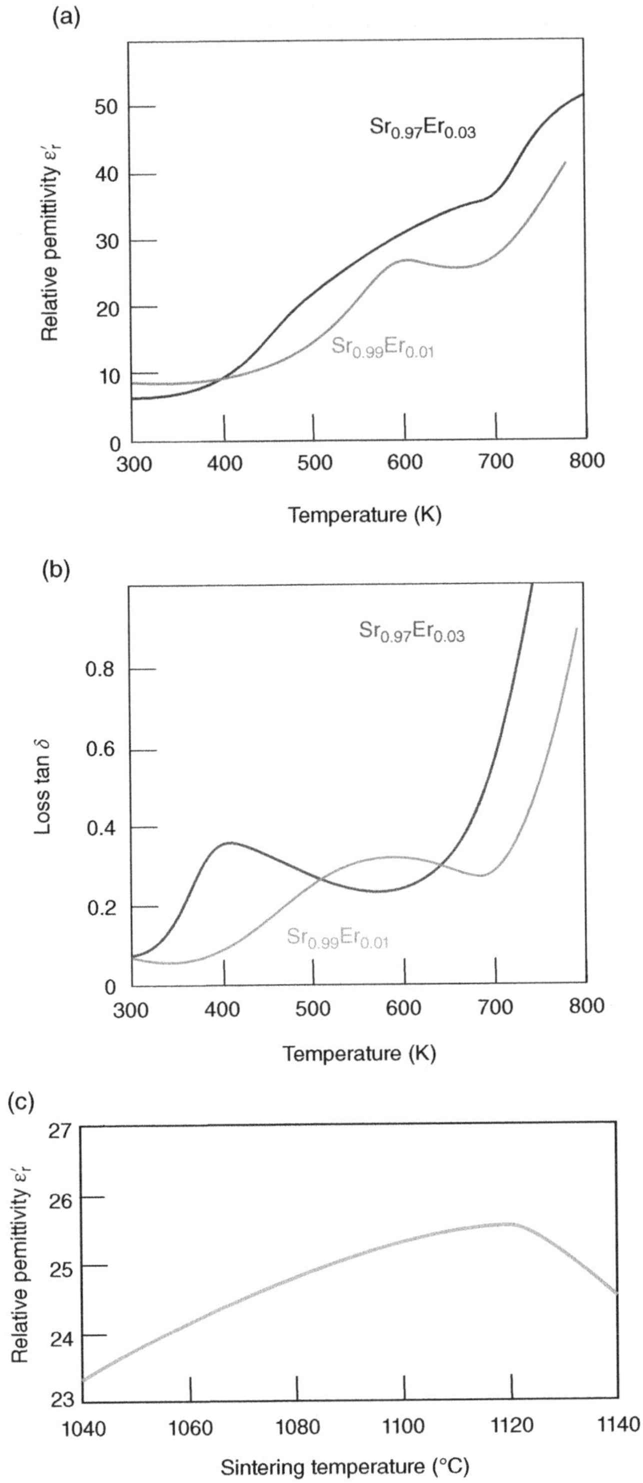

Figure 6.3 *Dielectric properties of ceramic perovskites: (a) the relative permittivity and (b) the loss tangent of* $Sr_{0.99}Eu_{0.01}SnO_{2.925}$ *and* $Sr_{0.93}Eu_{0.03}SnO_{2.735}$ *as a function of temperature; (c) the relative permittivity of the hexagonal 8H perovskite* $Ba_5Li_2W_3O_{15}$ *as a function of sintering temperature*

6.2.2 Colossal Dielectric Constant Materials

The interplay between preparation conditions, microstructure and properties is well illustrated by the ceramic oxides that have come to be called colossal dielectric constant (CDC) materials. The best known of these is the cubic perovskite phase $CaCu_3Ti_4O_{12}$ (CCTO) and the closely allied materials $SrCu_3Ti_4O_{12}$, $La_{2/3}Cu_3Ti_4O_{12}$, $Y_{2/3}Cu_3Ti_4O_{12}$, $Ce_{0.5}Cu_3Ti_4O_{12}$ and $Na_{0.5}Bi_{0.5}Cu_3Ti_4O_{12}$. These all adopt a cubic perovskite structure (Section 2.1.3). The relative permittivity of ceramic samples of these compounds is of the order of 10^5 at low frequencies and is almost independent of frequency and temperature over a broad temperature range of 100–500 K. The loss tangent is also very low and values close to 0.017 can be obtained in ceramic samples (Figure 6.4).

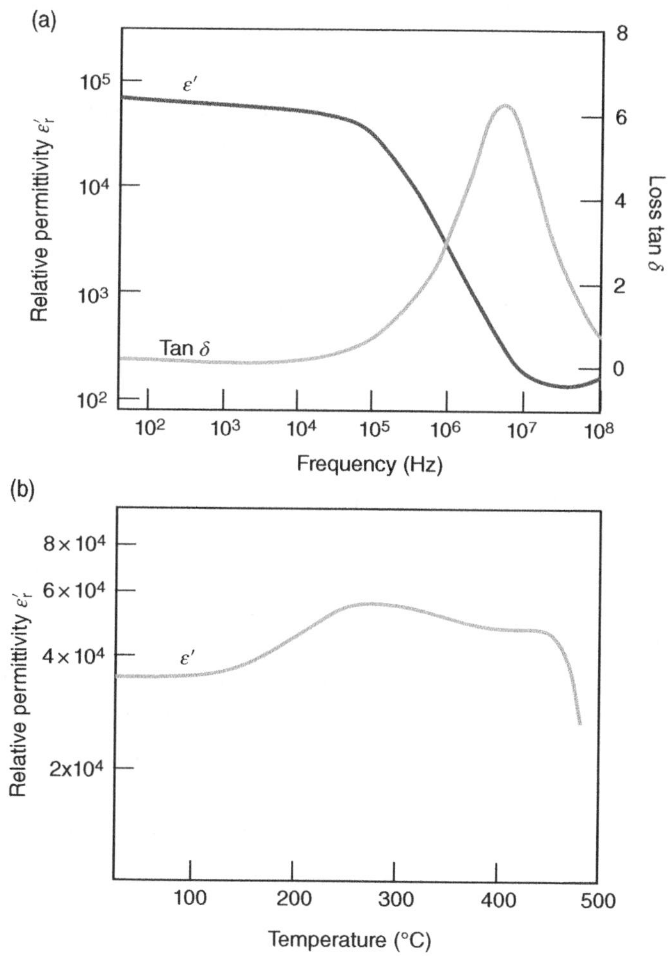

Figure 6.4 *Dielectric properties of $Na_{0.5}Bi_{0.5}Cu_3Ti_4O$ ceramics: (a) relative permittivity ε_r' and dielectric loss tan δ as a function of frequency at room temperature; (b) relative permittivity ε_r' as a function of temperature at a frequency of 1 kH*

These phases make ideal candidates for use in miniature electronic components and in energy storage applications. However, the measured relative permittivity of ceramic samples has been found to vary greatly, depending upon the preparation route, the sintering conditions and the ultimate grain size in the solid. The source of the high relative permittivity has still to be completely explored. The current indications are that the colossal effect may be the sum of two (or perhaps more) components. At least a significant contribution appears to arise in thin fully oxidised insulating grain boundaries enclosing large semiconducting or semiconducting/insulating grains containing a high population of mobile charge carriers. The high values of relative permittivity are frequently attributed to the Maxwell–Wagner (or Maxwell–Wagner–Sillars) effect, which describes the behaviour of two-phase materials when charges accumulate at the interfaces between the different grains. However, the conductivity of the semiconducting grains has also been implicated in the appearance of the colossal effect. Loss of oxygen during high-temperature sintering can result in the formation of oxygen vacancies together with mobile electrons. Control over the internal defect chemistry of the grains, as well as grain boundary geometry, is needed to confidently deliver reproducible commercial products.

6.3 Ferroelectric/Piezoelectric Perovskites

6.3.1 Spontaneous Polarisation and Domains

A ferroelectric perovskite, as distinct from a normal dielectric, has a permanent spontaneous polarisation $\mathbf{P}_s$ that can be switched in direction by the application of an external electric field. All ferroelectrics are also piezoelectrics and the majority of commercially important piezoelectric materials are ferroelectric perovskites.

The fundamental cause of ferroelectricity in oxide perovskites was originally attributed to the idea that small B-site cations could 'rattle around' inside BO_6 octahedra and off-centre configurations lent stability to the structure. The displaced B-site cation created an electric dipole within each octahedron. The direction of the dipole, or, equivalently, the position of the cation, could be switched from one off-centre position to another under the influence of an applied electric field.

The situation can be summarised with reference to the archetypal ferroelectric material, barium titanate, $BaTiO_3$ described in Section 1.5. The ideal cubic (paraelectric) form of $BaTiO_3$ is found above 398 K. The Ti^{4+} ions are situated at the centre of a regular octahedron of oxygen ions. Between 398 and 278 K the unit cell is tetragonal. The cubic cell contracts slightly along the **a**- and **b**-axes and expands slightly along the **c**-axis in transforming to tetragonal (Figure 6.5a). The change from cubic to tetragonal is accompanied by an off-centre movement of the octahedrally coordinated Ti^{4+} ions along the **c**-axis, accompanied by a slight change in octahedron dimensions so that two equatorial oxygen atoms move parallel to the **+c**-axis and two move in the opposite direction (Figure 6.5b). The O^{2-}—Ti^{4+} bond lengths parallel to the **c**-axis are now 0.22 and 0.18 nm, while the equatorial bond lengths remain at 0.2 nm. This results in the formation of a dipole within each octahedron, with a net dipole moment of the order of 26 μC cm^{-2}, each pointing along the **c**-axis (Figure 6.5c). The polar axis is therefore the **c**-axis. The off-centre Ti^{4+} position and the octahedral deformation can be changed in an electric field and hence tetragonal $BaTiO_3$ is a ferroelectric.

There is no preference as to which of the original cubic axes becomes the polar direction, and so this can take one of six equivalent orientations, parallel to ±**x**, ±**y** or ±**z**. On cooling a large crystal, any of these displacements is possible, and within different regions of the crystal different orientations occur, leading to the formation of domains in which all the dipoles are aligned internally (Figure 6.5d). The polarisation directions in a domain are related to those in neighbouring domains by the crystallography of the matrix.

Between the temperatures 278 and 183 K the tetragonal structure undergoes a further displacive transition resulting in an elongation along a face diagonal of the unit cell to give an orthorhombic phase. Compared to the cubic structure, there are now 12 equivalent directions for the dipole. Below 183 K the structure another displacive transition causes the unit cell to become rhombohedral with eight equivalent directions for the dipole orientation.

The deformations of the unit cell from the high-temperature cubic form are relatively small, and the structural relationships between the various crystallographic unit cells can be simplified by reference to a *pseudocubic* structure. Referred to such a pseudocubic unit cell, the polarisation directions $\mathbf{P}_s$ in $BaTiO_3$ are then as follows: tetragonal phase, $[001]_p$; orthorhombic phase, $[011]_p$; and rhombohedral phase, $[111]_p$.

The B-site displacement idea is too simplistic to explain the behaviour of a majority of ferroelectrics, and other features of the perovskite structure must be taken into account in explaining the formation of arrays of permanent switchable dipoles. Among these additional aspects are dipoles due to A-cation displacements, especially for lone pair cations such as Pb^{2+} or Bi^{3+}, octahedral tilting and irregularities in the positions of the oxygen ions that make up the BO_6 polyhedra.

Irrespective of the underlying cause of the ferroelectric spontaneous polarisation, many approximately cubic perovskite ferroelectrics show domain patterns similar to that encountered in $BaTiO_3$, with the spontaneous polarisation vector lying parallel to the pseudocubic $[001]_p$, $[011]_p$ or $[111]_p$ directions. In thin films, these domain patterns depend upon the orientation and misfit between the lattice parameters of the substrate and the overlying ferroelectric film. For example, in the important ferroelectric $PbZrO_3$–$PbTiO_3$ system, (100)-oriented films of the tetragonal ferroelectric phase have domains running perpendicular and parallel to the surface, while (111)-oriented films have domains running at 45° to the surface (Figure 6.6).

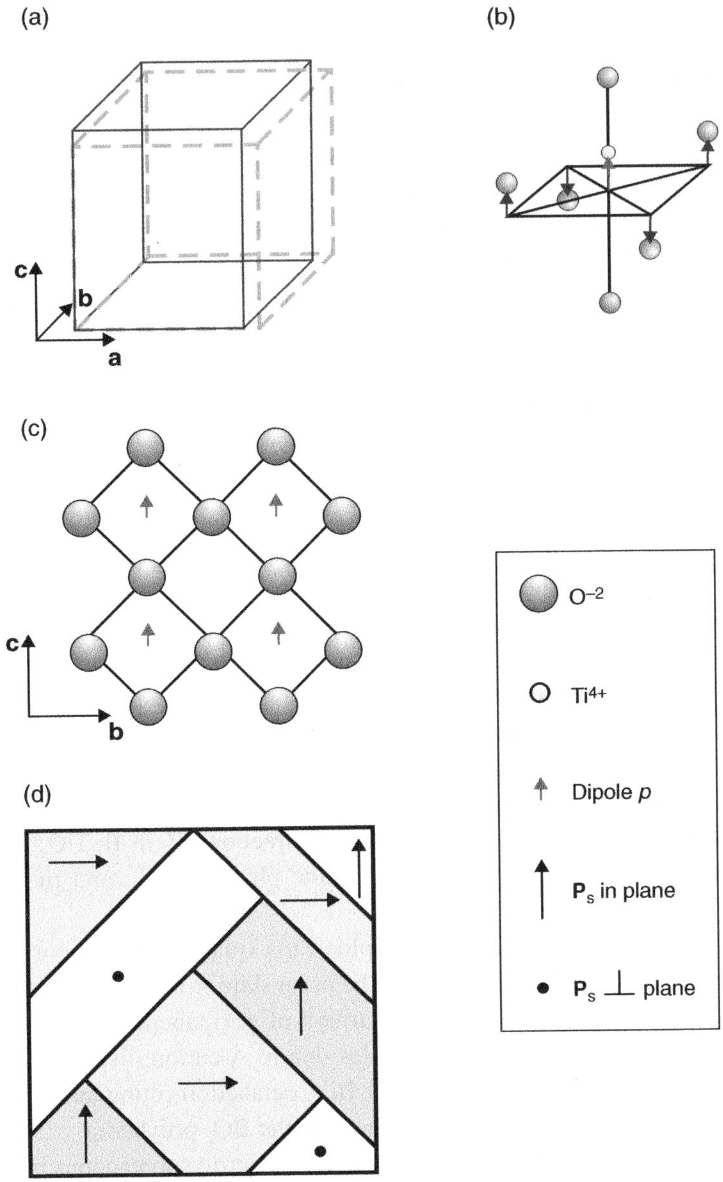

Figure 6.5 Tetragonal $BaTiO_3$: (a) the tetragonal unit cell compared to the cubic unit cell (dashed); (b) displacements found in tetragonal $BaTiO_3$; (c) dipole array in tetragonal $BaTiO_3$; (d) domain structure in a crystal slice. (All distortions are greatly exaggerated)

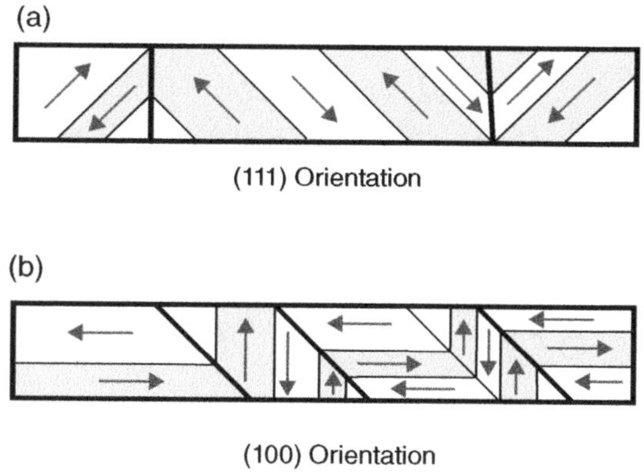

Figure 6.6 *Domain structures, schematic, in oriented thin films of tetragonal Pb(Zr,Ti)O$_3$: (a) (111) orientation; (b) (100) orientation*

From symmetry considerations, the ferroelectric layered perovskites, such as the Ruddlesden–Popper, Dion–Jacobson and Aurivillius phases in which the perovskite layers are an even number of octahedra in thickness, tend to have the spontaneous polarisation vector lying parallel to the perovskite sheets. In phases with an odd number of octahedra in thickness, the spontaneous polarisation can lie perpendicular to the sheets or at least have a component in this direction. Thus the $n=2$ Dion–Jacobson phase $RbBiNb_2O_7$ has the spontaneous polarisation vector lying parallel to the perovskite layers. This polarisation arises from the tilted NbO_6 octahedra and, more importantly, from the displaced lone pair Bi^{3+} cations that are found between the perovskite layers.

6.3.2 Ferroelectric Domain Switching

The normal state of a ferroelectric crystal is one in which the microstructure consists of a set of coherent domains, each of which has the internal electric dipoles parallel to each other, but not aligned with those of neighbouring domains. Domain walls are not atomically smooth but have a thickness of between 0.5 and 1 nm. Polarisation builds up at domain boundaries and a single domain can have surface charges of the order of 1.5×10^{14} electrons cm^{-2}, which can generate an internal electric field of 300 MV or more.

An applied electric field which is opposed to the internal dipoles such as those that can arise by octahedral cation displacement (Figure 6.7a) can eventually be strong enough to force the cations to flip to the opposite orientation (Figure 6.7b). The situation can be represented in terms of the potential energy of the crystal. Here each stable configuration is found at a minimum in the potential energy versus position curve, with a small energy barrier, ΔU, between them (Figure 6.7c). Provided

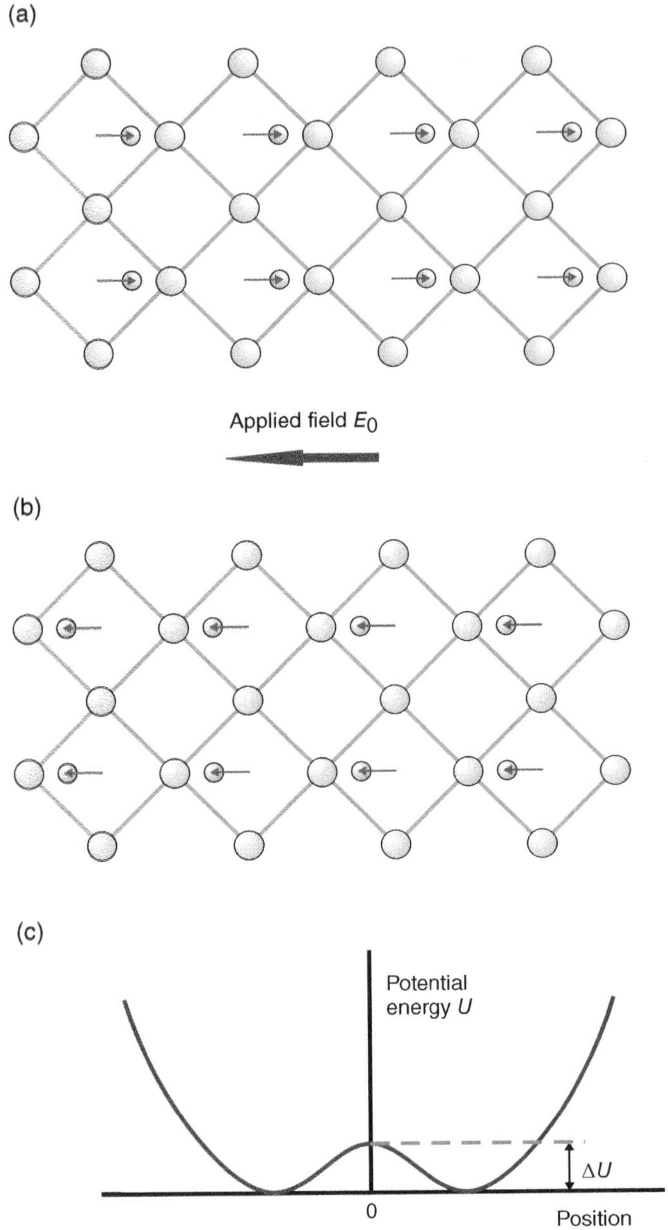

Figure 6.7 Switching of polarisation in a perovskite, schematic: (a and b) an array of parallel dipoles in a perovskite before and after switching. The orientation of the electric field required to switch the dipoles is shown. (c) Potential energy versus position for a dipole

that the applied field is strong enough to overcome the barrier height, the orientation of the dipoles will switch to the alternative orientation.

In recent years, the nucleation and growth of ferroelectric domains has been directly observed using advanced microscopy techniques. For example, ferroelectric domains in (001)-oriented single domain sheets of $BiFeO_3$ can be nucleated in response to the application of an electric field applied by a metallic probe (Figure 6.8a). Domains are nucleated at the lower film surface as arrow-shaped incursions into the film (Figure 6.8b and c), which amalgamate into a single domain as the field strength increases (Figure 6.8d). The polarisation vector in the starting sample, parallel to the pseudocubic $[111]_p$ direction, is switched to the $[11\bar{1}]_p$ direction under the electric field (Figure 6.8e).

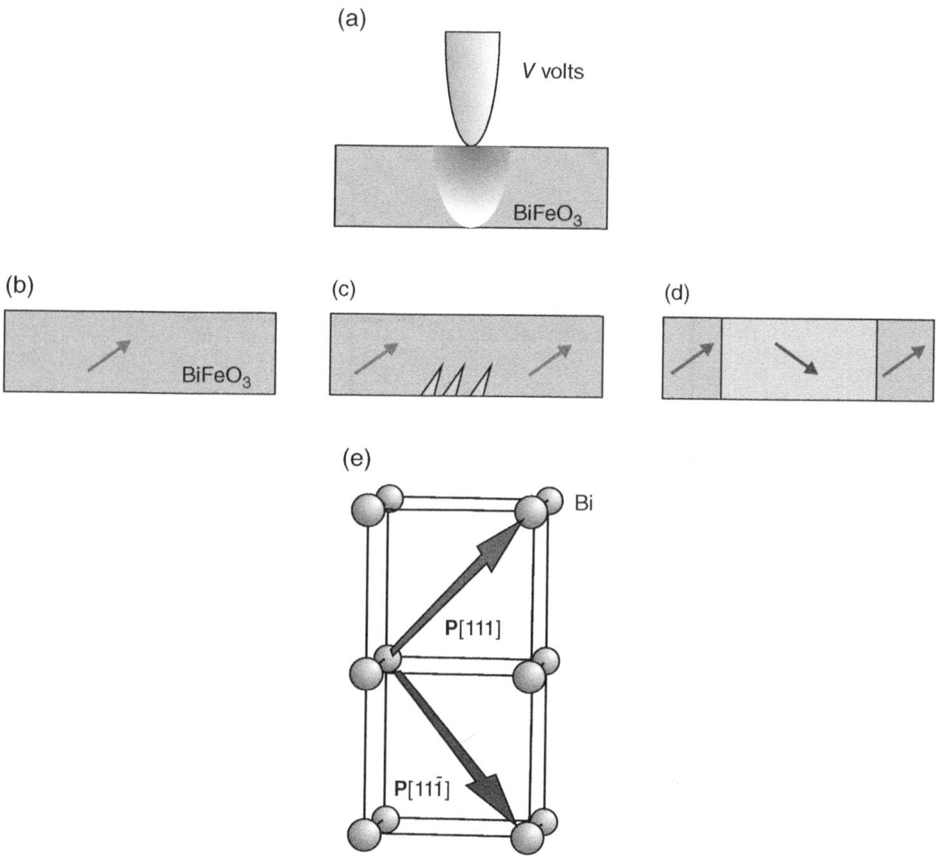

Figure 6.8 *Domain growth in ferroelectric $BiFeO_3$: (a) metal probe with voltage V applied to a single crystal film of $BiFeO_3$; (b–d) domain growth, initiated at the lower surface of the film; (e) the change in polarisation direction with respect to the pseudocubic unit cell of $BiFeO_3$ (Original data in Nekon* et al. *(2011))*

6.3.3 Ferroelectric Hysteresis Loops

In general a ferroelectric perovskite single crystal will be composed of a roughly equal number of domains oriented in all the equivalent directions allowed by the crystal symmetry. The overall polarisation of the crystal will be zero. The application of an electric field will cause a polarisation switch and lead to a classical hysteresis loop in which the important values are P_r, (the remanent or residual polarisation when the electric field is reduced to zero), and E_c, (the *coercive field*, which is the reverse field required to reduce the polarisation to zero). Extrapolation of the high-field portion of the curve to $E=0$ gives the value of the spontaneous polarisation P_s (Figure 6.9a).

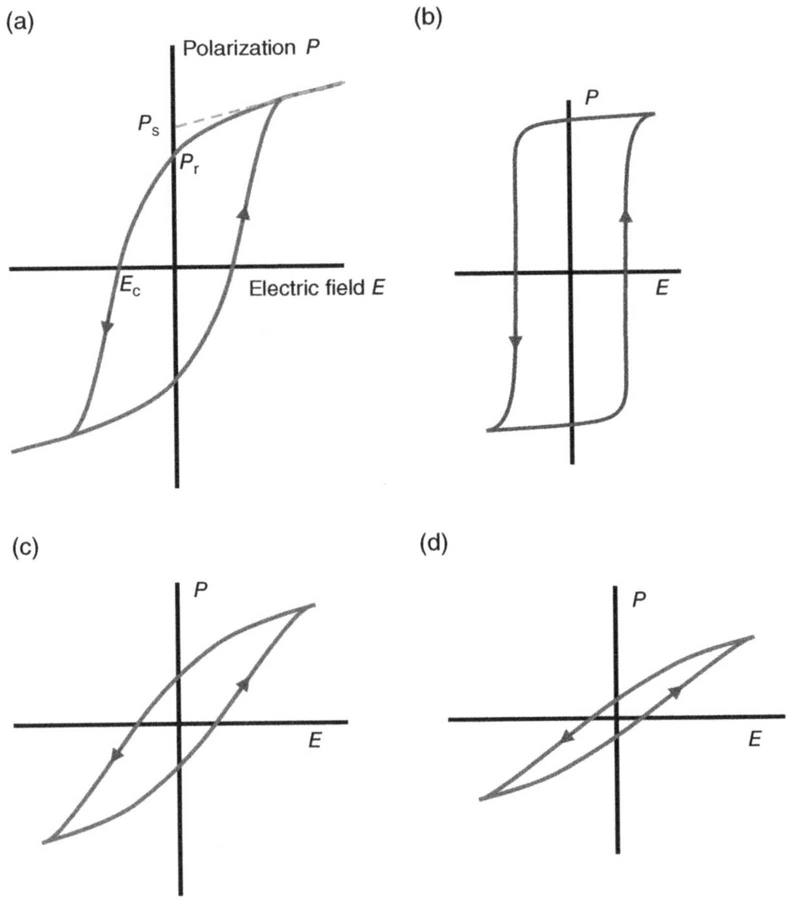

Figure 6.9 *Ferroelectric hysteresis loops, schematic: (a) typical loop indicating P_s, P_r and E_c; (b) single crystal ABO_3; (c) medium-grained ceramic ABO_3; (d) fine-grained ceramic ABO_3*

The form of the hysteresis loop for a perovskite depends upon a number of factors. In 'hard' ferroelectrics switching requires a high imposed field and is demonstrated by a high value of the coercive field, E_c. 'Soft' ferroelectrics are easily switched and show a low value of E_c. Thus hard ferroelectrics tend to show a broad hysteresis loop while soft ferroelectrics show narrow loops. In addition, the size and orientation of the loop changes with particle size. Single crystals tend to have rectangular hysteresis loops with rather square shoulders. Coarse-grained ceramic ferroelectrics have rotated and rounder loops, while in fine-grained ceramics the loops are rotated further and are much thinner (Figure 6.9b–d). Other aspects of microstructure may lead to asymmetric hysteresis loops.

6.3.4 Temperature Dependence of Ferroelectricity

With increasing temperature the thermal motion of the atoms will increase and eventually this alone can overcome the energy barrier separating the various orientations of the spontaneous polarisation. At sufficiently high temperatures the distribution of atoms becomes statistical and the crystal behaves as a normal (cubic) dielectric and is said to be in the *paraelectric* state. The temperature at which this change occurs is known as the transition temperature, Curie temperature or Curie point, T_C. The way in which the spontaneous polarisation of the crystal varies as the Curie temperature is approached depends upon whether the transformation to the paraelectric state can be classified as first or second order. In a first-order transition the change is an abrupt (discontinuous) change, whereas in a second-order transition the change is gradual and continuous (Figure 6.10a). In either case the relative permittivity rises to a sharp peak in the neighbourhood of T_C (Figure 6.10b).

The temperature dependence of the relative permittivity of many ferroelectric crystals in the *paraelectric state* can be described fairly accurately by the Curie–Weiss law:

$$\varepsilon_r = \frac{C}{(T - T_C)}$$

where C is a constant, T_C is the Curie temperature and T is the temperature (K). The value of C is determined by a plot of $1/\varepsilon_r$ versus T:

$$\frac{1}{\varepsilon_r} = \frac{T}{C} - \frac{T_C}{C}$$

Ideally the graph is linear, with a slope of $1/C$ and an intercept on the T-axis of T_C (Figure 6.11). Frequently the point of intercept, T_0, is slightly different from the measured value of T_C, and the Curie–Weiss equation is often written in the form

$$\varepsilon_r = \frac{C}{(T - T_0)}$$

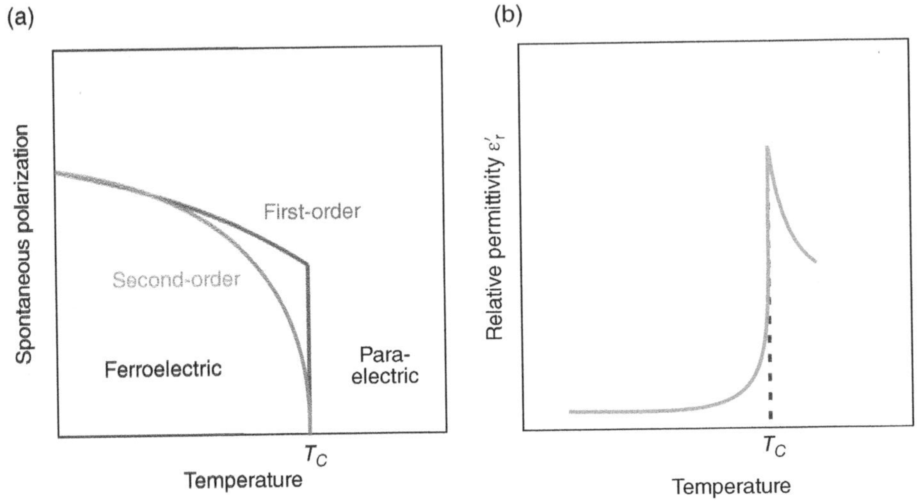

Figure 6.10 *Temperature dependence of ferroelectricity: (a) first-order and second-order transitions to the high-temperature paraelectric state; (b) the variation of relative permittivity across the Curie temperature* T_C

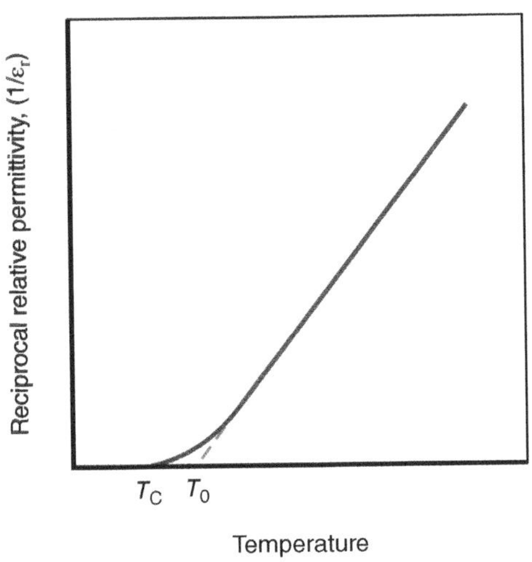

Figure 6.11 *Curie behaviour of an ideal ferroelectric in the paraelectric state*

where T_0 is the *extrapolated Curie temperature*. Note that the Curie temperature is the temperature at which the structural transition to or from the paraelectric state takes place. The value T_0 (also confusingly referred to as the Curie temperature) is derived from extrapolation and does not coincide with the phase transition.

6.3.5 Pyroelectrics, Piezoelectrics and Crystal Symmetry

A pyroelectric crystal possesses a permanent spontaneous polarisation, $\mathbf{P}_s$, at all times. (In ferroelectrics this direction can be switched, as described.) Due to symmetry constraints, only crystals from 10 non-centrosymmetric point groups permit spontaneous polarisation to occur and the vector is generally constrained to lie along a single crystallographic direction. The octahedral tilting, cation displacements and so on that produce the spontaneous polarisation in perovskites are all temperature sensitive, and a change in temperature will cause a change in the spontaneous polarisation that is manifested as the *pyroelectric effect*. For all crystal systems except triclinic and monoclinic, the change in the value of $\mathbf{P}_s$ with temperature, ΔP, the pyroelectric effect, is given by

$$\Delta P = \pi \Delta T$$

where π is the (single) pyroelectric coefficient of the phase and ΔT is the temperature change.

The direct and converse piezoelectric effects are electromechanical effects whereby an electrical and a mechanical property are linked together. In piezoelectric perovskites, the direct piezoelectric effect relates the change in polarisation of the solid following an applied stress. Here the stress is termed the *stimulus* and the resulting polarisation is termed the *response*. The application of a stress changes octahedral tilting, cation displacements and so on, leading to a change in polarisation. Both polarisation and stress are directional properties, but unlike pyroelectrics the relationship between these two is more complicated. Polarisation is a vector with three components. Stress, however, has to be represented by a second-rank tensor with nine components. In single crystals, the direct piezoelectric effect is usually written in matrix notation as

$$\mathrm{P}_i = d_{ij}\sigma_j \quad i = 1,2,3; \quad j = 1,2,3,4,5,6$$

where i are the components of the (response) vector $\mathbf{P}$ along the principal axes (OX, OY, OZ, parallel to the conventional crystallographic axes for crystals of symmetry orthorhombic or higher) and j are the stress (stimulus) components along the principal axes. There are 18 d_{ij} matrix coefficients, although in crystals of higher symmetry, many of these reduce to zero. They are not dimensionless numbers but take the units C N^{-1} with values of the order of 100–200 pC N^{-1}.

In the converse piezoelectric effect, a voltage (or electric field, $\mathbf{E}$, the stimulus) applied to a piezoelectric material will induce a change in shape (strain, ε, the response), given by

$$\varepsilon_j = d_{ij}E_i \qquad i = 1,2,3; \quad j = 1,2,3,4,5,6$$

The values of the coefficients d_{ij} are identical to those for the direct piezoelectric effect although they are usually quoted as $\mathrm{m\,V^{-1}}$.

6.3.6 Strain versus Electric Field Loops

The converse piezoelectric effect describes the relationship of the strain experienced by a ferroelectric when subjected to an applied electric field. A measurement of the strain resulting from an applied field takes the form of a loop. The shape of the loop can be correlated with the shape of the normal ferroelectric hysteresis loops described previously (Figure 6.9). For a classical symmetrical hysteresis loop (Figure 6.9a), the strain versus electric field loop has a symmetrical butterfly shape (Figure 6.12a). The large degree of hysteresis and the negative values of strain are due to domain wall switching. From the converse piezoelectric equation, it can be seen that the slope of the curve at small values of electric field will give a value for the appropriate piezoelectric coefficient d_{ij}. In practice, most samples that have been investigated are ceramic materials (Section 6.4). In these cases the slope of the graph may contain contributions from the microstructure or anisotropy of the material. However, these problems do not affect all ceramic materials and reliable values of the coefficient d_{33} (see Section 6.4) can be obtained. For example, a symmetrical strain versus electric field loop for ceramics of composition $0.95Bi_{0.5}Na_{0.5}TiO_3$–$0.05KNbO_3$ (Figure 6.12b) yields a value for the coefficient d_{33} of approximately 400 $\mathrm{pC\,N^{-1}}$ at room temperature.

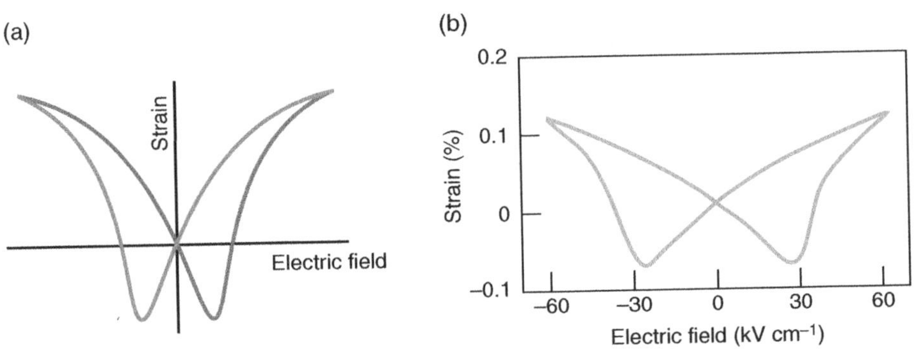

Figure 6.12 *Strain versus applied electric field loops: (a) symmetrical 'butterfly' loop for a ferroelectric with a symmetrical P–E hysteresis loop; (b) experimentally determined loop for a fine-grained ceramic* $0.95Bi_{0.5}Na_{0.5}TiO_3$–$0.05KNbO_3$

6.4 The Development of Ferroelectric/Piezoelectric Ceramic Bodies

6.4.1 Ceramic Piezoelectrics

The electronic properties of perovskites came to the fore in the early 1940s, when the search for materials with higher relative permittivity was intense. This resulted in the discovery of the important dielectric properties of the perovskite $BaTiO_3$. It was soon discovered that the high permittivity of this phase was due to the ferroelectric nature of the structure, with the highest values of permittivity occurring close to ferroelectric transitions. However, single crystal slices that could exploit this property were not an option for the majority of uses, not least because of the difficulty and cost of producing large crystals. The need was for polycrystalline shapes, often discs, tubes or cylinders, which preserved the advantages conferred by ferroelectricity. The resulting polycrystalline bodies are described as ceramics, the term 'ceramic' being synonymous with 'polycrystalline' for practical purposes.

An important point now emerges: the requirement that the piezoelectric, pyroelectric and ferroelectric effects are restricted to non-centrosymmetric crystals implies that these physical phenomena should not be observed in a polycrystalline solid. This is because the individual grains of a polycrystalline body will polarise in random directions that will cancel overall. This was changed by the discovery, in 1945, of a way to endow polycrystalline ceramic articles with ferroelectric properties.

In this process, called *poling*, the ceramic is cooled from a high temperature while subjected to a strong electric field. This acts so as to partially orient the spontaneous polarisation of each grain parallel to the poling axis, which is the direction of the electric field. Conventionally the poling direction is taken as the OZ-axis (Figure 6.13a). After poling the ceramic remains isotropic in directions normal to the poling axis, so that physically both OX and OY behave in exactly the same way and the axes OX and OY can be placed anywhere (as long as they remain at right angles to each other and to OZ).

The relationship between polarisation and the stress characterised by the direct piezoelectric effect is now greatly simplified. If the stress, σ, is applied in the same direction as the poling direction and the polarisation developed is also measured in this direction (Figure 6.13b), the relationship is given by

$$P = d_{33}\sigma$$

The coefficient d_{33} is called the longitudinal piezoelectric coefficient. If the faces over which the polarisation is measured remain the same, but the stress is applied in a perpendicular direction (Figure 6.13c), the relationship is

$$P = d_{31}\sigma$$

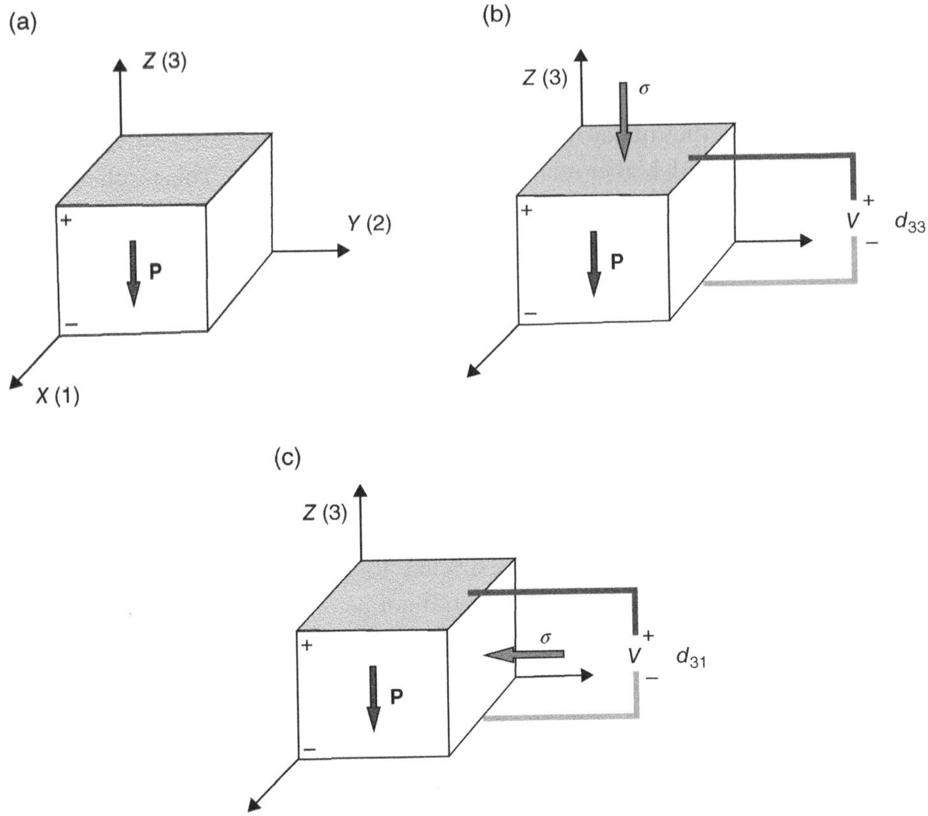

Figure 6.13 *Piezoelectric coefficients for ceramic samples: (a) poling direction; (b) polarisation and stress both parallel to OZ; (c) polarisation along OZ, stress perpendicular to OZ*

The coefficient d_{31} is called the transverse piezoelectric coefficient. The converse piezoelectric effect, relating strain, ε to the applied electric field E is similarly simplified to

$$\varepsilon = d_{33}E$$

or

$$\varepsilon = d_{31}E$$

depending upon the geometry of the arrangement.

The development of the poling process opened the door to large families of perovskite ceramic materials with properties superior to $BaTiO_3$ (BT) for many applications. The major perovskite materials to be utilised over the latter half of the twentieth century were lead-containing ferroelectric ceramic samples derived from lead zirconate $PbZrO_3$ (PZ) and lead titanate $PbTiO_3$ (PT), including phases in the system $PbTiO_3$–$PbZrO_3$ (PZT), lead lanthanum zirconate titanate (PLZT), lead stannate zirconate titanate (PSZT), lead magnesium niobate (PMN) and lead zinc niobate (PZN). These materials have effectively dominated the piezoelectric market since the 1960s.

In the current century, research into the dielectric properties of perovskite phases has continued at pace, stimulated by the growth of single crystals of many phases, especially of the relaxor ferroelectrics (Section 6.7). In addition, the search for lead-free phases to replace $PbZrO_3$–$PbTiO_3$ and related materials is intense. Here the potassium niobate, $KNbO_3$ (KN), sodium niobate, $NaNbO_3$ (NN) and perovskite phases in the $KNbO_3$–$NaNbO_3$ (KNN) system, including $K_{0.5}Na_{0.5}NbO_3$ (also called KNN), are being widely studied.

6.4.2 Electrostriction

Electrostriction, which is a change in sample dimensions in response to the application of an electric field to a dielectric, is a universal characteristic and provides another example of an electromechanical effect. Some materials get thinner while others get thicker in the direction of the electric field. This effect is not reversible and a deformation does not produce any polarisation. The effect is found in all materials, not just those that lack a centre of symmetry, including glasses and liquids. However, the electrostrictive effect is generally very small except for ferroelectric perovskites, especially relaxor ferroelectrics described in the following (Section 6.7).

Electrostriction is related to the converse piezoelectric effect. At modest electric field strengths, the piezoelectric equations given previously are adequate and there is a linear relationship between strain and electric field. However, at higher electric field strengths, these equations need to be extended to include a further term quadratic with respect to the electric field. The strain is now given by

$$\varepsilon_j = d_{ij}E_i + \gamma_{ij}E_iE_j \qquad i = 1,2,3; \quad j = 1,2,3,4,5,6$$

where γ_{ij} (sometimes written M_{ijkl}) is the fourth-rank electrostrictive tensor written in matrix form. Clearly, if the applied field is reversed, the sign of the electrostrictive strain does not change.

As with ceramic piezoelectric samples, the equation can be simplified. The order of the suffixes takes the same meaning as before, so that it is possible to write

$$\varepsilon_1 = \gamma_{31}E_3^2$$

where γ_{31} correlates a shape change ε_1, along OX with the square of the voltage E_3^2 applied along OZ. A similar equation correlates the strain with the polarisation of the ceramic:

$$\varepsilon_1 = Q_{31}P_3^2$$

Both γ (M) and Q are called electrostrictive coefficients.

6.5 Antiferroelectrics

Ferroelectricity is governed by two factors: (i) chemical bonds, which are short-range forces, and (ii) dipolar interactions, which are long range. Calculations of the energy of ferroelectric crystals indicate that a minimum energy results when the spontaneous polarisation in each unit cell is in a parallel (ferroelectric) configuration (Figure 6.14a) or in an antiparallel configuration (Figure 6.14b). The antiparallel arrangement is characteristic of antiferroelectrics. The balance between ferroelectric and antiferroelectric states is delicately poised, and some antiferroelectrics readily transform to ferroelectric states and vice versa. For example, antiferroelectric lead zirconate, $PbZrO_3$, is changed into a ferroelectric phase by the substitution of the order of five per cent Ti for Zr to form $PbTi_{0.05}Zr_{0.95}O_3$.

The unit cell of antiferroelectric $PbZrO_3$ is orthorhombic: space group *Pbam* (55), $a=0.58822$ nm, $b=1.17813$ nm, $c=0.82293$ nm. These parameters are related to the pseudocubic unit cell by $a=2a_p\cos45°$, $b=4a_p\sin45°$, $c=2a_p$, with $a_p=0.416$ nm. The real crystal structure is quite complex, with temperature-sensitive octahedral tilts and oxygen ion displacements present. However, the antiferroelectric array of dipoles appears to be principally associated with displacement of the lone pair Pb^{2+} ions along one of the pseudocubic [110] directions, parallel to the orthorhombic **a**-axis (Figure 6.15a). It is seen that there are *double rows* of

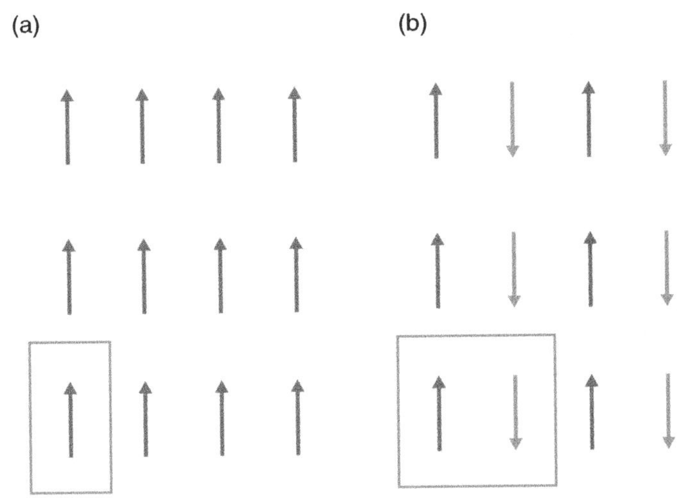

Figure 6.14 Antiferroelectric behaviour schematic: (a) parallel array of spontaneous polarisation dipoles characteristic of a ferroelectric crystal; (b) antiparallel array of dipoles in an antiferroelectric crystal

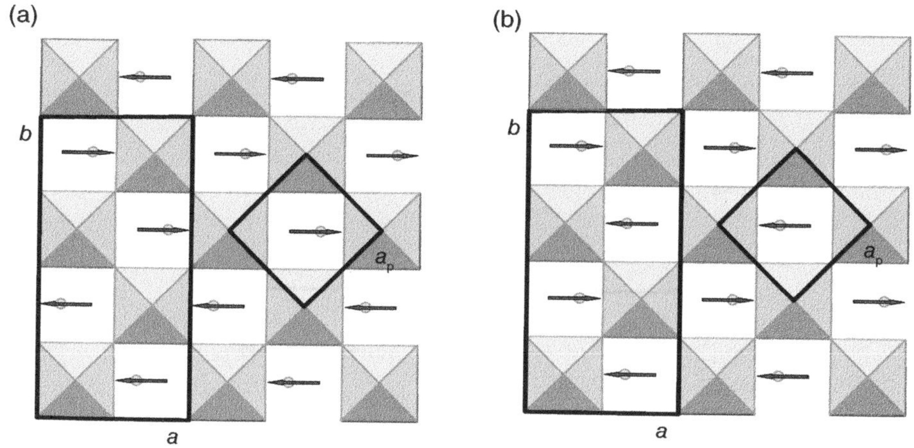

Figure 6.15 *(a) Idealised room temperature $PbZrO_3$ structure showing double rows of antiparallel dipoles due to Pb^{2+} ion displacement; (b) idealised intermediate $PbZrO_3$ structure at approximately 235°C with single rows of antiparallel dipoles due to Pb^{2+} ion displacement*

antiparallel aligned dipoles in the (001) plane of the structure, rather than single rows. Apart from these dipoles, there is also a displacement of the Pb^{2+} ions parallel to the **c**-axis, which leads to a local *ferroelectric* alignment of dipoles.

At a temperature of approximately 232°C, the structure changes to another antiferroelectric intermediate structure which is stable over a narrow temperature range of about 10°C. The antiferroelectric dipoles are again associated with Pb^{2+} ion displacements but with *single* rows of antiparallel aligned dipoles (Figure 6.15b). At slightly higher temperatures this intermediate phase transforms to a cubic paraelectric phase: space group $Pm\bar{3}m$ (221), $a = 0.41597$ nm at 250°C.

Clearly the spontaneous polarisation in an antiferroelectric perovskite ceramic body such as $PbZrO_3$ is zero in the absence of an electric field. If a field is now applied, there will be little change until the field reached a critical level E_f when it can switch one of the sets of dipoles. In this case the polarisation rapidly increases and the material is effectively transformed into a ferroelectric. Reversal of the high field will cause little change until it has reduced sufficiently, until, at E_a, the antiferroelectric state is again more stable, at which point the antiferroelectric structure reforms. The same applies to field applied in the opposite direction, which switch the other set of dipoles. The result is a double hysteresis loop, commonly regarded as a signature of antiferroelectric perovskite phases (Figure 6.16a). The strain versus electric field loops are also double, as would be anticipated from the polarisation versus electric field pattern (Figure 6.16b).

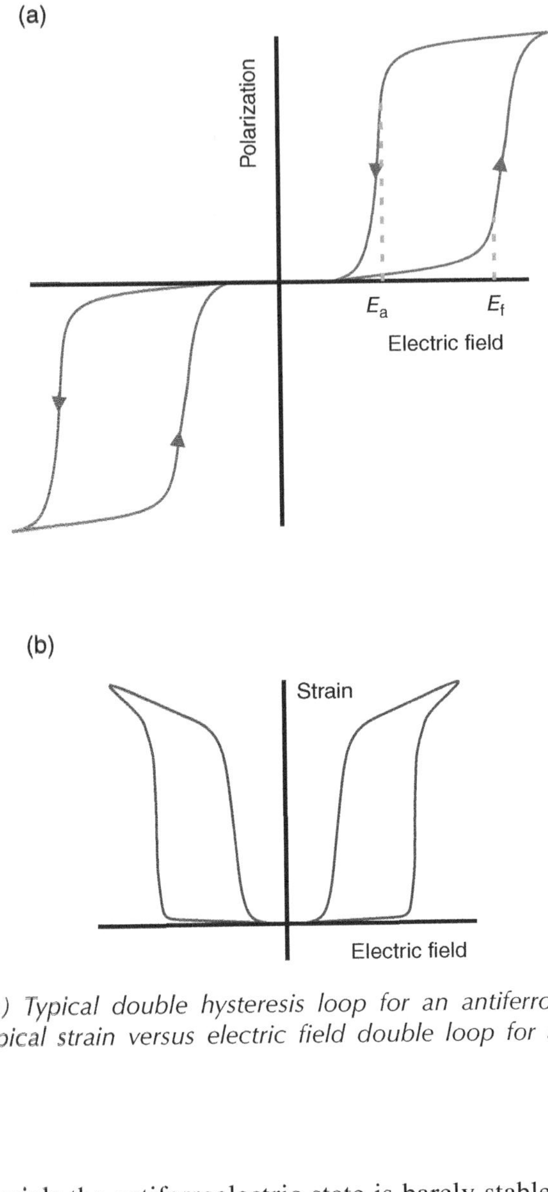

Figure 6.16 *(a) Typical double hysteresis loop for an antiferroelectric perovskite ceramic; (b) typical strain versus electric field double loop for an antiferroelectric perovskite*

In some materials the antiferroelectric state is barely stable or metastable. In such materials, application of an electric field will convert the phase to ferroelectric, as described, but removal of the field leaves the phase in a ferroelectric state. This material then behaves like a typical ferroelectric and displays a conventional hysteresis loop. Heating the material to a high temperature so as to form the paraelectric structure, followed by cooling, can reform the original antiferroelectric state.

The competing short-range and long-range forces can give rise to different local structures compared to the average structure determined by conventional diffraction methods. The perovskite $Bi_2Mn_{4/3}Ni_{2/3}O_6$ provides an example of this complexity. At room temperature the structure is incommensurate, with modulation vectors of $\mathbf{q}_1 = 0.4930\ \mathbf{a}^*$ along a pseudocubic $[1\bar{1}0]_p$ direction and $\mathbf{q}_2 = 0.4210\ \mathbf{a}^*$ along a pseudocubic $[001]_p$ direction. These modulations particularly affect the lone pair cation Bi^{3+} and induce a locally polarised structure with the polarisation vector along a $[1\bar{1}0]_p$ direction, that is, a local ferroelectric configuration. The long-range incommensurate periodicity, however, suppresses the overall polarisation and results instead in a general antiferroelectric behaviour.

6.6 Ferrielectrics

A feature analogous to ferrimagnetism, that is, ferrielectricity, has been found in the double perovskite Pb_2MnWO_6. The unit cell is orthorhombic: space group $Pmc2_1$ (21), $a = 0.80370$ nm, $b = 57857$ nm, $c = 1.16378$ nm. The BO_6 octahedra are tilted and somewhat distorted, but the polar dipoles arise from displacements of the lone pair Pb^{2+} cations, as in $PbZrO_3$ described previously. The dipoles from the two symmetry-independent Pb^{2+} cations are unbalanced and give an overall spontaneous polarisation parallel to the **c**-axis (Figure 6.17). It is seen that these dipoles are, very approximately, in an antiparallel, antiferroelectric, configuration, but each dipole is tilted away from a strict parallel alignment, leading to a residual ferrielectric configuration.

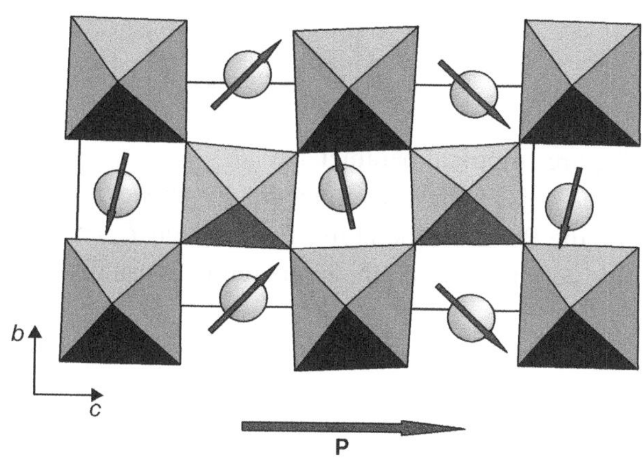

Figure 6.17 *The polar dipole arrangement in the ferrielectric double perovskite Pb_2MnWO_6: MnO_6 dark shaded, WO_6, light shaded, Pb^{2+} spheres (Original data from Orlandi* et al. *(2014)*

6.7 Relaxor Ferroelectrics

6.7.1 Macroscopic Characteristics of Relaxor Ferroelectrics

Relaxor ferroelectrics are a large group of ferroelectric perovskites with a general formula $Pb(B_1\ B_2)O_3$ in which B_1 is typically Mg^{2+}, Zn^{2+}, Ni^{2+}, Fe^{3+}, Sc^{3+} and In^{3+} and B_2 is typically Nb^{5+}, Ta^{5+} and W^{6+}. They are represented by $Pb(Mg_{1/3}Nb_{2/3})O_3$ (PMN), $Pb(Zn_{1/3}Nb_{2/3})O_3$ (PZN), $Pb(Mg_{1/3}Ta_{2/3})O_3$ (PMT), $Pb(W_{1/3}Fe_{2/3})O_3$ (PFW), $Pb(Sc_{1/2}Nb_{1/2})O_3$ (PSN), $Pb(Sc_{1/2}Ta_{1/2})O_3$ (PST), $Pb(In_{1/2}Nb_{1/2})O_3$ (PIN) and $Pb(Fe_{1/2}Nb_{1/2})O_3$ (PFN), all of which have mixed valence cations on the B-sites. In addition relaxor behaviour is found in a number of solid solutions including $(1-x)Pb(Mg_{1/3}Nb_{2/3})O_3$–$xPbTiO_3$ (PMN–PT) and $(1-x)Pb(Zn_{1/3}Nb_{2/3})O_3$, – $xPbTiO_3$ (PZN–PT) and the Ba-containing perovskites $Ba(Ti_{1-x}Zr_x)O_3$ and $Ba(Ti_{1-x}Sn_x)O_3$.

Relaxor ferroelectrics differ from the ferroelectric materials described previously in many ways. Of fundamental importance is the fact that the relative permittivity shows a wide diffuse peak which is approximately the same magnitude as in a normal ferroelectric. The broad peak is relatively temperature insensitive, but the transition is frequency dependent, with a maximum value at a temperature T_m, rather than a sharp transition at a temperature T_C (Figure 6.18a and b). The region under the curve represents a state called the ergodic relaxor (ER) state (see following text).

The relationship between frequency and T_m is often expressed in terms of the empirical Vogel–Fulcher–Tammann law (often just called the Vogel–Fulcher law, usually encountered in discussion of the viscous flow of amorphous liquids and glasses) and expressed for relaxors as

$$f = \left[\frac{1}{(2\pi\tau_0)}\right]\exp\left[\frac{-E_a}{(T_m - T_{VF})}\right]$$

where τ_0, E_a and T_{VF} are empirical parameters.

There is no Curie–Weiss behaviour immediately above T_m but only at a much higher temperature, the Burns temperature, T_d. Between T_m and T_d the relationship between the relative permittivity and the temperature is usually better described by a quadratic law:

$$\varepsilon_r = \frac{C'}{(T - T_0')^2}$$

Moreover, the crystal structure does *not* change significantly at T_m, while in a normal ferroelectric there is a noticeable change of symmetry, easily revealed by X-ray diffraction.

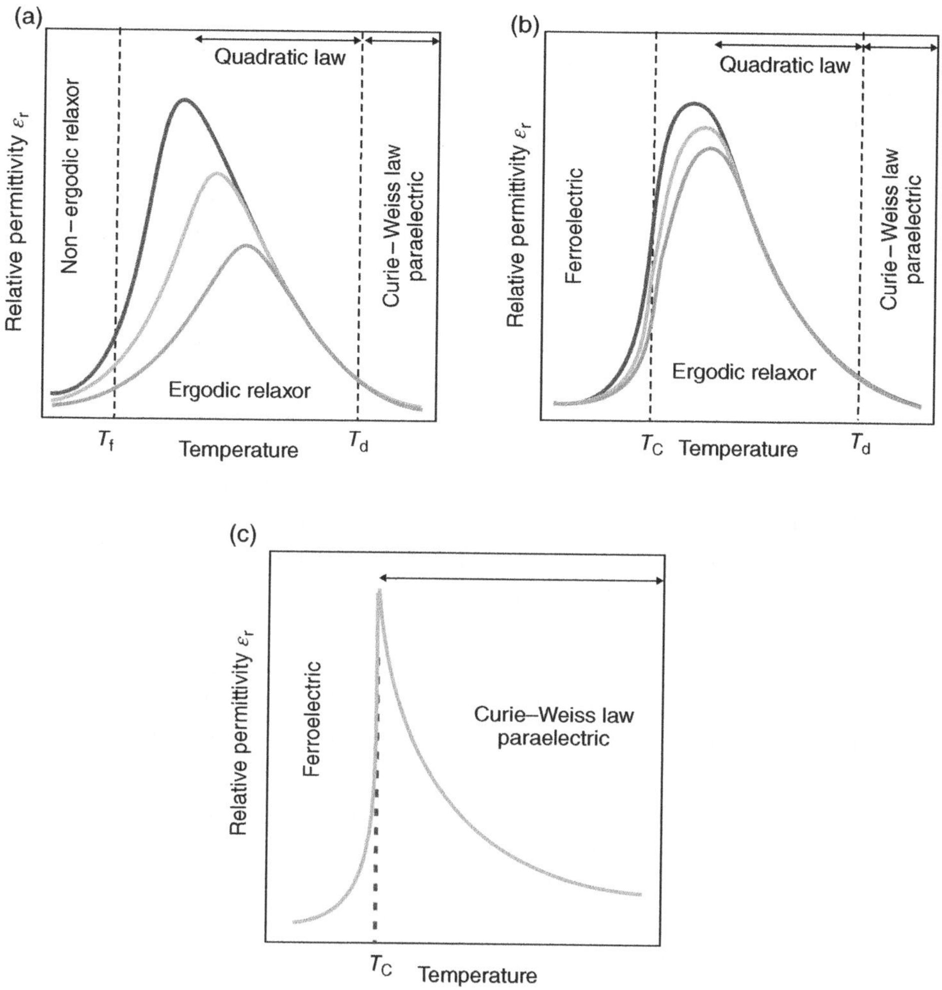

Figure 6.18 *Relative permittivity versus temperature behaviour: (a) a canonical relaxor; (b) non-canonical relaxor; (c) normal ferroelectric*

On cooling, relaxor ferroelectrics tend to roughly separate into two groups. In one group, called canonical relaxors, the ergodic state transforms to a 'frozen' *non-ergodic* state below a lower 'freezing temperature', T_f, some hundreds of degrees below T_m (Figure 6.18a). This state is not ferroelectric and the crystal structure of the material remains cubic. Heating the low-temperature non-ferroelectric state above the freezing temperature, T_f will transform it once again into the ergodic state.

In the second group, the diffuse peak in relative permittivity is steeper on the low-temperature side compared to the canonical relaxors, and at a temperature below T_C which is generally similar to the value of T_f for a canonical relaxor, the material transforms into a ferroelectric solid (Figure 6.18b). Heating the low-temperature ferroelectric state above the Curie temperature T_C will transform it once again into the ergodic state.

The sharpening of the dielectric permittivity peak that occurs between Figure 6.18a and b can be considered to be part of a continuum leading to classical ferroelectric behaviour (Figure 6.18c). This trend is often clear when solid solutions of perovskite phases are examined. For example, solid solutions between $BaTiO_3$ and the relaxor ferroelectric $BiFeO_3$ show an evolution from the classical sharp peak for $BaTiO_3$ to a characteristic relaxor peak with a sharp low-temperature margin as the $BiFeO_3$ content increases until a broad relaxor peak appears at highest concentrations.

It is clear that a sharp Curie temperature cannot be defined for relaxor materials in the same way as for simple materials such as $BaTiO_3$. To take account of this, a depolarisation temperature, T_d, is taken as the temperature at which a partial or complete loss of polarisation, and a consequent degradation in piezoelectric performance, occurs. In materials with relative permittivity behaviour similar to that in Figure 6.18b, T_d is often lower than the Curie temperature, T_C, and can often be well below this temperature. For classical materials as in Figure 6.18c, T_d coincides with T_C.

The relative permittivity and other dielectric behaviour of a relaxor ferroelectric as a function of temperature is often different depending upon whether the sample is cooled from higher temperatures in an electric field (FC) or in no electric field (ZFC). For example, the non-ferroelectric low-temperature state of a canonical relaxor can be transformed irreversibly into a ferroelectric state by the application of a sufficiently high electric field or if the material is cooled in the presence of an electric field. As with non-canonical relaxors, this state transforms to the ergodic state above T_f.

Finally we can note that the ferroelectric state of relaxors is characterised by an extremely narrow hysteresis loop with a low value for the remanent polarisation (Figure 6.19a). This sort of hysteresis loop can be considered to fall into the continuum described previously (Figure 6.9) and suggests that the microstructure of the phases is at an even more reduced scale than the fine-grained ceramic samples. Indeed, the behaviour when the materials are either cooled in or without an external electric field (FC or ZFC) is also taken as indicative of a complex microstructure. The strain versus applied electric field loop has a U shape, rather like the central portion of the strain curve for an antiferroelectric (Figure 6.19b).

6.7.2 Microstructures of Relaxor Ferroelectrics

As suggested, the defining characteristics of relaxor ferroelectrics are due to a complex microstructure. In normal materials the ferroelectric domain size is large, often identical to the grain size in a ceramic sample, and each domain is uniform. In relaxor ferroelectrics the B-sites are occupied by two cations, B_1 and B_2. These are not

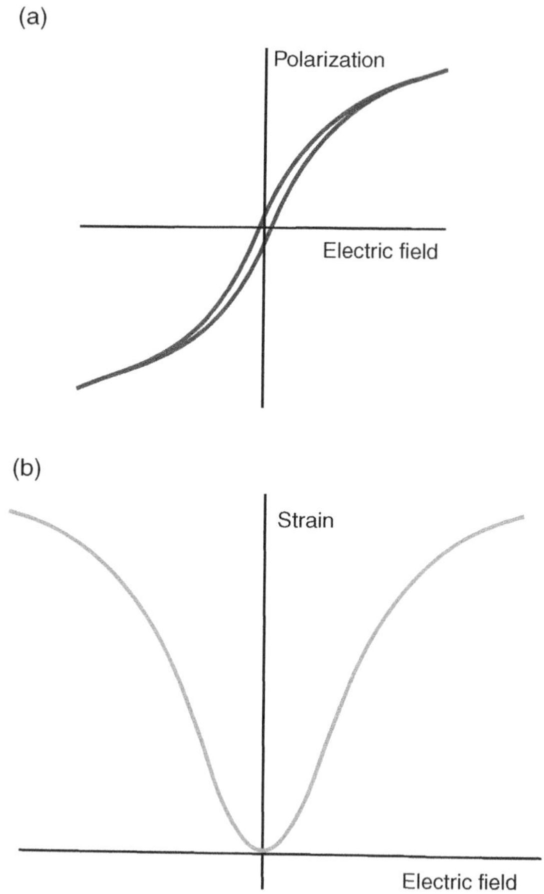

Figure 6.19 Relaxor ferroelectrics: (a) characteristically thin hysteresis loop; (b) U-shaped strain versus applied electric field 'loop'

homogeneously distributed so that an ordinary ferroelectric domain may be made up of microregions, each with slightly differing B_1 and B_2 amounts, leading to compositional disorder at an atomic scale similar to antisite defects. These inhomogeneous volumes are known as *nanodomains*, *microdomains*, *polar microregions* (PMRs) or *polar nanoregions* (PNRs). In effect they destroy long-range order and each microregion behaves as a slightly different ferroelectric. Every one of these will have a slightly different value of T_C to its neighbours, and this smears out the sharp transitions seen in normal ferroelectrics and gives rise to the broad maximum centred on T_m. Similarly, because of the spread of T_C values across the microdomains, a certain degree of spontaneous polarisation will persist above T_m in a relaxor ferroelectric.

The actual nature of the polar nanoregions remains an important question. At high temperatures, well above T_m, the cations are mobile and PNRs, in this case, are at best fleeting and continually dissolving and reforming. The material is in a non-polar paraelectric state similar to an ordinary ferroelectric and the Curie–Weiss law is obeyed (Figure 6.20a).

At the Burns temperature, T_d, nucleation and growth of stable polar nanoregions occurs.[1] This state is called the ER state, which implies that the PNRs, although still forming and reforming, form a quasi-equilibrium situation, so that over the course

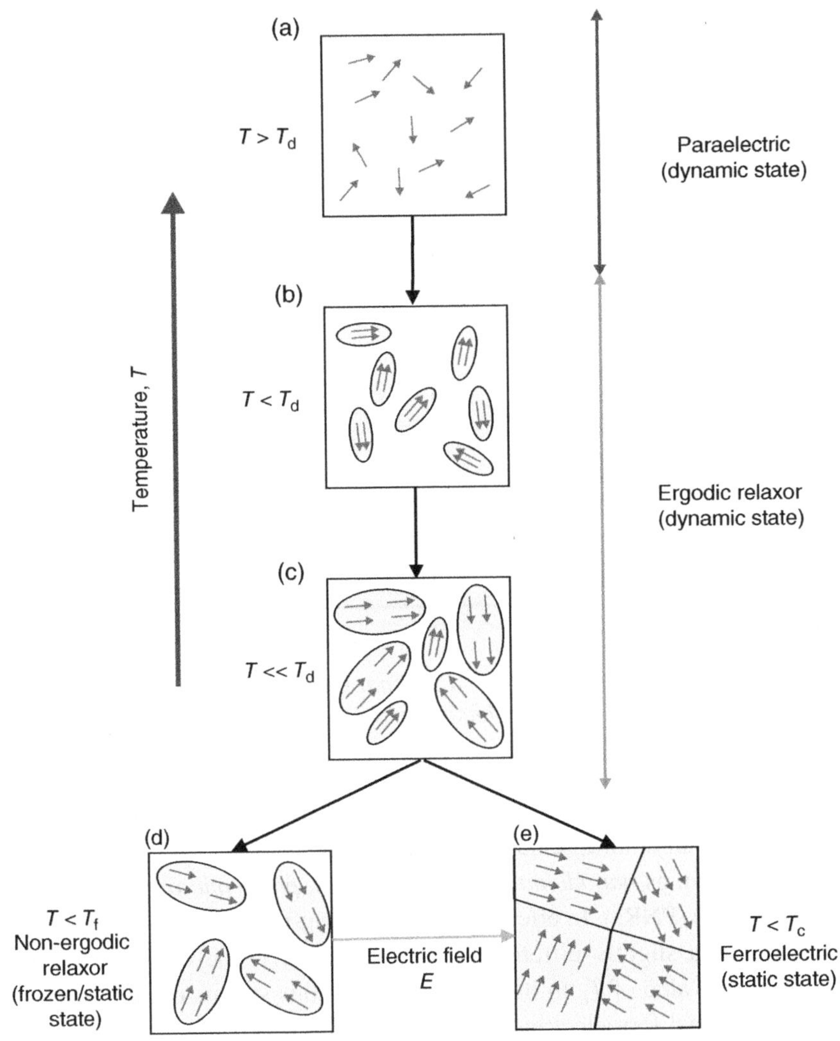

Figure 6.20 *Schematic development of relaxor ferroelectric microstructures with temperature*

---

[1] Note that in the literature the symbol T_d is used both for the decomposition temperature and the Burns temperature of relaxor ferroelectrics, although the two phenomena are totally different.

of time the properties measured remain the same. The system is to be viewed as a very dynamic state, and each PNR shows a polarisation dipole which can take directions related to the symmetry of the underlying structure, although the overall macroscopic symmetry remains cubic (Figure 6.20b). Statistical mechanics suggests that the domain sizes will follow a temperature-sensitive distribution with high temperatures giving rise to larger numbers of smaller volume microdomains. A decrease in temperature causes the dynamics to slow, and allowing for equilibrium to establish will tend to result in fewer and larger microdomains (Figure 6.20c). At T_m the PNRs are plentiful and have dimensions of the order of 0.1–0.2 nm and an estimated lifetime of 10^{-4}–10^{-5} s.

On further cooling, the behaviour diverges. In one group, the canonical relaxors, the ergodic state transforms to a 'frozen' *non-ergodic* state below a lower freezing temperature, T_f, some hundreds of degrees below T_m. In the non-ergodic state, the PNRs are now static and the phase does not show any ferroelectric properties (Figure 6.20d). The crystal structure of the material remains cubic and is believed to be similar to a magnetic spin glass (Section 7.6). However, this frozen state can be transformed irreversibly into a ferroelectric state by the application of a sufficiently high electric field. The same result can be obtained if the material is cooled in the presence of an electric field.

In the second group, the diffuse peak in relative permittivity is steeper on the low-temperature side compared to the canonical relaxors, and at a temperature T_C generally similar to the value of T_f for a canonical relaxor, the PNRs amalgamate into domains and form a ferroelectric solid (Figure 6.20e). In these materials, no non-ergodic state forms.

Relaxor behaviour can, to some extent, be tuned by varying processing temperature, cooling rate and composition. In addition, it is apparent that the imposition of an electric field will also influence the PNRs and potential ferroelectric ordering. Because of this, the canonical relaxor phase diagram requires three coordinates, composition and temperature plus electric field, to completely specify the situation. The most frequent diagrams are plots of the phase as a function of electric field versus temperature diagram, which are relevant to a single fixed composition. These may contain a field-induced ferroelectric state, an ergodic state and a non-ergodic state (Figure 6.21). The position of the boundaries between these phase fields varies considerably depending upon whether the temperature is held constant while the electric field is increased or decreased, or whether the electric field is held constant while the temperature is increased or decreased. The boundary position also depends upon whether the field or temperature is increasing or decreasing as the boundary is crossed. Moreover, the diagrams can look different when different experimental methods are employed, for example, X-ray diffraction as opposed to dielectric measurements. Figure 6.21 is an idealised figure derived from a number of sources.

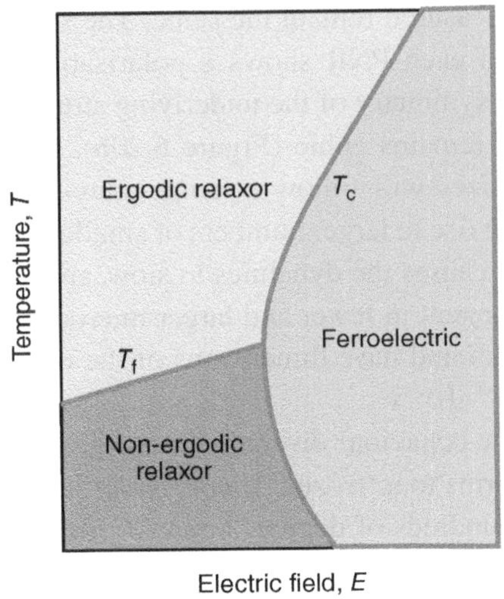

Figure 6.21 *Temperature versus electric field phase diagram for a typical relaxor ferroelectric. (T_C is the Curie temperature and T_f the freezing temperature)*

6.8 Improper Ferroelectricity

Ferroelectricity can arise in a number of ways other than that described previously, in which dipoles are generated in a crystal structure and combine to give a spontaneous polarisation if the crystal symmetry permits. These alternatives have been called improper ferroelectricity (perhaps better *extrinsic ferroelectricity*, see Lines and Glass 2001). In essence, improper ferroelectricity is ferroelectricity which is not due to the normal polarisation of the structure but arises from other interactions. Improper ferroelectricity has been considered to be rare in bulk materials and is a weak effect, usually rather difficult to detect, but the creation of artificial superlattices and studies of layered perovskites have changed this and now improper ferroelectrics are becoming widely studied.

Ferroelectric superlattices are alternating layers, a few unit cells thick, of two or three perovskite structure materials, grown sequentially so that a perfect crystal slab is produced. To achieve this, layers are built up atom by atom, thus allowing control of crystal composition and perfection as well as layer thickness. As anticipated, the ferroelectric properties of these superlattices can be tuned by varying the chemical constituents (ABO_3, $A'B'O_3$) of the layers and also by changing the relative proportions of each layer, *m* unit cells of ABO_3 and *n* unit cells of $A'B'O_3$ (Figure 6.22). However, tuning using strain can also be achieved. This possibility arises because the lattice parameters of the perovskite unit cells are close but not identical, so that degrees of strain exist at the interfaces that are related to the

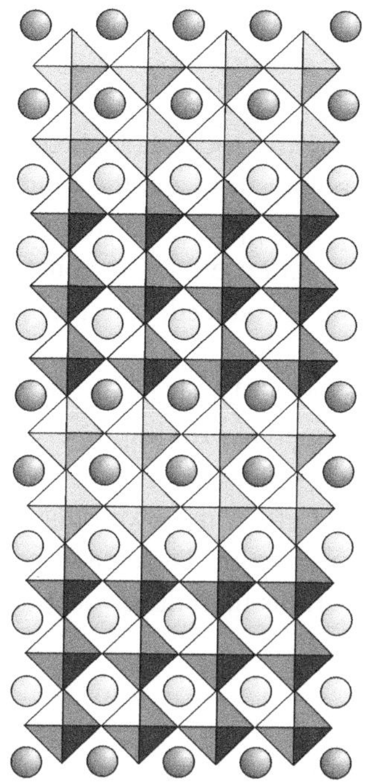

Figure 6.22 *A ferroelectric superlattice composed of m = 2 unit cells of $SrTiO_3$ and n = 3 unit cells of $BaTiO_3$*

difference between the two lattice parameters in the interface plane. This interfacial strain has an important role to play in the properties of such superlattices and a number of unique effects have been attributed to this cause. For example, a superlattice formed of the non-ferroelectric perovskite oxides $SrTiO_3$ or $CaTiO_3$ enclosing thin sheets of ferroelectric $BaTiO_3$ shows a much higher degree of polarisation than $BaTiO_3$ alone. This is attributed to the strain at the ferroelectric–non-ferroelectric interfaces which results in an enhanced polarisation of the TiO_6 octahedra in the $BaTiO_3$ component.

Interfacial strain in the superlattice series formed by *m* unit cells of ferroelectric $PbTiO_3$ and *n* unit cells of dielectric $SrTiO_3$ also produces curious behaviour. Normal ferroelectric behaviour is found when the layers are relatively thick. This diminishes as layer thicknesses fall but surprisingly, at the lowest values, ferroelectricity recovers. In bulk $PbTiO_3$ octahedral tilt is suppressed and in $SrTiO_3$ octahedral rotation is suppressed. However, in superlattices these distortions become possible, creating a strain between the two perovskite slices. As the slabs become thin, the strain component of the interfaces becomes relatively greater and ultimately, in the thinnest layers, is able to induce polarisation and an increased ferroelectric response.

A similar strain-induced feature is the development of ferroelectricity in superlattices formed by two non-ferroelectric components. This has been found, for instance, in the 1 : 1 superlattices $LaGaO_3/YGaO_3$ and $LaAlO_3/YAlO_3$. The bulk perovskite structures of these oxides contain slightly rotated octahedral GaO_6 or AlO_6 groups, but they do not lead to permanent polarisation and the solids are simply dielectrics. However, the construction of thin layer superlattices relaxes the bulk constraints so that in the interfacial regions the rotation effects become unbalanced to such an extent that permanent polarisation is introduced, resulting in measurable ferroelectricity of the superlattice.

In the layered Ruddlesden–Popper and Dion–Jacobson phases, improper ferroelectricity can also arise in a form termed *hybrid* improper ferroelectricity. Here, the ferroelectricity arises when two non-polar rotations of BO_6 octahedra combine to produce a polar structure. The same effect may well also occur in the Aurivillius phases although this has not yet been proven. The creation of improper ferroelectric polarisation due to magnetic interactions in multiferroic perovskites is described in Section 7.10.

6.9 Doping and Modification of Properties

Although simple ferroelectric materials show interesting and useful properties, there is always a desire to improve these or else to extend them to temperature or pressure regimes not accessible to existing materials. The notionally simplest and oldest established method of modification is the replacement of one or more of the constituents of the compound by A-site or B-site substitutions. Ferroelectric oxide perovskites, for example, are of interest as capacitor materials because of their high relative permittivity values, but usually the sharp maximum in dielectric constant at the Curie point must be broadened and moved to room temperature. For instance, $BaTiO_3$ has a high relative permittivity at the Curie temperature, about 393 K. The Curie temperature can be increased by A-site substitution in which of some of the Ba^{2+} ions are substituted with Pb^{2+} ions. These ions are 'softer' (i.e. more easily polarised) than the Ba^{2+} ions, as they have a lone pair of electrons, and so are more easily affected by an applied electric field. The phase $Ba_{0.6}Pb_{0.4}TiO_3$ has a Curie temperature of approximately 573 K, an increase of 200 K over that of the undoped material. In a similar way, the Curie temperature can be lowered by the substitution of Ba^{2+} ions by Sr^{2+}. These ions are smaller than Ba^{2+} ions and can be considered to be 'harder' and more difficult to polarise. The compound $Ba_{0.6}Sr_{0.4}TiO_3$ has a Curie temperature of 0°C. Similar changes can be brought about by B-site substitutions.

Much of the work on chemical modification is best understood in terms of phase diagrams. An important system, in this respect, is the $PbZrO_3$–$PbTiO_3$ (PZT) system

(Figure 6.23). This diagram shows that there are no two-phase regions and the consequence of changes in composition and temperature is simply to alter the symmetry of the perovskite phase. The major phase fields of interest are the ferroelectric trigonal *R*3*m* phase which has a rhombohedral unit cell in which the spontaneous polarisation lies along the pseudocubic $[111]_p$ direction and the neighbouring tetragonal *P*4*mm* phase which has the spontaneous polarisation lying along the pseudocubic $[001]_p$ direction. In neither of these phases is there evidence of octahedral tilt.

Phases with compositions near to boundaries between two symmetry types are structurally somewhat unstable, resulting in enhanced piezoelectric and ferroelectric properties. These symmetry-sensitive boundaries, which are composition dependent, but relatively temperature insensitive, are known generally as morphotropic phase boundaries (MPBs), but in recent years the term has become more specialised and now tends to refer solely to a boundary between phases with tetragonal unit cells and rhombohedral unit cells. For example, the boundary between the rhombohedral and tetragonal phases in the $PbZrO_3$–$PbTiO_3$ (PZT) system, occurring at about 0.48 mol% $PbTiO_3$, is regarded as the MPB in this system.

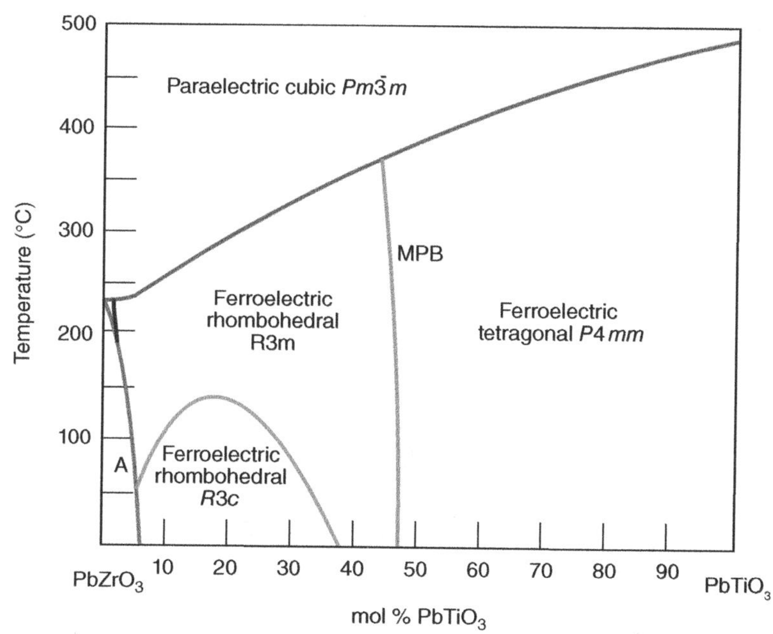

Figure 6.23 *The $PbZrO_3$–$PbTiO_3$ phase diagram showing the morphotropic phase boundary (MPB) separating the rhombohedral and tetragonal phases. Region A is antiferroelectric: orthorhombic space group* Pbam.

One strategy in the search for enhanced piezoelectric and ferroelectric properties is to form solid solutions between two or more systems, each of which contains an MPB, in order to map out larger composition regions over which superior properties might occur. Illustrative examples include the relaxor ferroelectric containing systems $Pb(Mg_{1/3}Nb_{2/3})O_3$–$PbTiO_3$ (PMN–PT), $Pb(Mg_{1/3}Nb_{2/3})O_3$–$PbZrO_3$ (PMN–PZ) and $Pb(Zn_{1/3}Nb_{2/3})O_3$–$PbTiO_3$ (PMN–PT). More complex pseudoternary systems that have been investigated include $Pb(Mg_{1/3}Nb_{2/3})O_3$–$PbZrO_3$–$PbTiO_3$ (PMN–PZT), $Pb(Y_{1/2}Nb_{1/2})O_3$–$PbZrO_3$–$PbTiO_3$ (PYN–PZT), $Pb(In_{1/2}Nb_{1/2})O_3$–$Pb(Mg_{1/3}Nb_{2/3})O_3$–$PbTiO_3$ (PIN–PMN–PT) and $Pb(Lu_{1/2}Nb_{1/2})O_3$–$Pb(Zn_{1/3}Nb_{2/3})O_3$–$PbTiO_3$ (PLZNT). For example, in this latter system MPBs exist at approximately 50 mol% $PbTiO_3$ in the $Pb(Lu_{1/2}Nb_{1/2})O_3$–$PbTiO_3$ system and 10 mol% $PbTiO_3$ in the $Pb(Zn_{1/3}Nb_{2/3})O_3$–$PbTiO_3$ system. The ternary phase field between these two MBPs would then be considered as a potentially rich area to explore for attractive relaxor properties (Figure 6.24).

Lead zirconate titanate (PZT)-based ceramics are widely used in piezoelectric devices, but in view of the toxicity of lead, much effort is now being placed in the attempt to discover lead-free perovskite phases with similar or improved properties to those of the lead-based phases. The systems currently under investigation include the archetypal $BaTiO_3$ together with $Bi_{0.5}Na_{0.5}TiO_3$, $NaNbO_3$, $KNbO_3$ and especially solid solutions $(K_xNa_{1-x})NbO_3$ (KNN). (The phase $K_{0.5}Na_{0.5}NbO_3$ itself is also called KNN.)

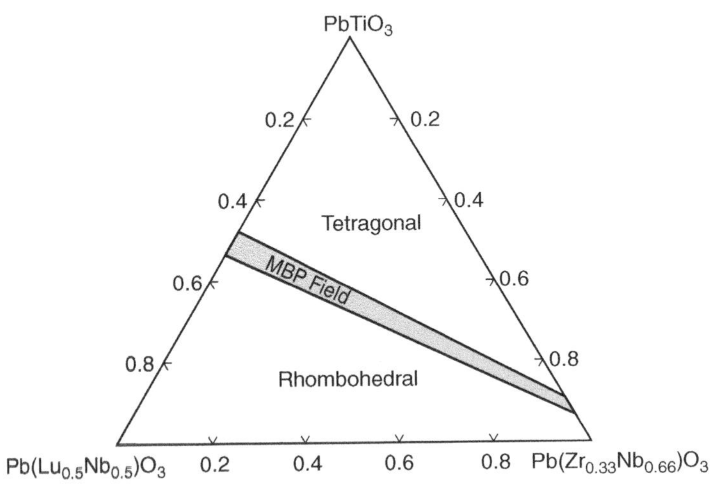

Figure 6.24 *The pseudoternary phase diagram $Pb(Lu_{1/2}Nb_{1/2})O_3$–$Pb(Zn_{1/3}Nb_{2/3})O_3$–$PbTiO_3$ (PLZNT) system. The region between the tetragonal and rhombohedral phase fields is the MBP field. (Data from Li and Long (2014). © Elsevier)*

The symmetry forms of $KNbO_3$ mirror those of $BaTiO_3$, with three ferroelectric phases occurring. Below approximately −10°C the symmetry is rhombohedral. Above this temperature the structure is orthorhombic, the normal room temperature form, tetragonal at approximately 225°C and cubic at temperatures of approximately 435°C and above. All of these are first-order transitions, mainly due to B-site cation displacements, which are parallel to the pseudocubic directions $[001]_p$ in the tetragonal phase $[011]_p$ in the orthorhombic phase and $[111]_p$ in the rhombohedral structure. The perovskite $NaNbO_3$ adopts more complex symmetry changes due to considerable octahedral tilting, and these have still not been completely clarified. At high temperature, above 640°C the structure is cubic. As the phase cools, transitions occur at 575, 520, 480, 370 and −100°C. The lowest-temperature phase is an orthorhombic ferroelectric structure. At room temperature, the structure is also orthorhombic but antiferroelectric. The antiferroelectric form can be transformed into an orthorhombic ferroelectric by the application of high electric fields.

The rather complex phase diagram of the $NaNbO_3$–$KNbO_3$ system contains mainly single perovskite phase fields, each differentiated from its neighbours by symmetry changes. At the highest temperatures a cubic phase field occurs: space group $Pm\bar{3}m$ (221), $a = 0.39928$ nm for $K_{0.5}Na_{0.5}NbO_3$ at 401°C (Figure 6.25). Above approximately 200°C a broad tetragonal phase field exists: space group $P4mm$ (94), $a = 0.39767$ nm, $c = 0.40190$ nm, for $K_{0.5}Na_{0.5}NbO_3$ at 201°C. However, the transition

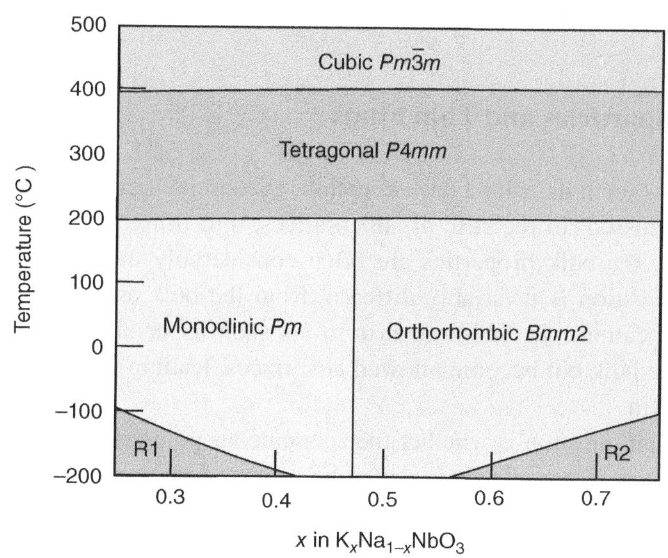

Figure 6.25 *Approximate phase diagram for part of the $NaNbO_3$–$KNbO_3$ (KNN) system. The phase R1 has a rhombohedral unit cell, space group* R3c*, and the phase* R2 *has a rhombohedral unit cell with space group* R3m

temperatures between the cubic to tetragonal and tetragonal to phases stable at lower temperatures depend on whether the sample is being cooled or heated, and the boundaries drawn in Figure 6.25 are only approximate. The structures that form below the tetragonal phase region are all very similar and are differentiated by changes in octahedral tilt and B-site cation displacement, with a result that some uncertainty in the position of phase boundaries still exists. The best characterised of these phases are an orthorhombic phase (space group *Bmm*2 (38), $a=0.56573$ nm, $b=0.39551$ nm, $c=0.56717$ nm for $K_{0.5}Na_{0.5}NbO_3$ at 22°C) and a monoclinic phase (space group *Pm* (6), $a=0.564304$ nm, $b=0.393187$ nm, $c=0.561260$ nm, $\beta=89.91°$ for $K_{0.3}Na_{0.7}NbO_3$ at 0°C). A boundary between these two forms occurs close to $K_{0.475}Na_{0.525}NbO_3$. The piezoelectric properties peak near to this phase boundary, and materials near to this composition are the ones being most actively investigated at present.

The properties of phases in the $KNbO_3$–$NaNbO_3$ system can be adjusted following the classical methods described previously, namely, A- and B-type doping, and solid solution formation with other perovskites such as $BaTiO_3$ and Ba(Zr, Ti)O_3. A-site and B-site doping has a considerable effect on the position of the phase boundaries. For example, the substitution of 5% of the A-site ions with Li is sufficient to drop the orthorhombic to tetragonal phase boundary from close to 200°C to room temperature. The same is true for reaction with other perovskite phases, such as Ba(Ti, Zr)O_3, which not only modify transition temperatures but also the spontaneous polarisation and piezoelectric coefficients.

6.10 Nanoparticles and Thin Films

In the previous sections, with a few exceptions (Section 6.8), mostly bulk properties are being discussed. In the case of superlattices, thin films, exposed surfaces and nanoparticles, the bulk properties are often considerably modified. For instance, surface composition is invariably different from the bulk as are the coordination polyhedra of cations. In addition, distortions such as octahedral tilting may be blocked in the bulk but become allowed at surfaces, leading to significant changes in surface strain.

An important question is whether the spontaneous polarisation of a ferroelectric will vanish at small particle sizes, as the long-range forces giving rise to order are no longer supported. In part this query is motivated by the need for ever smaller components as electronic circuits are miniaturised. In this respect, barium titanate is probably the most studied of the ferroelectric perovskites. Evidence now suggests that ferroelectric properties persist in small particles prepared by high-energy milling of bulk material. At dimensions of 100 μm $BaTiO_3$ particles are multidomain, and this drops to single domain when particle size is of the order of 30 μm. However, different methods of nanoparticle preparation can give somewhat different particle morphology and crystal structure for the $BaTiO_3$. Generally nanoparticles tend to be tetragonal, but evidence suggests that the *c/a* ratio falls as the particle size decreases, implying that the desirable ferroelectric properties weaken. In addition, a sharp transition from tetragonal to cubic is frequently absent. These features may be due to the fact that although the ionic displacements do not change, surface relaxation, resulting in structural disorder, contributes a larger amount to the observations while the particle cores, which may be likened to bulk material, contribute relatively less. Despite this uncertainty, barium titanate samples show polarisation and switching in particles down to approximately 10 unit cells, say, 5 nm diameter. As the dimensions fall below this, the values of T_C and polarisation begin to diminish. It is beginning to appear that a dimension of the order of 5 nm represents a critical size for ferroelectric phenomena in this material.

A similar picture seems to emerge for other nanoparticles. In this respect nanoparticles of $BiFeO_3$ approximately 54 nm size are apparently ferroelectric, as are particles of composition $Bi_{0.9}In_{0.1}FeO_3$ (A-site substitution), $BiFe_{0.95}Ti_{0.05}O_3$ (B-site substitution) and $Bi_{0.9}In_{0.1}Fe_{0.95}Ti_{0.05}O_3$ (A- and B-site substitution), these latter giving hysteresis loops even for particles of 14 nm size. Homogeneous tetragonal perovskite nanoparticles of composition $(KNbO_3)_{0.8}(PbTiO_3)_{0.2}$ 50 nm in size have also been found to be ferroelectric at room temperature and to exhibit a tetragonal–paraelectric cubic phase transition at 500 K.

The question also arises as to whether ferroelectricity and domain structures are still preserved in very thin films. Films of lead titanate appear to remain ferroelectric down to 1 or 2 unit cells (0.4–1 nm) in thickness, although the Curie temperature gradually falls for the thinnest films suggesting that below this critical size matters may change. Many thin films appear to have poorer electrical properties than single crystals, attributed to polar defects introduced during film deposition. For example, oxygen vacancies can form during deposition of $BiFeO_3$ films together with a balancing population of electrons. In this case each oxygen vacancy ($V_O^{\bullet\bullet}$) is balanced by the formation of two electrons (2e′). Some or all of the electrons can reside on the Fe^{3+} cations, changing the nominal charge state to Fe^{2+} (Fe'_{Fe}). These in turn can form electric dipoles by association with oxygen vacancies to give ($Fe'_{Fe}-V_O^{\bullet\bullet}$). In more complex compositions such as $(Bi_{0.9}La_{0.1})(Fe_{0.97}Co_{0.03})O_3$, the electrons produced by the loss of oxygen can reside on either the Fe^{3+} ions or the Co^{3+} ions. In this latter case, dipolar associated defects ($Co'_{Fe}-V_O^{\bullet\bullet}$) can also form. These additional dipoles appear to be the source of the degradation in properties of the film compared to the bulk response.

Mechanical stress will deform a solid and the question arises as to whether it is possible to switch polarisation in a ferroelectric by the application of stress. The connection between these physical properties, which describes the coupling between strain and polarisation, is called flexoelectricity. This is of potential importance because if the polarisation state of a ferroelectric film can be changed mechanically, then the process can be used to create a permanent domain array that will serve as a memory store that can be mechanically written but read or erased electrically. In particular, if the switched domains are of nanometre dimensions, the amount of data that can be stored will be significant.

Computation shows that the crucial feature to enable mechanical switching is the creation of a *strain gradient* in the ferroelectric. A homogeneous stress simply shifts the double potential well associated with ferroelectricity (Figure 6.7c) without changing the shape of the curve. On the other hand, a strain gradient alters the form of the curve so that one configuration is preferred over the alternative (Figure 6.26a). The switching of the polarisation direction can then be envisaged to occur in a thin film if a force can be applied to a point location on the film surface. This is because, in a thin film, the strain gradient will be sufficiently extended to penetrate the film and hence completely switch the strained volume. (In a bulk sample the strain field will only penetrate a small way into the crystal and on removal of the strain the surroundings will force the polarisation to revert to its original direction.) Flexoelectric domain writing has been achieved in $BaTiO_3$ films approximately 4.8 nm (12 unit cells) thickness, oriented with the polarisation vector perpendicular to the surface. The initial polarisation was reversed by forcing the probe of a scanning probe microscope into the crystal surface, creating a domain with dimensions of the order of 30 nm diameter (Figure 6.26b). The resultant mechanically written domain pattern can be both read and erased electrically, thus generating a mechanically written permanent memory store.

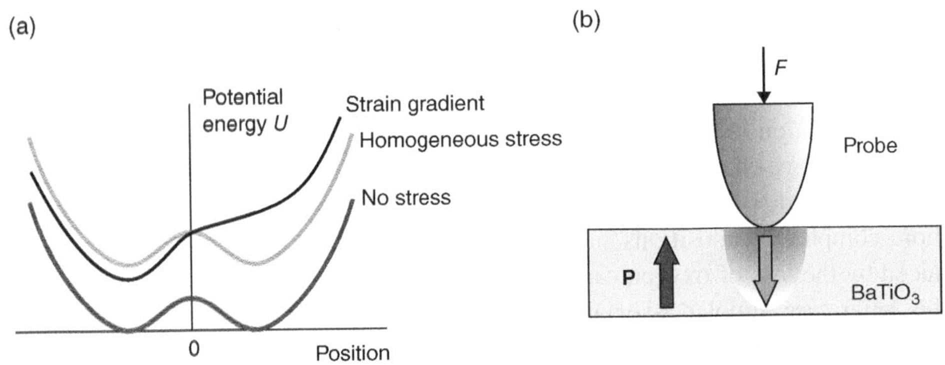

Figure 6.26 *The flexoelectric effect in* $BaTiO_3$*: (a) the evolution of the potential energy curve under a homogeneous stress and in a strain gradient; (b) domain switching via mechanical stress imposed by a probe (Original data from Lu* et al. *(2012))*

References

T. Li, X. Long, *Mater. Res. Bull.*, **51**, 251–257 (2014).
M. E. Lines, A. M. Glass, *Principles and Applications of Ferroelectrics and Related Materials*, Oxford Classics, Oxford University Press, Oxford (2001).
H. Lu *et al.*, *Science*, **336**, 59–61 (2012).
X. Lv *et al.*, *J. Am. Ceram. Soc.*, **96**, 1188–1192 (2013).
C. T. Nekon *et al.*, *Science*, **334**, 968–971 (2011).
F. Orlandi *et al.*, *Inorg. Chem.*, **53**, 10283–10290 (2014).

Further Reading

For an introduction to dielectric properties and the relationship with crystal structures see:
R. J. D. Tilley, *Understanding Solids*, 2nd edition, John Wiley & Sons, Ltd, Chichester (2013), particularly Chapter 11.
R. J. D. Tilley, *Crystals and Crystal Structures*, John Wiley & Sons, Ltd, Chichester (2006), particularly Chapter 4.

For the tensor description of the dielectric physics of crystals see:
J. F. Nye, *Physical Properties of Crystals*, Oxford Science Publications, Oxford (1985).

For a description of earlier studies of ferroelectric materials see;
M. E. Lines, A. M. Glass, *Principles and Applications of Ferroelectrics and Related Materials*, Oxford Classics, Oxford University Press, Oxford (2001).

A great deal of information on perovskite ceramics, including numerical values for the properties of components, is to be found on manufacturer's websites.
For an entry into the extensive literature of ceramic perovskite materials, it is useful to consult a few recent issues of the Journal of the American Ceramic Society. The papers found, together with the references cited to each, will rapidly form a comprehensive data source.
The following review articles provide entry points into the extensive literature of ferroelectric perovskites:
G. H. Haertling, *J. Am. Ceram. Soc.*, **82**, 797–818 (1999).
J.-F. Li *et al.*, *J. Am. Ceram. Soc.*, **96**, 3677–3698 (2013).
L. Ji, F. Li, S. Zhang, *J. Am. Ceram. Soc.*, **97**, 1–27 (2014).
A. A. Heitmann, G. A. Rossetti Jr., *J. Am. Ceram. Soc.*, **97**, 1661–1683 (2014).
L. Li, J. R. Jokisaari, X. Pan, *Mater. Res. Soc. Bull.*, **40**, 53–61 (2015).

The Maxwell–Wagner effect is described in detail in:
M. Iwamoto, The Maxwell-Wagner Effect, in: Encyclopedia of Nanoscience and Technology, ed. B. Bhushan (Ed.), *Encyclopedia of Nanoscience and Technology*, Springer Reference, (2013).

Domain switching in $BiFeO_3$ is described by:
C. T. Nekon *et al.*, *Science*, **334**, 968–971 (2011).

The crystal structures of the various forms of $PbZrO_3$ are described by:
S. Teslic, T. Egami, *Acta Crystallogr.*, **B54**, 750–765 (1998).
N. Zhang *et al.*, *Acta Crystallogr.*, **B67**, 461–466 (2011).

The January issue of *Journal of Materials Science* is devoted to "Frontiers of Ferroelectricity" (J. Mater. Sci., **41** [1] (2006), including a review of relaxor ferroelectric behaviour:
A. A. Bokov, Z.-G. Ze, *J. Mater. Sci.*, **41**, 31–52 (2006).

The properties arising at interfaces between perovskite slabs is discussed in:
H. Y. Hwang *et al.*, *Nat. Mater.*, **11**, 103–113 (2012).

Hybrid improper ferroelectrics are defined in:
N. A. Benedek, C. J. Fennie, *Phys. Rev. Lett.*, **106**, 107204 (2011).
N. A. Benedek, *Inorg. Chem.*, **53**, 3769–3777 (2014).

7

Magnetic Properties

7.1 Magnetism in Perovskites

Magnetic materials can be approached from the point of view of free electron band theory or from a modification of the familiar ionic model. Those perovskites that can be treated best in free electron terms form a group of metallic phases analogous to the metallic elemental magnets. Interaction of the electron spins, usually important only at lower temperatures, may be strong enough to lead to a parallel (ferromagnetic) alignment. At higher temperatures these transform into metal-like Pauli paramagnetic compounds, with almost no temperature dependence for the magnetic susceptibility.

Many perovskites, though, are magnetic due to the incorporation of paramagnetic cations within the perovskite structure, that is, cations that display a magnetic moment themselves, and these can frequently be considered in ionic terms. The most important magnetic species are transition metal and lanthanoid cations which have incompletely filled d and f shells. The geometry of the crystallographic site occupied by these cations is important as it governs the d and f cation orbital energy level positions, which are the result of crystal-field/ligand-field interactions. In particular, for the 3d transition metal cations, the magnetic moment arises principally from the electron spin (Russell–Saunders quantum number S) alone and is often just called the spin. Depending upon the strength

Perovskites: Structure–Property Relationships, First Edition. Richard J. D. Tilley.

of the crystal-field interaction, the total spin can often take one of several values, high spin (HS), intermediate spin (IS) or low spin (LS). Magnetic behaviour is then associated with the organisation between the magnetic moments or spin on these cations. At higher temperatures the magnetic moments are disordered, leading to paramagnetic compounds with a temperature-dependent magnetic susceptibility obeying the Curie or Curie–Weiss laws. At lower temperatures magnetic moments can interact, leading to antiparallel (antiferromagnetic) or parallel (ferromagnetic) alignment, as well as a host of other magnetic configurations.

The interaction that has been proposed to account for the antiferromagnetic ordering takes place via an intermediate anion in a process called *superexchange*. This mechanism relates the magnetic pairing between two cations to the degree of overlap between the relevant orbital lobes on the cations and those on the intermediate anion. However, no actual electron exchange takes place between cations and anion, and the exchange is said to be virtual. There are two possibilities, 180° and 90° superexchange, both described in more detail later. The magnetic outcome of linking two paramagnetic cations via superexchange is predicted by the Goodenough–Kanamori–Anderson (GKA) rules, which take into account the cation d-energy level occupancies, which vary across the transition metal series and depend upon the strength of any crystal-field/ligand-field splitting. In 180° superexchange, the cation–anion–cation link is more of less linear, depending on the degree of octahedral tilt and distortion, and in this geometry, the GKA rules indicate that the two paramagnetic cations will have a strong tendency to align in an antiferromagnetic fashion. In the case of 90° superexchange, the bond angle is square, and the GKA rules indicate that magnetic moment alignment is ferromagnetic, but the tendency is much weaker than in the preceding case.

An important aspect of superexchange is the fact that no actual electron transfer takes place. Superexchange is therefore possible between cations in both octahedral and tetrahedral sites and between cations in other coordination polyhedra, such as square pyramidal. This means that antiferromagnetic ordering can take place in brownmillerites or other perovskites in which the octahedral coordination polyhedron has been significantly changed. (This is not so with interactions in which electron exchange actually occurs.)

The energy difference between these various spin alignments is delicately balanced and is a function of composition and temperature. Cation substitution readily transforms one magnetic phase to another. For example, in the $La_{1-x}Sr_xMnO_3$ system, a ferromagnetic metallic phase is found as well as four different ordered antiferromagnetic insulating structures, which evolve with the Sr content (Figure 7.1).

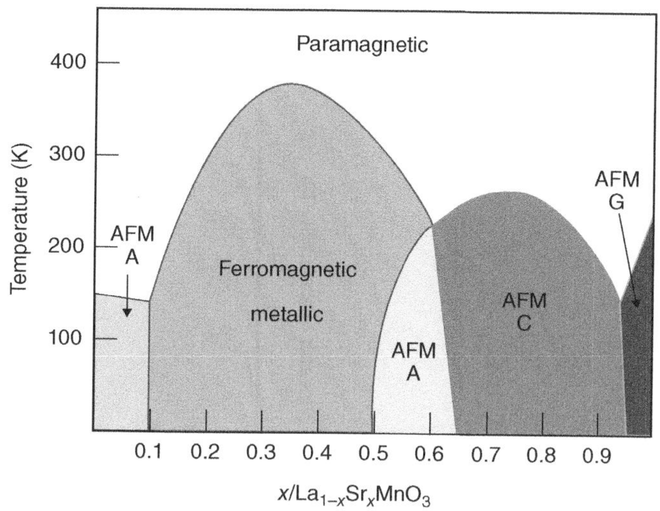

Figure 7.1 *Simplified magnetic phase diagram of the* $La_{1-x}Sr_xMnO_3$ *perovskites. AFM A, C and G refer to antiferromagnetic insulating phases (Section 7.3)*

7.2 Paramagnetic Perovskites

For perovskites containing paramagnetic cations, at higher temperatures the magnetic moments in the structure are arranged at random. The arrangement is dynamic, and orientation of the dipoles is continually changing, due to thermal effects, to constitute the paramagnetic state. In a magnetic field, these magnetic dipoles will attempt to orient themselves parallel to the magnetic flux density in the solid, which will be opposed by thermal agitation. The paramagnetic susceptibility, χ, will thus vary with temperature. For a simple paramagnetic, this dependence will follow the Curie law:

$$\chi = \frac{C}{T}$$

where C is the Curie constant and T is the temperature (Figure 7.2a and b). Ferromagnetic perovskites transform to the paramagnetic state above the Curie temperature T_C, and antiferromagnetic perovskites transform to the paramagnetic state above the Néel temperature T_N. Well above these transition temperatures, in both cases, the paramagnetic solid obeys Curie–Weiss law:

$$\chi = \frac{C}{(T - \theta)}$$

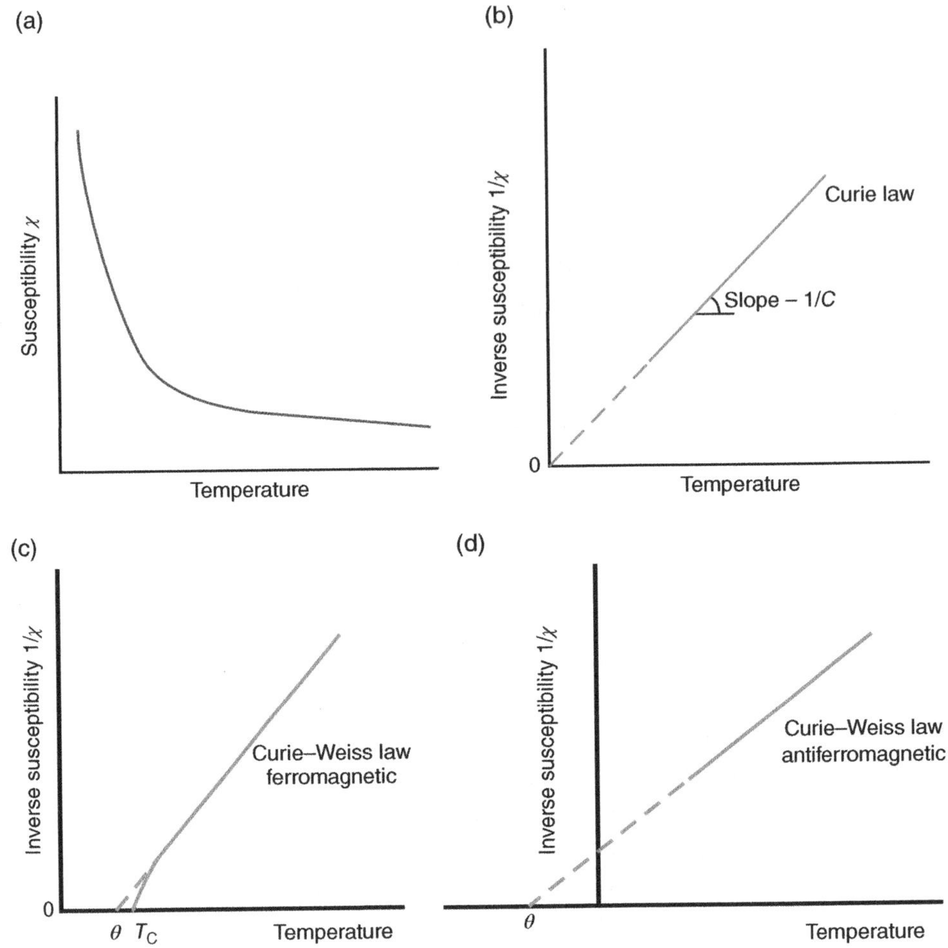

Figure 7.2 *Paramagnetic susceptibility: (a) the form of the susceptibility (χ) versus temperature curve; (b) Curie law plot of (1/χ) versus temperature; (c) Curie–Weiss law plot for a ferromagnetic phase; (d) Curie–Weiss law plot for an antiferromagnetic phase*

where θ is the Curie–Weiss constant (Figure 7.2c and d). The Curie–Weiss constant is positive and close to the Curie temperature for ferromagnetic materials and negative for antiferromagnetic materials.

The magnitude of the molar Curie constant is proportional to the square of the magnetic moment on the magnetic cations. The expression for the Curie constant can be written as

$$C_{\text{molar}} = \frac{\left[N_A \mu_0 g_J^2 \mu_B^2 J(J+1)\right]}{3k_B} = \left[\frac{N_A \mu_0 \mu_B^2}{3k_B}\right]\mathbf{m}^2$$

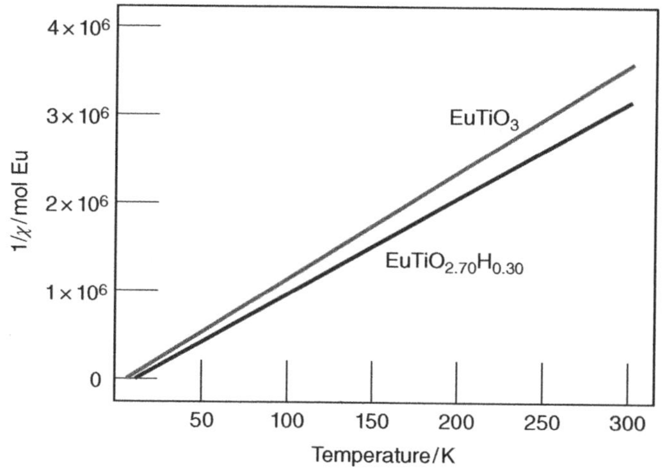

Figure 7.3 *Inverse susceptibility of the perovskites $EuTiO_3$ and $EuTiO_{2.70}H_{0.30}$ (For original data see Yamamoto* et al. *(2015))*

where C_{molar} is the molar Curie constant, N_A is Avogadro's constant, μ_0 is the vacuum permeability, μ_B is the Bohr magneton and **m** is the magnetic moment of the paramagnetic ions. Inserting values for the constants,

$$C_{\text{molar}} = 1.572 \times 10^{-6} \mathbf{m}^2 \text{ mol}^{-1}$$

A plot of the inverse paramagnetic susceptibility against temperature will then give a value for the magnetic moment giving rise to the magnetic behaviour, **m**, which can be used to ascertain the magnetic state of the paramagnetic cation. For example, $1/\chi$ versus T plots for the perovskites $EuTiO_3$ and $EuTiO_{2.93}H_{0.7}$ (Figure 7.3) yield values of 8.37×10^{-5} and 9.21×10^{-5} K mol^{-1}, respectively, for the Curie constants. The magnetic moment of the paramagnetic cation is then 7.3 μ_B ($EuTiO_3$) and 7.65 μ_B ($EuTiO_{2.93}H_{0.7}$). These values correspond to Eu^{2+} and confirm that in both perovskites, the Ti is in the non-magnetic Ti^{4+} state.

This method is widely used to give information about the nominal valance state of the cations in paramagnetic materials. The perovskite $SrCrO_3$ prepared at high pressure and temperature (75 kbar, 1073 K) forms a cubic perovskite phase: space group $Pm\bar{3}m$, $a = 0.52911$ nm. Above 150 K a plot of inverse susceptibility versus temperature gives the effective magnetic moment of the cations as 2.98 μ_B. This is in agreement with the expected spin-only value for d^2 Cr^{4+} cation. In the form of Mn_2O_3 that adopts the $AA'_3B_4O_{12}$ structure (Section 2.1.3), the effective magnetic moment of the nominally Mn^{3+} cations, from an inverse susceptibility plot, is found to be 5.51 μ_B. The theoretical value for HS Mn^{3+} is 4.90 μ_B. The difference suggests that some Mn^{3+} cations have disproportionated into Mn^{2+} and Mn^{4+} cations, as Mn^{2+} has a spin-only magnetic moment of 5.92.

Although the form of the inverse susceptibility versus temperature curve is a good guide to low-temperature ferromagnetic or antiferromagnetic properties, it can be mimicked by other physical processes, and the magnetic ordering of the phase needs to be verified by separate techniques to avoid errors. For example, the double perovskites $Cs_2NaYbCl_6$, Sr_2YbNbO_6, Sr_2YbTaO_6 and Sr_2YbSbO_6 show negative values of the Curie–Weiss constant, suggesting an antiferromagnetic state at lower temperatures. However, the magnetic behaviour originates in the lanthanoid Yb^{3+} cation, in octahedral coordination. This cation has a ground state $^2F_{7/2}$ that is split by the crystal field of the surrounding anions. In the lanthanoids this splitting is rather small compared to that experienced by the 3d transition metal cations, as the 4f electron orbitals are well shielded. However it remains significant in that it divides the ground state into two closely spaced energy levels. As the temperature is increased from a low value, the population of the upper energy level is increased by thermal excitation. The resulting curve mimics the behaviour of a Curie–Weiss antiferromagnetic phase, although no antiferromagnetic spin ordering is ever present.

Magnetic perovskites that are best treated in terms of delocalised electrons in a conduction band give rise to Pauli paramagnetic behaviour above any magnetic moment ordering temperature. In this state, the spin-up and spin-down electron states are slightly altered so as to give rise to a weak, temperature-dependent paramagnetic state unlike the situation outlined earlier.

7.3 Antiferromagnetic Perovskites

7.3.1 Cubic Perovskite-Related Structures

In these phases, the B-site cations are in apex-shared octahedra, and the A-site cations in the cage sites between the octahedra and the cation–anion–cation geometry is linear so that the superexchange mechanism suggests that these phases should favour the antiferroelectric (antiparallel) alignment of the magnetic moments. Schematically (Figure 7.4), a cation with an overall spin-up configuration can interact with a spin-down electron in a filled oxygen p orbital, with a consequence that the other (spin-down) electron in the oxygen p orbital must then induce a favourable exchange to enforce an overall spin-down configuration in the other cation.

This result is broadly true, especially for perovskites in which a single magnetic cation occurs, such as $LaFeO_3$ (Table 7.1). In these compounds the phases are usually insulators or fairly high-resistivity semiconductors. For many, but not all, there is an abrupt transition from an antiferromagnetic arrangement to a paramagnetic array of spins that takes place at the Néel temperature, T_N. In some antiferromagnetic perovskites the transition from long-range antiferromagnetic order to paramagnetic

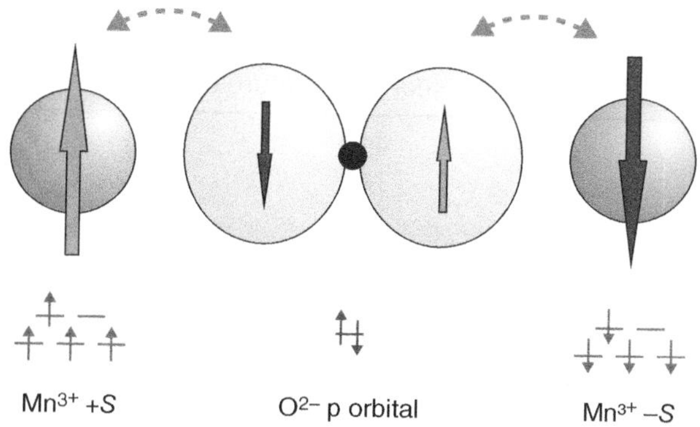

Figure 7.4 Schematic of 180° superexchange between two paramagnetic cations via the filled p orbitals of an intermediate oxygen ion, illustrated by Mn^{3+}

disorder traverses an intermediate region with short-range order and an unusual magnetic susceptibility signature. In these cases the termination of the antiferromagnetic order is characterised by a critical temperature, often written as T_M, T_m or T_c (not to be confused with the Curie temperature T_C, for ferromagnetic phases).

The ordering energy in antiferromagnetic materials is delicately balanced, and a number of different ordering patterns are known. A-type ordering has ferromagnetic ordering on $(001)_p$ planes (referred to an ideal perovskite unit cell), and each $(001)_p$ plane is aligned in the opposite direction (Figure 7.5a). Each cation has four neighbours with parallel spins and two with antiparallel spins. The C-type alignment is similar but with $(110)_p$ ferroelectric sheets (Figure 7.5b). Each cation has four neighbours with antiparallel and two with parallel spins. The G-type alignment has $(111)_p$ ferroelectric sheets (Figure 7.5d). Each cation has six antiparallel neighbours. In the E-type arrangement (Figure 7.5c) the cation ordering is more complex. Here each cation has four neighbours with antiparallel spins and two with parallel spins. The ordering type, A, C, E and G, is sometimes given a subscript x, y and z, which indicates the spin orientation along the **a**-, **b**- or **c**-axes of the unit cell. A more complex pattern of antiferromagnetic ordering, called CE order, is found in mixed valence manganites with compositions close to $La_{0.5}A_{0.5}MnO_3$ (A = alkaline earth cation). In these oxides, the critical components are the Mn^{3+} and Mn^{4+} cations, and the interactions between them lead not only to an intricate antiferromagnetic spin ordering consisting of a regular arrangement of the C- and E-type patterns but also to charge and orbital ordering (Section 8.6).

The perovskite $LaCrO_3$ is a G-type antiferromagnetic phase, $T_N \sim 300$ K, as are $CrCrO_2H$, $T_N \sim 380$ K and $BaFeO_2F$, $T_N \sim 645$ K. The phase $BiCoO_3$ is a C-type antiferromagnetic with $T_N = 470$ K, containing Bi^{3+} and Co^{3+} cations in the HS $S = 2$ state, occupying square pyramidal coordination polyhedra. Above the Néel temperature

Table 7.1 *Antiferromagnetic perovskites and perovskite-related phases*

Perovskite	Néel temperature (K)	Comments
$NaOsO_3$	410	Metal
$LaTiO_3$	146	G-type
$CeTiO_3$	136	G-type
$PrTiO_3$	115	G-type
$NdTiO_3$	94	G-type
$SmTiO_3$	54	G-type
$EuTiO_3$	5.5	—
$LaCrO_3$	290	—
$LaMnO_3$	140	A-type
$CaMnO_3$	—	G-type
$LaFeO_3$	750	—
$BiCoO_3$	470	C-type
$Ca_2Fe_2O_5$	—	G_x
Ca_2FeMnO_5	—	G_y
Ca_2GaMnO_5	—	G_y
$Ca_2Fe_{1.5}Mn_{0.5}O_5$	—	$G_y \rightarrow G_x$
$CaSrFe_{1.5}Mn_{0.5}O_5$	—	$G_y \rightarrow G_x$
Sr_2MgOsO_6	110	—
$SrCrO_2H$	380	G-type
$CaMn_3V_4O_{12}$	54	—
$LaMn_3V_4O_{12}$	44	—
$CaCu_3Ti_4O_{12}$	25	G-type
$Ce_{0.5}Cu_3Ti_4O_{12}$	24	G-type
Ba_2MnWO_6	45	—
Ba_2FeWO_6	40	—
Ba_2CoReO_6	40	—
Ba_2MnWO_6	45	—
Ba_2CoWO_6	18	—
Sr_2MnMoO_6	12	—
Sr_2MnWO_6	13	—
Sr_2FeWO_6	40	—
Sr_2CoReO_6	65	—
Sr_2NiWO_6	54	—
Ca_2MnWO_6	16	—
Ca_2CoReO_6	130	—
Sr_2FeOsO_6	108 and 70	—
La_2NaRuO_6	30	—
$LaFe_{0.5}V_{0.5}O_3$	299	G
$NdFe_{0.5}V_{0.5}O_3$	304	G
$EuFe_{0.5}V_{0.5}O_3$	304	G
$YFe_{0.5}V_{0.5}O_3$	335	G
$YBa_2Fe_3O_{8+\delta}$	660	G
$NaFeF_3$	90	G

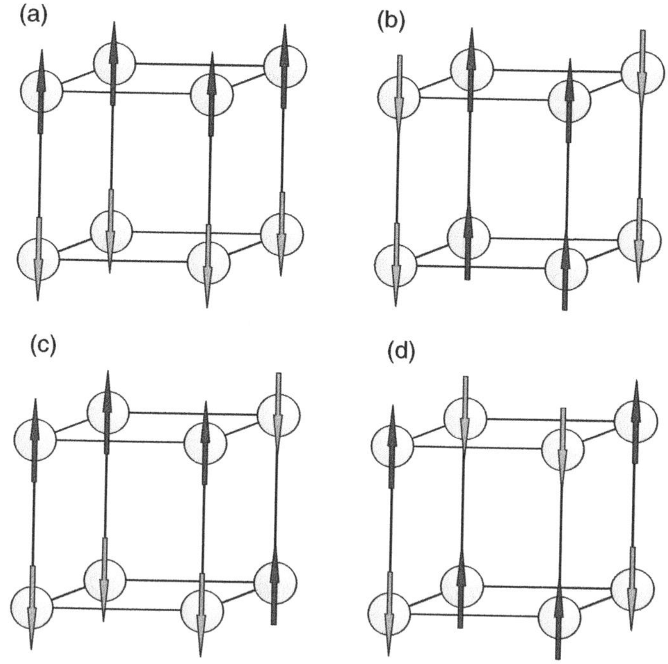

Figure 7.5 Antiferromagnetic spin ordering patterns, schematic, referred to an idealised cubic unit cell; (a) A-type $(001)_p$ alignment; (b) C-type $(110)_p$ alignment; (c) E-type; (d) G-type $(111)_p$ alignment

the unit cell is tetragonal: space group *P4mm* (99), $a = 0372937$ nm, $c = 0.472382$ nm. The C-type antiferromagnetic ordering of spins on the Co^{3+} cations at lower temperatures doubles the a- and b-parameters so that in the antiferromagnetic form $a \approx 0.7459$, $c \approx 0.4724$. The magnetic moments lie parallel to the tetragonal **c**-axis (Figure 7.6). The measured magnetic moment on the Co^{3+} cations, via inverse susceptibility plots, is 3.24 μ_B. This is somewhat lower than the theoretical spin-only value, 4.9 μ_B, which indicates that other interactions cannot be neglected.

Double perovskites with only one paramagnetic cation behave in a similar fashion except that antisite disorder between the two B cations can complicate matters. The phase Sr_2MgOsO_6 becomes antiferromagnetic at a Néel temperature of 110 K. When two (or more) magnetic cations are present, the ordering possibilities increase, although in many cases the two magnetic sublattices maintain an independent configuration. In the double perovskite phase Sr_2CoOsO_6 two different antiferromagnetic arrangements are found, the first with a T_N of 108 K and the second with a T_N of 70 K. At the first of these transitions, the magnetic moments on both the Co^{2+} (d^7, $S = 3/2$) and Os^{6+} ($5d^2$, $S = 1$) fluctuate dynamically but have an average long-range antiferromagnetic order. At the lower transition temperature, the Co^{2+} spins

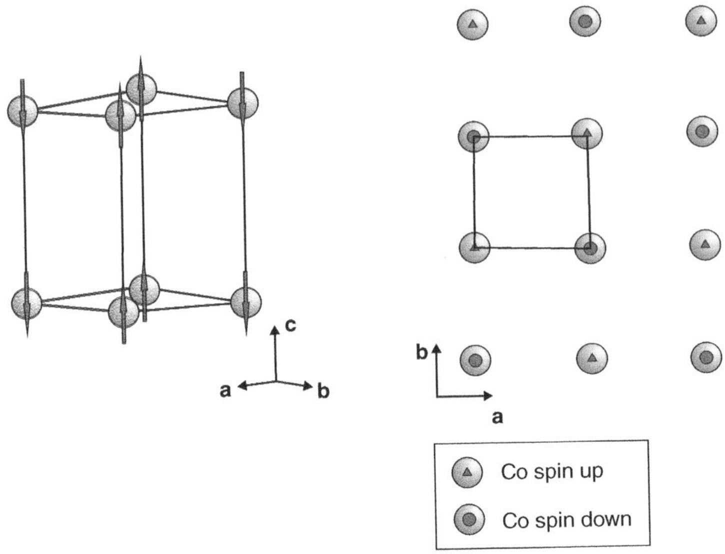

Figure 7.6 The C-type antiferromagnetic ordering in $BiCoO_3$. Only the Co^{3+} cations are shown

are frozen into a spin-canted state (Section 7.7). In these materials, the antiferromagnetic ordering is mediated by a superexchange between three cations, Os-O-Co-O-Os or Co-O-Os-O-Co.

The brownmillerite structure phases have either a G_x-type antiferromagnetic structure with the spin orientation along the shortest (usually **a**-) axis or G_y-type ordering with the spin orientation along the longest (usually **b**-) axis. The balance between these ordering patterns is delicately poised. The brownmillerite $Ca_2Fe_2O_5$, with a single magnetic cation, shows G_x-type ordering at all temperatures. The closely related brownmillerite Ca_2FeMnO_5, with two magnetic cations, shows G_y-type ordering at all temperatures up to a critical temperature of approximately 407 K (Figure 7.7). Above this temperature, short-range order is found until a temperature of 500 K is reached; after which typical paramagnetic behaviour ensues. The mixed phase $Ca_2Fe_{1.5}Mn_{0.5}O_5$ with an orthorhombic unit cell, space group *Pmna* (62), $a = 0.536319$ nm, $b = 1.50508$ nm, $c = 0.552838$ nm, adopts G_y-type ordering below approximately 400 K which changes to G_x-type ordering above 465 K. In this phase, the Mn^{3+} cations prefer octahedral coordination, while the tetrahedra contain only Fe^{3+} cations, and the antiferromagnetic arrangement occurs between both cation types. The phase $Ca_2Fe_{1.33}Mn_{0.67}O_5$ phase only adopts the G_y-type structure. The phase $SrCaFe_{1.5}Mn_{0.5}O_5$ also shows the same trend, with the transformation from low-temperature G_y-type changing to high-temperature G_x-type at approximately 400 K.

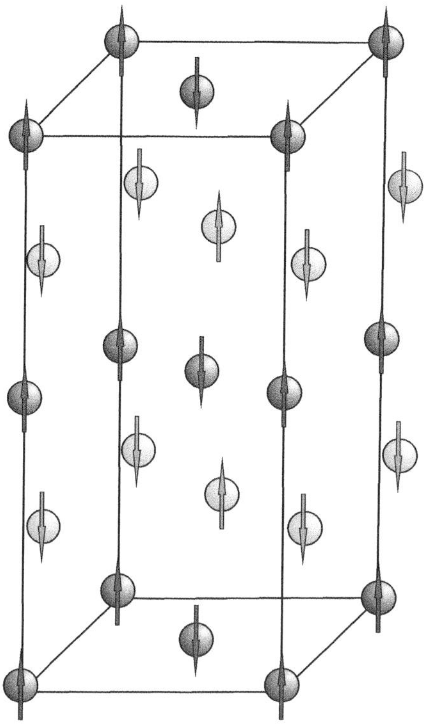

Figure 7.7 *G_y-type ordering in the brownmillerite Ca_2FeMnO_5. Only the B-site cations are shown. Each layer of cations parallel to the cell basal plane contains only one cation type, Mn or Fe*

The cubic $AA'_3B_4O_{12}$ ordered perovskites are of interest as the antiferromagnetic behaviour concerns the A′ cations even when other magnetic cations are present (Figure 7.8). The two phases $Ca_2Mn_3^{2+}V_4^{4+}O_{12}$ and $LaMn_3^{2+}V_4^{3.75+}O_{12}$ show Curie–Weiss behaviour suggestive of antiferromagnetic order at temperatures below approximately 54 K (Table 7.1). In both phases, the antiferromagnetic behaviour is solely due to the Mn^{2+} cations. The electrons on the V^{4+} cations are delocalised and do not correlate with those on the Mn^{2+} at all, causing the phases to show a low, almost metallic, resistivity. Similar behaviour is demonstrated by $LaMn_3^{3+}Mn_4^{3+}O_{12}$ and $LaMn_3^{3+}Cr_4^{3+}O_{12}$. In both of these phases, antiferromagnetic behaviour affects both the A′- and B-sites independently, and there is no correlation between the spins on either sublattice. In $LaMn_3Mn_4O_{12}$ the B-site Mn^{3+} spins order antiferromagnetically at 78 K, and the A′-site Mn^{3+} cations order independently in an antiferromagnetic arrangement at 21 K. In $LaMn_3Cr_4O_{12}$ the B-site Cr^{3+} cations order antiferromagnetically at 150 K, and the A′-site Mn^{3+} cations order antiferromagnetically at 50 K. The phase $CaCu_3Ti_4O_{12}$ shows G-type antiferromagnetism

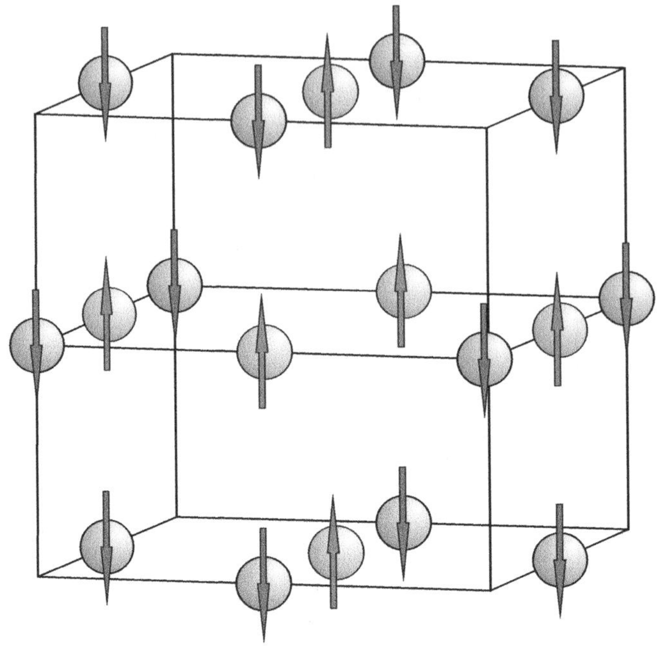

Figure 7.8 *The G-type antiferromagnetic order of the A′ cations in many $AA'_3B_4O_{12}$ cubic phases including $CaCu_3Ti_4O_{12}$*

with $T_N \sim 25$ K, as do the $Ln_{2/3}Cu_3Ti_4O_{12}$ phases, but with lower T_N values. The Ce member of this group is slightly anomalous because the Ce takes a 4+ valence giving a formula $Ce_{0.5}Cu_3Ti_4O_{12}$. In all of these copper phases, the Cu^{2+} cations are ordered antiferromagnetically in a G-type arrangement (Figure 7.8). In this material the A cations, Ce^{4+} and oxygen vacancies can order or take on a random arrangement. In the ordered version, below the Néel temperature the two Cu^{2+} sublattices differ slightly in their antiferromagnetic coupling to show a small ferrimagnetic moment between the temperatures of 24 and 10 K. This behaviour is termed L-type ferrimagnetism. Other ferrimagnetic $AA'_3B_4O_{12}$ phases are described in Section 7.5.

The lanthanoid titanates all adopt the $GdFeO_3$ structure and form a series which changes from G_x-type antiferromagnetic ordering for the larger cations, La to Sm, to ferromagnetic (Section 7.4) with the magnetic moment directed along the **c**-axis for Gd to Yb plus Y (Figure 7.9). Despite the fact that there are two magnetic cations present in most of these phases, the magnetic effects are mainly due to the Ti^{3+} $3d^1$ cations, as the lanthanoid cations have magnetic moments that order only at a much lower temperature than the Ti^{3+} cations. This means that the phases show two values of T_N, the lower associated with Ln spins and the higher with Ti spins.

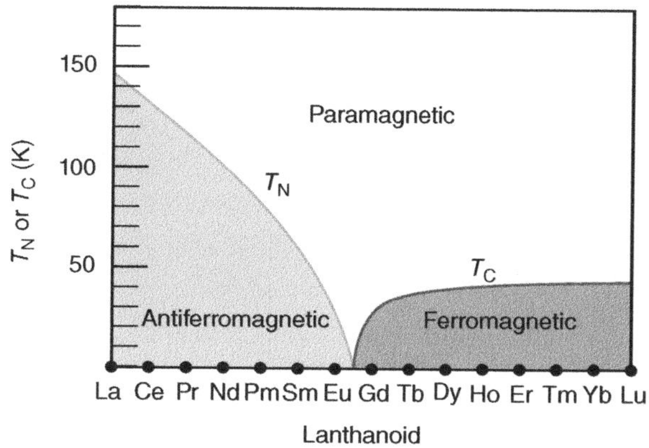

Figure 7.9 *Schematic magnetic phase diagram for the lanthanoid titanates, $LnTiO_3$. The values of the Curie temperature, T_C, and the Néel temperature, T_N, are somewhat variable and depend on sample stoichiometry and microstructure*

7.3.2 Hexagonal Perovskites

The structures of the hexagonal perovskites consist of columns of face-sharing octahedra or octahedra and trigonal prisms, usually linked by layers of apex-sharing octahedra or tetrahedra. In addition, the cation–anion–cation linkage is different than in the cubic-related structures so that both 90° and 180° superexchange geometries are possible. Any magnetic ordering can now occur between cations in any column (intra-column) or between columns (inter-column) or between the polyhedra that link the columns together. The ordering temperatures for these alternatives can be quite different. Thus, it is possible to have intra-column order at one temperature, while long-range order between the columns is lacking and only occurs at a second temperature. This may well give rise to two magnetic anomalies as the sample is cooled. For example, 4H-$SrMnO_3$ shows two antiferroelectric transitions, one at $T_{N1} = 350$ K, in which the magnetic moments between the face-sharing octahedra dimers align, and T_{N2} at 270 K, in which a three-dimensional order occurs as the face-centred units couple through the apex-sharing octahedra.

In the hexagonal ABX_3 polytypes, the 2H structure is the simplest. In 2H-$BaCoO_3$ the magnetic moments in the columns are ferromagnetically ordered, as expected from 90° superexchange, but there is an antiferromagnetic arrangement between adjacent columns, giving an overall antiferromagnetic structure (Figure 7.10a). This arrangement is not, however, universal. Despite the fact that the room temperature structure of 2H-$BaMnO_3$ is isostructural with that of 2H-$SrCoO_3$, the ordering of the magnetic moments on the Mn^{4+} cations is quite different than that shown by Co^{4+}. The low-temperature unit cell, applicable to the antiferromagnetic structure, is larger than that of the room temperature form, and the magnetic moments are

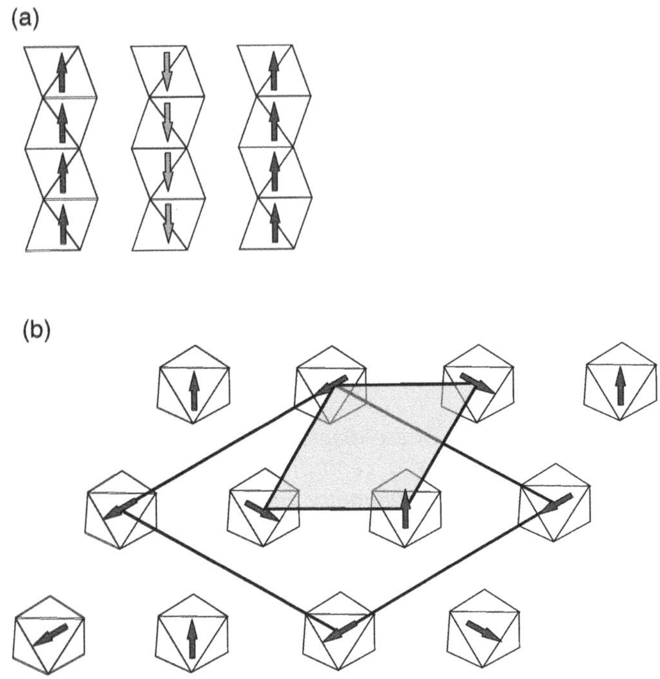

Figure 7.10 Antiferromagnetic alignment of magnetic moments in (a) 2H-$BaCoO_3$ and (b) 2H-$BaMnO_3$. In (b) the magnetic moments lie in (001) at z = 0; the magnetic moments on the cations at z = ½ are antiferromagnetically arranged. The unit cell of the room temperature 2H unit cell is shown shaded, and the unit cell of the low-temperature antiferromagnetic phase is outlined

constrained to lie in the (001) plane. The moments in each column are arranged antiferromagnetically, and the spin direction between each column is rotated by 120° (Figure 7.10b).

The ABX_3 polytypes that contain short columns of face-shared octahedra seem to prefer an antiferromagnetic intra- and inter-column ordering, exemplified by 4H-$BaMnO_3$ with pairs of face-sharing octahedra (Figure 7.11a), 9R-$Ba_{0.875}Sr_{0.125}MnO_3$ with triplets of face-sharing octahedra (Figure 7.11b) and 8H-$BaMnO_{2.95}$ with quartets of face-sharing octahedra (Figure 7.11c). Exactly the same type of antiferromagnetic ordering appears to occur in the hexagonal perovskites in which the face-sharing columns are linked by apex-sharing octahedra, as in 6H-$BaFeO_2F$ with pairs of face-sharing octahedra (Figure 7.12a) and 15R-$BaFeO_2F$ with double pairs of face-sharing octahedra (Figure 7.12b). In these compounds the preferred direction of the magnetic moment is apparently parallel to the basal (001) plane, although further studies are needed to confirm this definitively. These phases have high T_N values of 300–700 K.

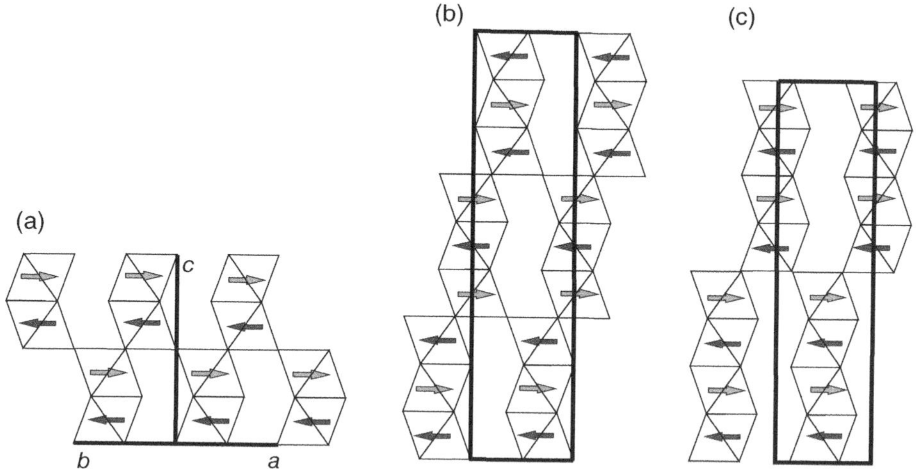

Figure 7.11 *Antiferromagnetic ordering in (a) 4H-$BaMnO_3$; (b) 9R-$Ba_{0.875}Sr_{0.125}MnO_3$; (c) 8H-$BaMnO_{2.95}$*

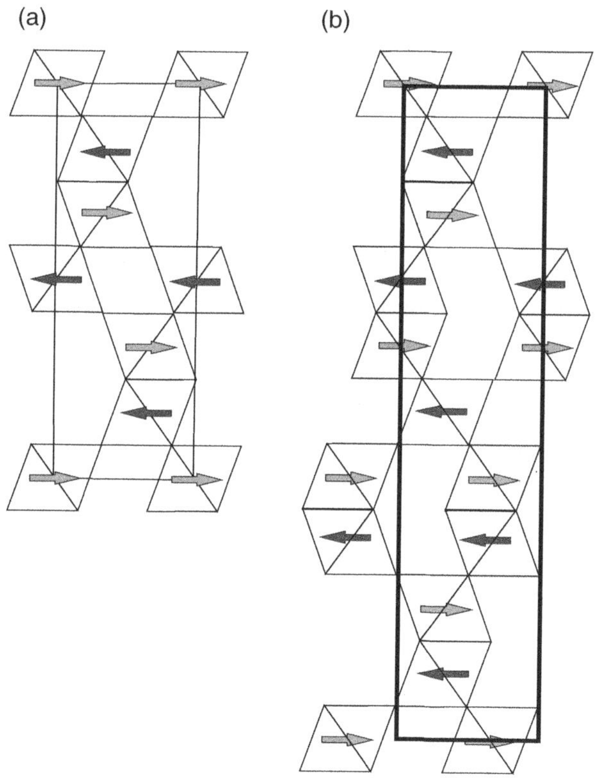

Figure 7.12 *Antiferromagnetic ordering in (a) 6H-$BaFeO_2F$; (b) 15R-$BaFeO_2F$*

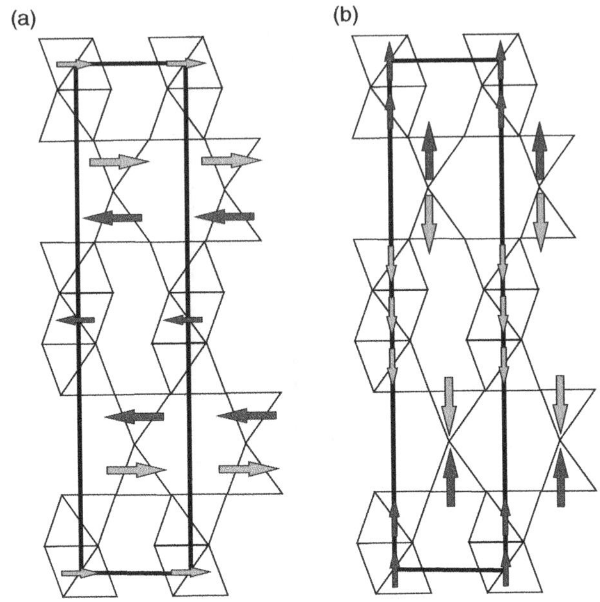

Figure 7.13 *Antiferromagnetic ordering in (a) $Ba_5Co_5ClO_{13}$; (b) $Ba_5Co_5FO_{13}$*

The hexagonal perovskite structures linked by tetrahedra or with wider spaces between the columns of face-sharing octahedra are equally diverse. Slight changes in stoichiometry or composition will easily alter the magnetic ordering of the phase. This aspect can be illustrated by the phases 10H-$Ba_5Co_5FO_{13}$ and 10H-$Ba_5Co_5ClO_{13}$ (Figure 7.13). In these structures, the Co is able to occupy octahedral and tetrahedral sites. In the phase containing F, the magnetic moments are perpendicular to the **c**-axis, and the main contribution comes from the Co in the tetrahedral sites. These contain Co^{4+} IS cations with a measured magnetic moment of 2.50 μ_B. The central octahedron of the triplets contains Co^{3+} LS cations with a measured magnetic moment of approximately 0.24 μ_B, while the outer pair of octahedra contains Co^{3+} LS cations with no measured magnetic moments. The phases with Cl as the X-site substituent show quite different ordering in which the magnetic moments align along the **c**-axis and all of the ions contribute to the magnetic structure. The tetrahedra contain Co^{4+} IS cations with a measured magnetic moment of 2.27 μ_B. The central octahedron of the column of triplets contains Co^{3+} LS cations with a measured magnetic moment of approximately 0.38 μ_B. The outer pair of octahedra in the column contains Co^{3+} HS cations with a measured magnetic moment of approximately 0.61 μ_B. The phase $Ba_6Co_6ClO_{15.5}$ with columns of four face-sharing octahedra is similar, with the two inner octahedra in the column enclosing Co^{3+} LS cations, while the Br-containing analogues are not antiferromagnetic at all.

7.4 Ferromagnetic Perovskites

Ferromagnetic perovskites are characterised by magnetic hysteresis loops. These are similar to those for ferroelectric materials (Figure 6.9) with the difference that the electric field is replaced by the magnetic field and the electric polarisation by the magnetic flux density. A number of ferromagnetic perovskites show metallic band-like ferromagnetism. The perovskite $SrRuO_3$ falls into this group and is a conduction band ferromagnetic phase below a Curie temperature, T_C, of 150 K. Above this temperature the phase becomes a Pauli paramagnetic compound, with almost no temperature dependence for the magnetic susceptibility. The phases $CaRuO_3$ and $LaNiO_3$ are also metal-like Pauli paramagnets but do not appear to order into a ferromagnetic phase at low temperatures (but see Section 7.8). The phase Sr_2FeMoO_6 is a more complex ferromagnetic, or possibly ferrimagnetic, phase described as a half-metal, with T_C approximately 415 K. Here the band structure of the 'spin-up' electrons is subtly different to that of the 'spin-down' electrons so that an energy gap at the Fermi level exists in just one of the two spin-polarised bands.

The first ferromagnetic metallic perovskites reported were the mixed valence $La_{1-x}A_xMnO_3$ phases, where A is an alkaline earth metal, especially Ca^{2+}. This behaviour is typified by the ferromagnetic metals $La_{0.7}Sr_{0.3}MnO_3$, T_C approximately 370 K, containing the cation pair Mn^{3+}/Mn^{4+}, and $La_{0.7}Sr_{0.3}CoO_3$, T_C approximately 220 K, containing Co^{3+}/Co^{4+}. $LaMnO_3$ itself, containing only HS Mn^{3+} (t_{2g}^3 e_g^1), is an antiferromagnetic insulator. In the substituted material, each Ca^{2+}-substituted La^{3+} is balanced by the formation of one Mn^{4+} in order to maintain charge balance. This necessity is confirmed by the fact that oxygen-rich $LaMnO_{3.12}$ is ferromagnetic with a Curie temperature T_C of approximately 170 K. As in the case of the alkaline earth substituted phases, each additional oxygen added to the structure is balanced by the introduction of two Mn^{4+} cations to give a nominal formula of $LaMn^{3+}_{1-x}Mn^{4+}_{2x}O_{3+x}$.

The puzzling appearance of metallic ferromagnetic behaviour was explained by the *double exchange* mechanism, devised by Zener in 1951. The key in double exchange, as in superexchange, is the mediation of the intervening oxygen anion which lies between two cations, of different valence, typically HS Mn^{3+} ($3d^4$, t_{2g}^4) and HS Mn^{4+} ($3d^3$, t_{2g}^3). In double exchange, one Mn^{3+} ion *transfers* an electron to an adjacent O^{2-} ion, but as this has filled p orbitals, to make this possible, the O^{2-} ion simultaneously transfers one of its electrons to a neighbouring Mn^{4+} ion. In essence, the electron hops from Mn^{3+} to Mn^{4+} via the intermediate oxygen ion (Figure 7.14). In the process the charges on the Mn ions are reversed and because all of the orbitals on the oxygen ion are full and electrons are spin paired, a spin-up electron moving from Mn^{3+} will displace a spin-up electron from O^{2-} onto Mn^{4+}. This double exchange is only favourable if there are parallel spins on the two cations involved, in which case it leads to ferromagnetic alignment. Note that in double exchange,

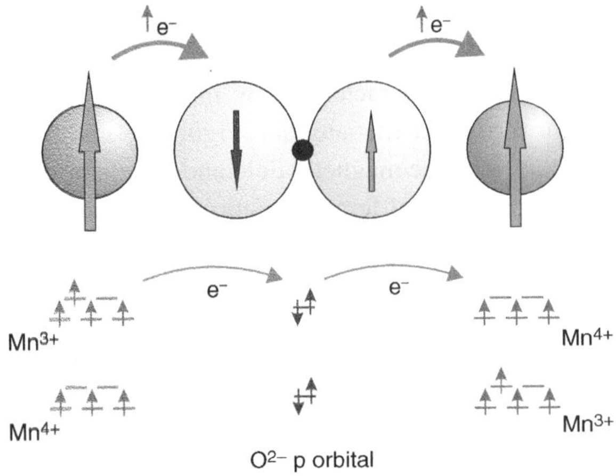

Figure 7.14 Double exchange, schematic: a cation, left, transfers a spin-up electron to an oxygen p orbital, centre (arrow), which simultaneously transfers a spin-up electron, right (arrow), to an empty orbital on the receiving cation, forcing ferromagnetic alignment of spins. The interaction is illustrated for HS Mn^{3+} ($3d^4$, t_{2g}^4) and HS Mn^{4+} ($3d^3$, t_{2g}^3) cations

electron transfer from one cation to another takes place (quite unlike the situation in superexchange), and ferromagnetic alignment always results.

In general, double exchange can operate in compounds that have a cation in two valence states. Many crystalline structures potentially fulfil these requirements, including the ion pairs already noted (Mn^{4+}/Mn^{3+} and Co^{2+}/Co^{3+}). However, pairs of cations are not a sufficient criterion for ferromagnetic order. In the alkaline earth doped manganites, when the composition reaches $Ln_{0.5}A_{0.5}MnO_3$ (where Ln is a lanthanoid and A is an alkaline earth cation), the Mn^{3+} and Mn^{4+} states can order. This charge-ordered (CO) state is generally found to be an antiferromagnetic insulator containing a chessboard type of arrangement of undistorted Mn^{3+} octahedra and Mn^{4+} Jahn–Teller elongated octahedra. The charge ordering interactions are temperature sensitive, and the CO state is lost above a charge ordering temperature.

Double exchange requires similar energies between the pair of cations involved and so is found not to operate between cations in tetrahedral sites and those in octahedral sites. The reason for this separation lies with the d-orbital energies, which are controlled by the crystal-field splitting. The d-orbital energy of a cation in an octahedral site is quite different from that of a cation in a tetrahedral site, and this difference prevents double exchange. However, the balance between antiferromagnetic and ferromagnetic ordering is delicate, and the small lanthanoid titanates $LnTiO_3$ are ferromagnetic, while the larger lanthanoids give rise to antiferromagnetic phases (Figure 7.9). However, the magnetic ordering in these antiferromagnetic phases is easily disrupted. This is well illustrated by $EuTiO_3$, which is a low-temperature antiferromagnetic phase. Electron doping produced by replacing O^{2-} by H^- in $EuTiO_{3-x}H_x$ will cause the material to become metallic and ferromagnetic with $T_C \sim 12$ K. The same effect is observed by electron doping via aliovalent substitution at the A-sites as in $Eu_{1-x}La_xTiO_3$ or by oxygen removal in $EuTiO_{3-\delta}$. The transformation to a ferromagnetic metal appears to be due to the interaction of the spins on nearest neighbour Eu^{2+} cations, mediated by itinerant conduction electrons.

A number of ferromagnetic double perovskites have fairly high Curie temperatures. The magnetic interactions are then between the two B- and B′-site cations with intermediate O^{2-} anions. The double and superexchange mechanisms do not totally explain this feature. In the double perovskite Sr_2FeMoO_6 it has been suggested that electron hopping between Fe^{3+} ($3d^5$, HS)–oxygen–Mo^{5+} ($4d^1$)–oxygen–Fe^{3+} all with parallel spins, is brought about by the spin-polarised energy band structure, resulting in a ferromagnetic metal (see also Chapter 8). However these interactions are delicately balanced and many 3d transition metal-containing double perovskites are paramagnetic or antiferromagnetic rather than ferromagnetic (Table 7.2).

Table 7.2 *Ferromagnetic perovskites and perovskite-related phases*

Perovskite	Curie temperature (K)[a]
$GdTiO_3$	36
$YTiO_3$	29
$La_{0.667}Sr_{0.333}CoO_3$	227
$La_{0.8}Ca_{0.2}MnO_3$	230
$La_{0.7}Ca_{0.3}MnO_3$	220
$La_{0.7}Sr_{0.3}MnO_3$	370
$SrCoO_3$	266
$Sr_{0.8}Ca_{0.2}CoO_3$	286
$SrRuO_3$	165
Ba_2MnReO_6	110
Ba_2FeMoO_6	308
Ba_2FeReO_6	303
Sr_2CrMoO_6	420
Sr_2CrReO_6	620
Sr_2CrWO_6	458
Sr_2FeMoO_6	420
Sr_2FeReO_6	400
Ca_2CrMoO_6	148
Ca_2CrReO_6	360
Ca_2CrWO_6	160
Ca_2MnReO_6	110
Ca_2FeMoO_6	365
Ca_2FeReO_6	522

[a]The measured Curie temperature depends upon the exact composition, defect population and sample microstructure. Literature values vary quite widely.

7.5 Ferrimagnetic Perovskites

The magnetic moments in a number of perovskites take on the more complex ferrimagnetic ordering, which is similar to antiferromagnetic order but with the magnetic moments in each direction unbalanced. This ordering pattern is pronounced in a number of $ACu_3Fe_4O_{12}$ perovskites, with A as an alkaline earth cation. In the phase $CaCu_3Fe_4O_{12}$, the high-temperature (metallic) phase has the nominal valence states of Cu^{2+} and Fe^{4+}. Cooling below a Curie temperature of 210 K results in disproportionation of the charges from Fe^{4+} to Fe^{3+}/Fe^{5+} and the formation of a semiconducting ferrimagnetic phase with the spin arrangement [Cu_3^{2+} (↓) Fe_2^{3+} (↑) Fe_2^{5+} (↑)]. A similar ferrimagnetic structure has also been reported for $CaCu_3Mn_4O_{12}$.

The lanthanoid perovskites $LnCu_3Fe_4O_{12}$ are similar but show a change in behaviour dependent upon the size of the lanthanoid cations. At high temperatures, above approximately 250 K, all are paramagnetic metals. The nominal valence states of the transition metal cations are Cu^{2+} and $Fe^{3.75+}$. Below this temperature compounds containing the larger lanthanoids, La, Pr, Nd, Sm, Eu, Gd and Tb, all show antiferromagnetic order with sharp T_N values of approximately 370 K (La) decreasing to approximately 240 K (Gd). The change is due to an intersite charge transfer between the transition metal cations:

$$3Cu^{2+} + 4Fe^{3.75+} \rightarrow 3Cu^{3+} + 4Fe^{3+}$$

Phases containing the smaller lanthanoids, Dy, Ho, Er, Tm, Yb and Lu, transform to a ferrimagnetic state at approximately $T_C \sim 240$ K for all compounds. This is due to a charge disproportionation between the transition metal cations:

$$8Fe^{3.75+} \rightarrow 5Fe^{3+} + 3Fe^{5+}$$

The ferrimagnetic ordering proposed is Cu^{2+} (↓), Fe^{5+} (↑), Fe_1^{3+} (↑), Fe_2^{3+} (↓). This ordering pattern is also found in $YCu_3Fe_4O_{12}$, which has a similar charge disproportionation at 250 K and undergoes the same ferrimagnetic and metal–semiconductor transition. Although the magnetic behaviour of these materials is complicated by the presence of magnetic lanthanoid ions, the electronic and magnetic phase fields occupied by the $LnCu_3Fe_4O_{12}$ compounds are clearly differentiated (Figure 7.15). The mixed B-site cation phases such as $CaCu_3Fe_2Nb_2O_{12}$ also show ferrimagnetic behaviour as do a number of other materials with this structure, such as $Ce_{0.5}Cu_3Ti_4O_{12}$ at low temperature.

A number of double perovskites are ferrimagnetic, a feature arising when there are two magnetic cations on the B-sites. Examples include the half-metallic Sr_2FeMoO_6, $T_C \sim 415$ K (also described as ferromagnetic; Section 7.4); Sr_2CrWO_6, $T_C \sim 450$–500 K; Sr_2CrReO_6, $T_C \sim 625$ K; and the insulating Sr_2CrOsO_6, $T_C \sim 725$ K, and Ca_2FeOsO_6, $T_C \sim 320$ K.

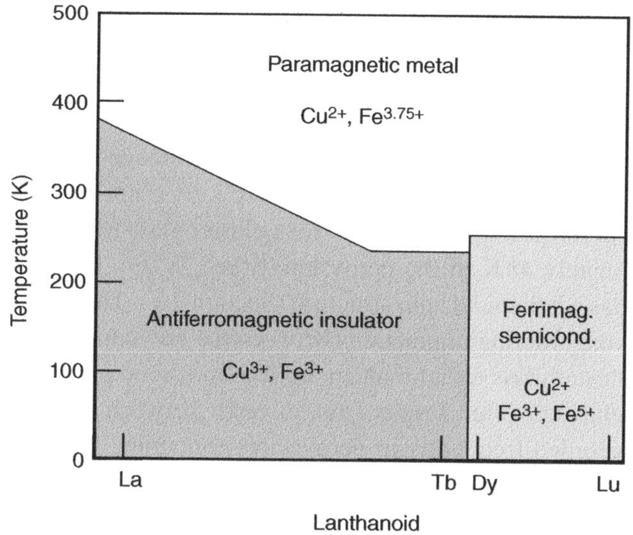

Figure 7.15 *Electronic and magnetic phase fields of* $LnCu_3Fe_4O_{12}$ *compounds*

7.6 Spin Glass Behaviour

A paramagnetic perovskite has the magnetic moments of the ions arranged at random in a dynamic fashion so that the orientation of any particular cation dipole is always changing. The balance between dynamic reorientation and magnetic interaction is often delicate, and for a number of systems, the result of this undecided conflict is the formation of a spin glass. In this form, the magnetic moments on the ions are oriented at random, as in a paramagnetic phase. However, they are no longer dynamically changing but are frozen in place below a temperature called the freezing or glass-forming temperature, T_f or T_g. If the microstructure of the phase is not homogeneous but has small volumes of spin glass-like structure, the material is described as a cluster glass or a spin cluster glass. That is, in a cluster glass the frozen disordered magnetic moments are not distributed evenly throughout the crystal but form more or less discrete clusters within it.

The low resistivity, almost metallic, phase $NaMn_3^{2.33+}V_4^{4+}O_{12}$ behaves as a spin glass at low temperatures. At moderate temperatures the inverse susceptibility curve shows Curie–Weiss behaviour, but at the lowest temperatures a spin glass state forms due to the complex magnetic interactions between the A′-site cations that are in mixed valence states Mn^{2+}, Mn^{3+} and Mn^{4+}. When only two valence states are present, ferromagnetic or antiferromagnetic ordering via double exchange or superexchange is probable, but with three competing valence states, the balance between the two forms is unresolved and a spin glass forms. Similar spin glass behaviour is encountered in $LaMn_3Ti_4O_{12}$ in which the Mn cations are forced to take charges of Mn^{3+}, Mn^{2+} and Mn^{+}. Note that it is the interactions between the A′-site cations that give rise to the spin glass state and there is no correlation between the cations on the A′- and B-sites in these cases.

The double perovskite Sr_2MgOsO_6 becomes antiferromagnetic at a Néel temperature of 110 K. Curiously the similar Ca_2MgOsO_6 does not form a magnetically ordered phase but forms a spin glass at 19 K. A cluster glass forms below a temperature of approximately 48 K in the perovskite $SrFe_{0.9}Ti_{0.1}O_3$. Here the parent compound $SrFeO_3$ has a helicoidal spin structure (Section 7.7). The replacement of Fe^{4+} by Ti^{4+} disrupts the ordering interactions between the Fe^{4+} cations resulting in a low temperature a cluster glass rather than an antiferromagnetic phase.

Spin glass behaviour often results from disorder between paramagnetic cations and has been recorded frequently in perovskites containing both Mn and Ru. For example, although the hexagonal 6H-$BaRuO_3$ is paramagnetic at all temperatures, phases in the solid solution 6H-$BaRu_{1-x}Mn_xO_3$ with x between 0.1 and 0.5 all show spin glass behaviour, with the freezing temperature, T_f, increasing to a maximum of approximately 50 K at $x=0.5$.

7.7 Canted Spins and Other Magnetic Ordering

A number of other possible magnetic ordering patterns are also known. In some antiferromagnetic phases, the colinearity of the magnetic dipoles does not persist to low temperatures, and the spins misalign by a small amount, 1° or so, to give a canted arrangement. The antiferromagnetic double perovskite Sr_2CoOsO_6 shows two low-temperature Néel temperatures; the first, at approximately 108 K, signals the antiferromagnetic ordering of the average effective magnetic moments in the structure, while the second, at approximately 67 K, is due to the magnetic moments on the Co^{2+} freezing into a canted antiferromagnetic state. Phases exhibiting low-temperature canting usually show some degree of weak ferromagnetism due to this misalignment.

The perovskites $TbCoO_3$ and $DyCoO_3$ show antiferromagnetic ordering due to the moments on the lanthanoids, Tb^{3+} ($4f^8$) and Dy^{3+} ($4f^9$). These perovskites both adopt the $GdFeO_3$ structure at all temperatures. At the lowest temperatures, $T_N=3.3$ K for $TbCoO_3$ and 3.6 K for $DyCoO_3$, the magnetic moments order into a complex canted antiferromagnetic pattern which can be broken down into the combination of both A-type and G-type antiferromagnetic ordering, in which the magnetic moments lie in the (001) plane and are at an angle to the **a**- and **b**-axes (Figure 7.16).

Canted spin alignment is not restricted to oxides. The perovskites $NaMnF_3$, $NaFeF_3$, $NaCoF_3$ and $NaNiF_3$ all show G-type antiferromagnetic arrangement of the

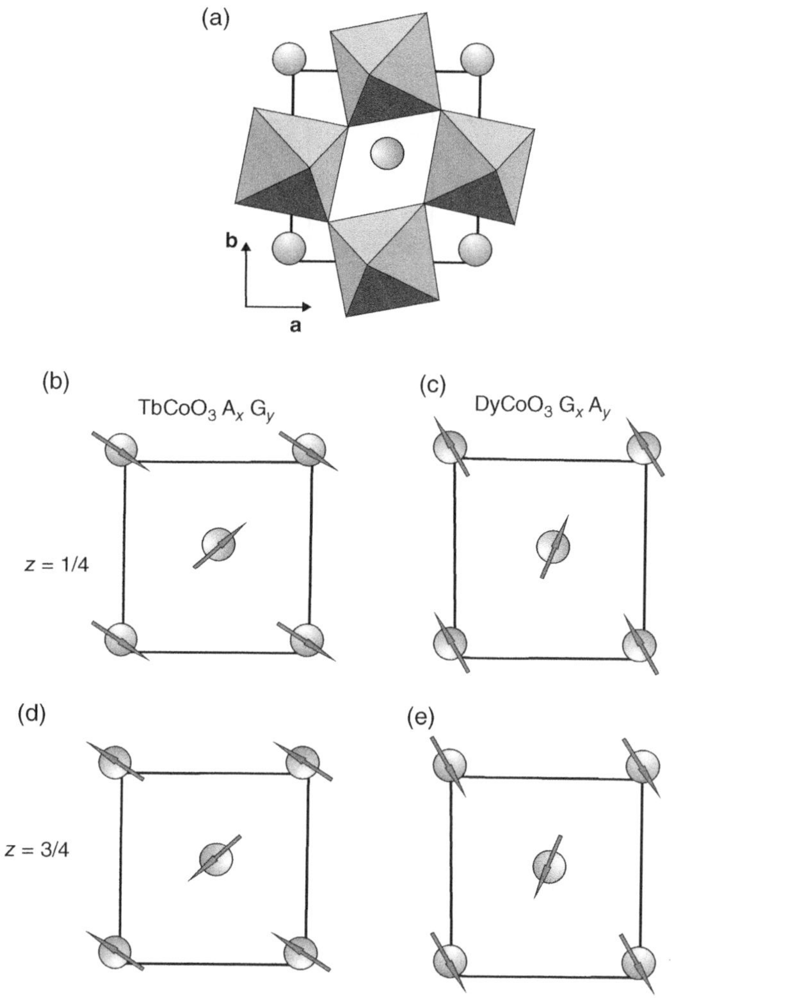

Figure 7.16 *Canted spin moments in $TbCoO_3$ and $DyCoO_3$: (a) structure projected down (001); (b and d) canted spin orientation on Tb^{3+} at z = ¼, ¾; (c and e) canted spin orientation ion Dy^{3+} at z = ¼, ¾ (Data from Knizek* et al. *(2014))*

spins. However, canting of the magnetic moments results in a weak ferromagnetic order at low temperatures. The hexagonal perovskite $K_4Fe_3F_{12}$, which has the $La_4Ti_3O_{12}$ structure (Figure 3.17a), has two sets of misaligned spins, one on the Fe^{2+} cations and one on the Fe^{3+} cations. These form two independent ferromagnetically aligned arrays below a magnetic transition temperature of approximately 120 K. This ordering has been described as pseudoferrimagnetic (Figure 7.17). The organic–inorganic perovskite $(C_2H_5NH_3)_2FeCl_4$ is also antiferromagnetic $T_N = 98.9$ K, with spins canted at 0.6° towards the **c**-axis.

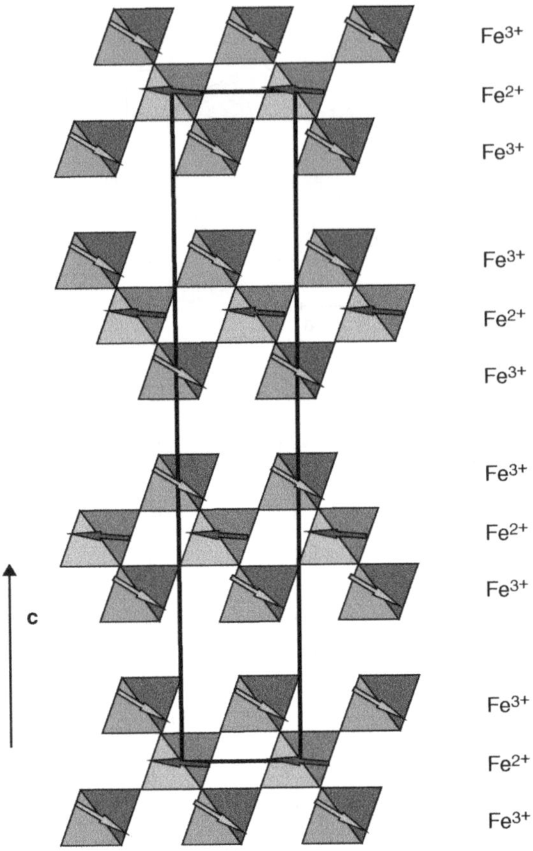

Figure 7.17 *The spin arrangement in* $K_4Fe_3F_{12}$

If the canting is not the same for each magnetic cation but varies in a regular way, then a number of commensurate or incommensurate spin structures can arise, including helical, helicoidal, cycloidal and sinusoidal. The helicoidal and sinusoidal ordering patterns are illustrated by the magnetic structure of $TbMnO_3$ (Section 7.10).

7.8 Thin Films

The balance between the ordering forces involved in various magnetic alignments is very delicately poised, and one form can be destabilised with respect to one another in many ways. The situation is modified in thin films and will depend upon the film thickness, the substrate carrying the film, the interfacial region between film and substrate and, in multilayer films, the number and spacing of the layers. In instances where these interfaces can be controlled to give sharp boundaries, the interfaces can be manipulated to give rise to many potentially useful (and frequently unexpected) properties. In particular, the properties of an epitaxial grown single crystal thin film of a perovskite will be strongly affected by the lattice parameter mismatch at the interface, leading to epitaxial strain, which can be engineered by changing both the film thickness and the substrate. A few illustrative examples follow.

The orthorhombic $GdFeO_3$-type perovskite $LuMnO_3$ is an E-type antiferromagnetic perovskite. Thin films of this phase grown epitaxially on the non-magnetic perovskite $YAlO_3$ can, however, show coexisting ferromagnetic and antiferromagnetic regions. The reason for this is lattice strain at the interface between the $YAlO_3$ and $LuMnO_3$ phases. For films grown on (110) planes of the substrate (Figure 7.18a), the lattice parameters of $YAlO_3$ are smaller than those of $LuMnO_3$. The cell diagonal of $YAlO_3$ is 0.742 nm, while the same diagonal in $LuMnO_3$ is 0.774 nm (Figure 7.18b). The $LuMnO_3$ film is thus in compression at the interface with $YAlO_3$ when the two are constrained to fit together. The compressive strain has the effect of inducing a monoclinic distortion in the unit cell of $LuMnO_3$, which causes changes in the Mn—O bonding, the Mn orbital ordering and the octahedral tilting in the region of the film closest to the substrate. The result is a thin ferromagnetic region, of the

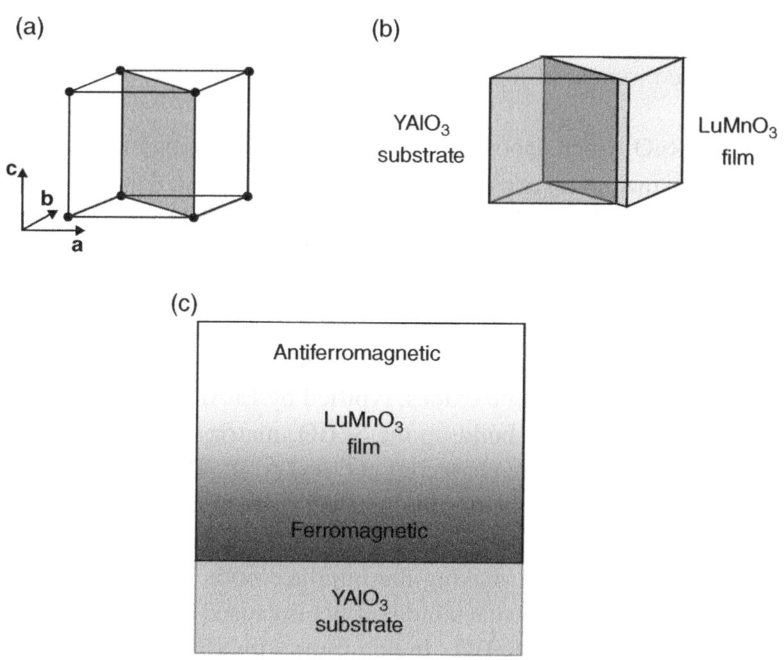

Figure 7.18 $LuMnO_3$ thin films: (a) the (110) plane of $LuMnO_3$ and $YAlO_3$; (b) misfit along the (110) planes of $LuMnO_3$ and $YAlO_3$; (c) magnetic properties of a $LuMnO_3$ film on $YAlO_3$

order of 10 nm in depth that successively transforms to antiferromagnetic ordering on moving away from the join (Figure 7.18c).

A similar situation exists in films of $La_{0.67}Ca_{0.33}MnO_3$. In the bulk the phase is a ferromagnetic metal with a Curie temperature of approximately 370 K. Thin films, though, have properties that depend in part upon the substrate. A thin film grown on $LaAlO_3$ is under a compressive stress, which makes the film a CO insulator at low temperatures. The application of a magnetic field of several tesla will cause this to revert to the ferromagnetic metallic form again. The transformation comes about by the formations of islands of metal within the CO insulating matrix so that regions experiencing high stress remain insulating while regions of low stress are ferromagnetic and metallic. Films of the same material which are less than approximately 4 nm in thickness on a $SrTiO_3$ substrate show canted spins and a Curie temperature T_C of approximately 540 K. The same material grown on $BaTiO_3$ suffers a greater degree of strain, the consequence of which is that the Curie temperature T_C rises to 650 K.

A number of perovskite films show ferromagnetism at the interface with the substrate. The metallic perovskite $CaRuO_3$ is paramagnetic, but thin films grown on $SrTiO_3$ (100) faces are subjected to tensile strain in the epitaxial layers, leading to ferromagnetism. The magnetic moment increases with increasing strain. Films of antiferromagnetic (001) $TbMnO_3$ grown on (001) $SrTiO_3$ are under interfacial strain. In this case, the strain is relieved by the formation of ferromagnetic domain walls. The net magnetic moment of the films, due to these domain walls, decreases with film thickness.

The phase $CaRuO_3$ when deposited on insulating antiferromagnetic $CaMnO_3$, the antiferromagnetic insulating perovskites $LaMnO_3$ and $LaFeO_3$ deposited on $LaCrO_3$ and antiferromagnetic $LaMnO_3$ films on diamagnetic $SrTiO_3$ all show ferromagnetic regions close to the substrate interface. In these cases the causes of this phenomenon are thought to be superexchange between cations in the boundary or charge transfer across the boundary rather than interfacial strain.

Rather more surprising is the generation of magnetic interfaces between thin films of non-magnetic perovskite oxides, typified by $LaAlO_3$ deposited on $SrTiO_3$. The SrO and TiO_2 layers that build up the $SrTiO_3$ matrix have no charge, but the equivalent layers in $LaAlO_3$ are charged: LaO^+ and AlO_2^-. This imbalance results in electron transfer of 0.5 e per perovskite unit cell from $LaAlO_3$ to Ti 3d levels in $SrTiO_3$. This gives rise to a ferromagnetic interface, which is also associated with a two-dimensional electron gas, making the interface both ferromagnetic and electronically conducting. A somewhat similar effect is caused by oxygen deficiency in surface layers of the insulator $SrTiO_3$. In thin films with a TiO_2 termination, oxygen loss creates surface vacancies which have Ti^{3+} neighbours. These spins are aligned in an antiferromagnetic way in the surfaces, due to superexchange coupling via neighbouring oxygen anions and anion vacancies.

In superlattices of differing perovskite oxides, these effects are frequently magnified and, dependent upon layer thickness, can produce a variety of magnetic ordering patterns.

7.9 Nanoparticles

As described in the previous section, decreasing dimensions can have important consequences for magnetic properties. In the case of nanoparticles, the surface starts to dominate the bulk as the particle size decreases. The surface differs from the core in that cation polyhedral coordination changes, there is often a composition gradient, and surface defects are commonly present. All of this results in the exchange interactions that are vital for magnetic order being significantly weakened. Compared to the bulk, nanoparticles display a large coercivity, low field saturated magnetisation and a lowering of the Curie temperature. This often results in the particle being made up of two regions, a core and a shell. The magnetic structure of the shell is invariably different from that of the core and can include magnetic disorder, spin glass behaviour, superparamagnetism or relatively well-aligned magnetic moments. It can be appreciated that preparation techniques may well influence ordering in the shell region of the particles.

Nanoparticles of the bulk ferromagnetic phase $La_{0.6}Sr_{0.4}MnO_3$ develop a core–shell structure when particle diameters drop to 40 nm or less, with the shell thickness increasing as particle diameter falls. Both the core and the shell of the nanoparticles contain 60% Mn^{3+} (t_{2g}^4, HS) and 40% Mn^{4+} (t_{2g}^3, HS). All appear to have a ferromagnetic single domain core structure similar to the bulk, but the surface 'shell' adopts an antiferromagnetic alignment. As the particle size decreases the relative shell thickness, hence the antiferromagnetic proportion increases. In the closely related $La_{0.8}Sr_{0.2}MnO_3$, it appears that the nanoparticle core is ferromagnetic, but the shell now contains disordered spins. Nanoparticles of the hexagonal perovskite 4H-$SrMnO_3$, with an (hchc) layer packing are found to contain increasing numbers of extra c layers as oxygen is lost. The magnetic transitions observed in the bulk are broadened, and extra transitions are noted from shell octahedra that behave differently with respect to magnetic ordering than the core octahedra.

7.10 Multiferroic Perovskites

Ferroic is a generic term for crystals that have an internal structure that can be switched from one stable orientation to another that is equally stable, by the application of a suitable driving force along an appropriate direction. Multiferroics are single phases that exhibit more than one ferroic property simultaneously. A number of perovskites, especially $BiFeO_3$, $BiMnO_3$, $TbMnO_3$ and $HoMnO_3$, are multiferroic. They have been most explored with respect to ferroelectric/ferromagnetic behaviour, in which spontaneous electrical polarisation is switched by a magnetic field and spontaneous magnetisation is switched by an electric field. This is, in fact, an example of the magneto-electric effect, which more generally refers to the initiation of magnetisation in a body by the application of an electric field or the generation of a ferroelectric polarisation in a body by the application of a magnetic field.

At first sight multiferroic properties would be thought to be incompatible with one another. Generally a ferroelectric polarisation is associated with cation displacement in a (frequently) distorted octahedral coordination polyhedron, and this is favoured by cations with an empty d shell, typically Ti^{4+}, as in $BaTiO_3$ and $PbTiO_3$. On the contrary, ferromagnetism requires partly filled d orbitals, typically Co^{4+} as in $SrCoO_3$. The general solution to marrying these disparate properties in a single phase is to use two different cations, one for each role. Thus $BiFeO_3$ employs Bi^{3+}, with a lone pair of electrons, as the cation responsible for the ferroelectric component, while Fe^{3+} is responsible for the magnetic moment. Albeit that this approach is a useful working guide, in practice, the interactions are far more complex. This is well illustrated by the multiferroic perovskite $TbMnO_3$.

The perovskite $TbMnO_3$ adopts the $GdFeO_3$ structure with $a \approx 0.5302$ nm, $b \approx 5856$ nm, $c \approx 0.7400$ nm (Figure 7.19a). The phase undergoes a transition from a paramagnetic state to an ordered magnetic state at approximately 42 K. In this magnetic phase the magnetic moments on the Tb^{3+} ($4f^8$) are arranged in an incommensurate sinusoidal wave along the **b**-axis (Figure 7.19b), although the magnetic moments have components along both [100] and [010]. At a temperature of approximately 28 K, the Mn^{3+} spins adopt a cycloidal geometry, in which the spins rotate around [100] (Figure 7.19c and d). At the same temperature, a spontaneous electric polarisation, E, arises, with a direction along [001].

The perovskite $BiFeO_3$ is a ferroelectric phase above a temperature T_C of approximately 1100 K due to the Bi^{3+} lone pair of electrons, which gives rise to polarisation along <111>. This phase becomes a G-type antiferromagnetic phase with $T_N \sim 640$ K. The magnetic moments, however, are not strictly antiparallel but, as in $TbMnO_3$, adopt an incommensurate spin cycloid configuration. The closely related $BiMnO_3$ is ferroelectric below 450 K and becomes ferromagnetic below 105 K. The low-temperature magnetic perovskite $BiCoO_3$ is not a multiferroic phase although it generates a spontaneous electric polarisation at low temperatures, because the direction of the polarisation cannot be switched. It is, therefore, classed as a pyroelectric material (Chapter 6).

Multiferroic behaviour is not confined to simple perovskite structures. Thin films of the more complex layered ferroelectric perovskite-derived Aurivillius phase

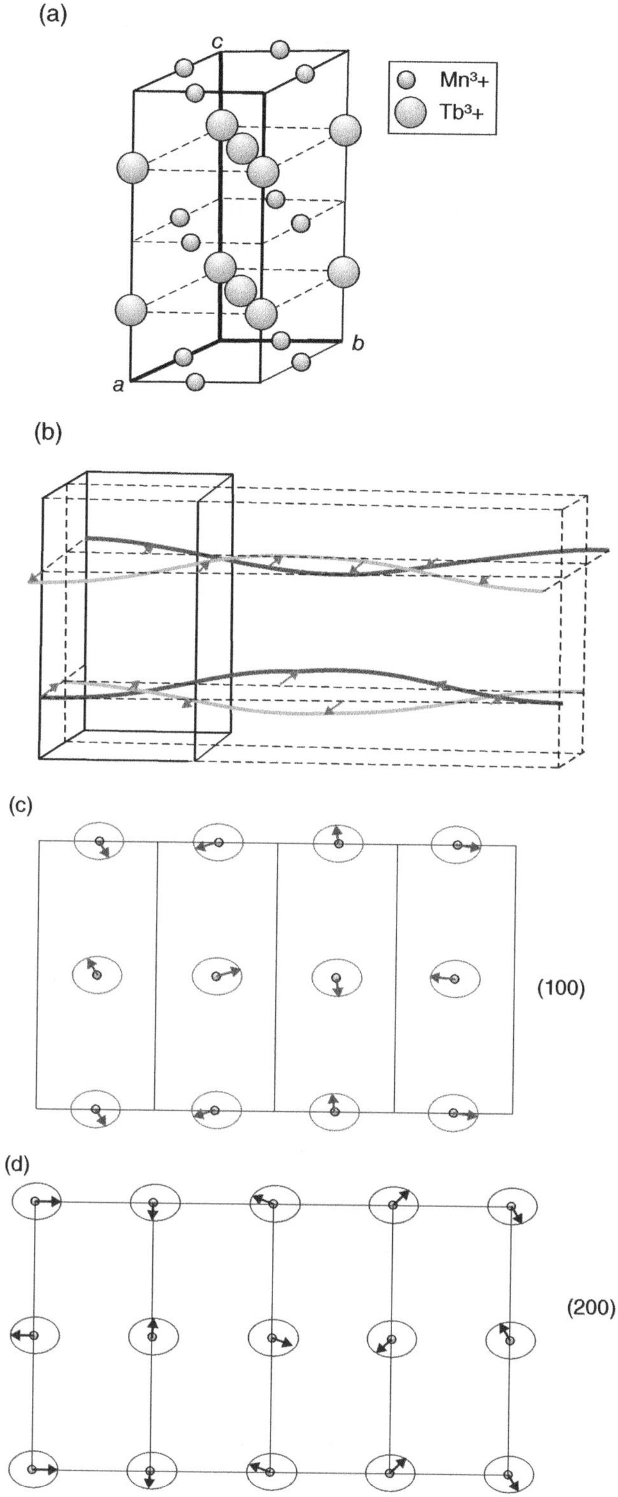

Figure 7.19 *$TbMnO_3$: (a) crystal structure; (b) schematic depiction of the incommensurate sinusoidal spin ordering on Tb^{3+} lying on (100) and (200) below approximately 42 K; (c and d) schematic depiction of the cycloidal spin geometry of Mn^{3+} projected onto (100) and (200)*

$Bi_6Ti_{2.8}Fe_{1.52}Mn_{0.68}O_{18}$ also show multiferroic behaviour and undergo ferroelectric domain switching following the application of a magnetic field at room temperature. In this phase, the magnetic component appears to reside in the Mn^{4+} species rather than the Fe^{3+} cations. The Aurivillius phase $Sr_{0.25}Bi_{6.75}Fe_{1.5}Co_{1.5}Ti_3O_{12}$ is also multiferroic, showing ferromagnetic and ferroelectric behaviour at room temperature.

References

K. Knizek *et al.*, *Solid State Sci.*, **28**, 26–30 (2014).
T. Yamamoto *et al.*, *Inorg. Chem.*, **54**, 1501–1507 (2015).

Further Reading

For a concise introduction to magnetic materials, see:
R. J. D. Tilley, Understanding Solids, 2nd edition, John Wiley & Sons, Ltd, Chichester (2013), Chapter 12.

For more details see:
J. M. D. Coey, Magnetism and Magnetic Materials, Cambridge University Press, Cambridge (2010).
N. Spaldin, Magnetic Materials, Cambridge University Press, Cambridge (2003).

For an introduction to magnetochemistry and the method of obtaining the magnetic moments of paramagnetic cations in solids, see:
A. F. Orchard, Magnetochemistry: Oxford Chemistry Primers, Oxford University Press, Oxford (2003).

For a review of double perovskites which are ferromagnetic above room temperature, see:
D. Serrate, J. M. de Teresa, M. R. Ibarra, *J. Phys. Condens. Matter*, **19**, 023201 (2003).

For the effect of interfaces on magnetic properties of thin films, see:
H. Y. Hwang *et al.*, Emergent phenomena at oxide interfaces, *Nat. Mater.*, **11**, 103–113 (2012).
Materials Research Society Bulletin, **38**, December (2013), various authors.
S. Farokhipoor *et al.*, *Nature*, **515**, 379–383 (2014).

For information on multiferroic perovskites, see:
N. Hill, *J. Phys. Chem.*, **B104**, 6694–6700 (2000).
L. Keeney *et al.*, *J. Am. Ceram. Soc.*, **96**, 2339–2357 (2013).
Y. Tokura, *Science*, **312**, 1481–1482 (2006).

8
Electronic Conductivity

In this chapter the focus is upon electronic conductivity in perovskites. The electrons in perovskites are believed to be strongly correlated; that is, they do not behave as a classical electron gas, but are the subject to electron–electron interactions. This leads to considerable modification of the collective electron behaviour of the conduction electrons, resulting in metal–insulator transitions, high-temperature superconductivity, half-metals and colossal magnetoresistance (CMR). The effects of strong correlation are important for the 3d, 4d and 4f elements. In many ways the topics described here are thus a continuation of the previous chapter on magnetic perovskites, and in truth the two subject areas cannot be separated in a hard and fast manner.

8.1 Perovskite Band Structure: Metallic Perovskites

In the earliest theories of the solid state, electrons were perceived as being free and non-interacting – an electron gas. In this model, the atomic orbitals of the component atoms are spread out into energy bands, the detailed form of which depends upon the crystal structure of the phase. An upper energy band which is only partly filled with electrons characterises a metal with itinerant (freely moving) non-interacting electrons.

As most perovskite studies have focussed on d-electron-containing oxides, these will mainly feature here. The band structures of these compounds can be calculated

Perovskites: Structure–Property Relationships, First Edition. Richard J. D. Tilley.

using the standard methods of solid-state physics, especially density functional theory. The resulting diagrams are often complex, but in the main it is possible to interpret the conductivity in a broad-brush fashion by using a simplified generic band structure diagram.

The band structure of insulating cubic $SrTiO_3$ (Figure 8.1a) may be regarded in this light. The lower group of bands, constituting the valence band of the crystal, consists of nine components mainly derived from the 2p orbitals of the three O atoms in the unit cell. The Fermi level lies just at the top of this energy band. The next set of bands, making up the bottom of the conduction band, is formed from the 3d orbitals on the octahedrally coordinated Ti atoms. The lower energy bands consist of a triply degenerate set mainly derived from the 3d t_{2g} orbitals. Above and overlapping with this is a doubly degenerate pair of bands largely derived from the 3d e_g orbitals. There is a band gap between the top of the valence band and the bottom of the conduction band. Finally at highest energies are bands derived from the Sr 5s and 5p orbitals, and also, not shown, there may be bands that derive from the 4s and 4p orbitals on the Ti atoms. By far the most important bands from the point of view of electronic properties are the O-derived valence band and the Ti 3d-derived conduction band. In $SrTiO_3$ the valence band is filled, while the t_{2g} and e_g components of the conduction band are empty as Ti^{4+} is a d^0 ion.

The density of states $N(E)$ as a function of energy across the bands (Figure 8.1b) gives an appreciation of the spread of energy levels in the various bands. This can be converted into the more severe 'flat band' approximation of the band structure and density of states (Figure 8.1c and d) that is familiar from discussions of elemental semiconductors.

In $SrTiO_3$, the filled valence band and the empty conduction band, separated by an appreciable band gap, give rise to an insulating phase. This picture also applies to other related phases such as $BaTiO_3$ and $PbTiO_3$. However, in these terms, any B-site cation with partly filled d orbitals would be expected to be metallic, with itinerant electron conductivity, as the d-electrons will partly fill the t_{2g} and e_g conduction bands. This is indeed true of many perovskites – including $CaVO_3$, $SrVO_3$, $CdVO_3$ and $MnVO_3$, with V^{4+} ($3d^1$); $SrFeO_3$, with Fe^{4+} ($3d^4$); $LaNiO_3$ with Ni^{3+} ($3d^7$); $NaOsO_3$, with Os^{5+} ($5d^3$); and $CaCu_3Ir_4O_{12}$, with Ir^{4+} ($5d^5$) – and many cobaltites containing Co^{2+}, ($3d^7$), Co^{3+} ($3d^6$), Co^{4+} ($3d^5$) and manganites containing Mn^{2+} ($3d^5$), Mn^{3+} ($3d^4$) and Mn^{4+} ($3d^3$).

The scheme described must be modified when layered perovskites are considered. In the layered phases, the band structure gradually changes from that of a two-dimensional layer structure, say, a Ruddlesden–Popper A_2BO_4 composition, to that of a three-dimensional ideal perovskite as the layer thickness increases. However, the simple model described is still useful even in these cases, because the band structure of many layered phases is dominated by the octahedral bonding of the BX_6 groups in the perovskite-like layers.

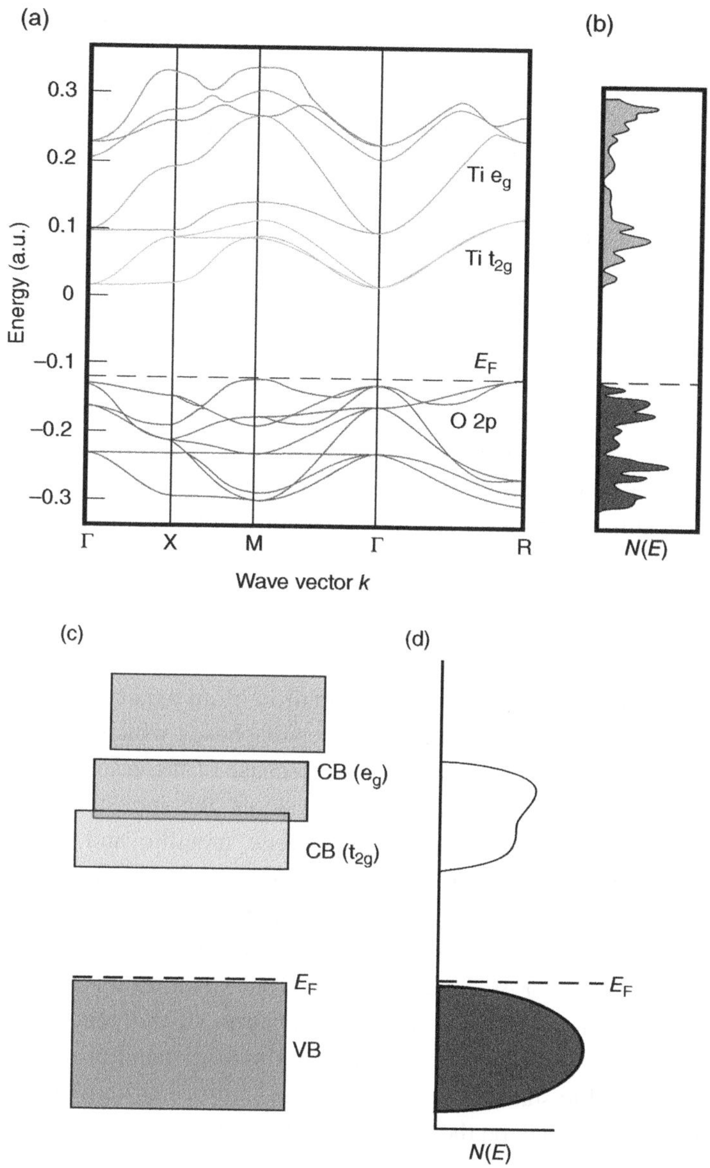

Figure 8.1 *(a) The (approximate) band structure of cubic* $SrTiO_3$*; (b) density of states* N(E) *corresponding to (a); (c) 'flat band' impression of the band structure (CB, conduction band; VB, valence band); (d) simplified density of states* N(E) *corresponding to (c). All figures are drawn to the same energy scale. The Fermi energy is represented by the dashed line just above the valence bands (More complete information is in Piskunov* et al. *(2004))*

Despite the success of this simple picture, not all B-site cations with partially filled d orbitals give rise to metallic perovskites, and many are antiferromagnetic (AFM) insulators. This dilemma was resolved in the middle of the twentieth century with respect to transition metal oxides by assuming that electron correlation, taking the form of Coulomb repulsion between electrons, split the partly filled 3d band into a filled and an empty sub-band, with a band gap between the two. This gap is the Hubbard or Mott–Hubbard gap, denoted by U, and the compounds that display this type of behaviour are called Mott insulators.

Loss of the conducting state can come about by mechanisms other than Coulomb repulsion. In perovskites, the conditions that break the conducting band structure can be described in terms of octahedral tilting; octahedral distortion, particularly Jahn–Teller distortion; cation ordering; cooperative ordering of cation orbitals; ordering of cation displacements, giving rise to ferroic structures; ordering between bonds of different lengths; and ordering of the spin distribution between different orbitals, called spin polarisation (see Goodenough 2014).

8.2 Metal–Insulator Transitions

The structural features that prevent the formation of an expected metallic perovskite are rather finely balanced, and a number of phases which lie on the cusp of metallic/non-metallic behaviour are known. Because of the ease with which A-, B- and X-site substitutions can be made in perovskites, the spin states and electronic conductivity of phases can be juggled between metallic and semiconducting/insulating, allowing these metal–insulator transitions (MIT) to be investigated. Depending upon the nature of the correlation that prevents metallic conductivity, it is found that the metallic phase can be found at low or high temperatures compared to the insulting phase. For example, if a Jahn–Teller distortion is a principal factor in preventing metallic conduction, high temperatures, which tend to suppress this distortion, allow the metallic state to form. On the other hand, if magnetic interactions lead to a metallic state, these may well be disrupted at high temperatures and thus the metallic state may be the low temperatures norm.

8.2.1 Titanates and Related Phases

The simplest conceptual way in which an insulating perovskite such as the archetypal alkaline earth titanates $CaTiO_3$, $SrTiO_3$ and $BaTiO_3$ can be turned into a conductor is to introduce electrons into the empty 3d t_{2g} band formed from the orbitals on the Ti^{4+} cations. This will occur if the compound can be appropriately doped with aliovalent ions or alternatively reduced chemically.

When $BaTiO_3$ loses oxygen, the resistivity falls considerably. The reduced materials show semiconducting/insulating behaviour until the composition approaches

$BaTiO_{2.75}$, at which point the phase becomes metallic and the now low resistivity increases with temperature[1] (Figure 8.2a). There are two factors at work that bring about this change. At high degrees of oxygen reduction, the oxygen vacancy population evens out the structural distortions that lead to ferroelectricity in the fully oxidised phase, and the oxide adopts a cubic structure in which the Ti—O bonds are

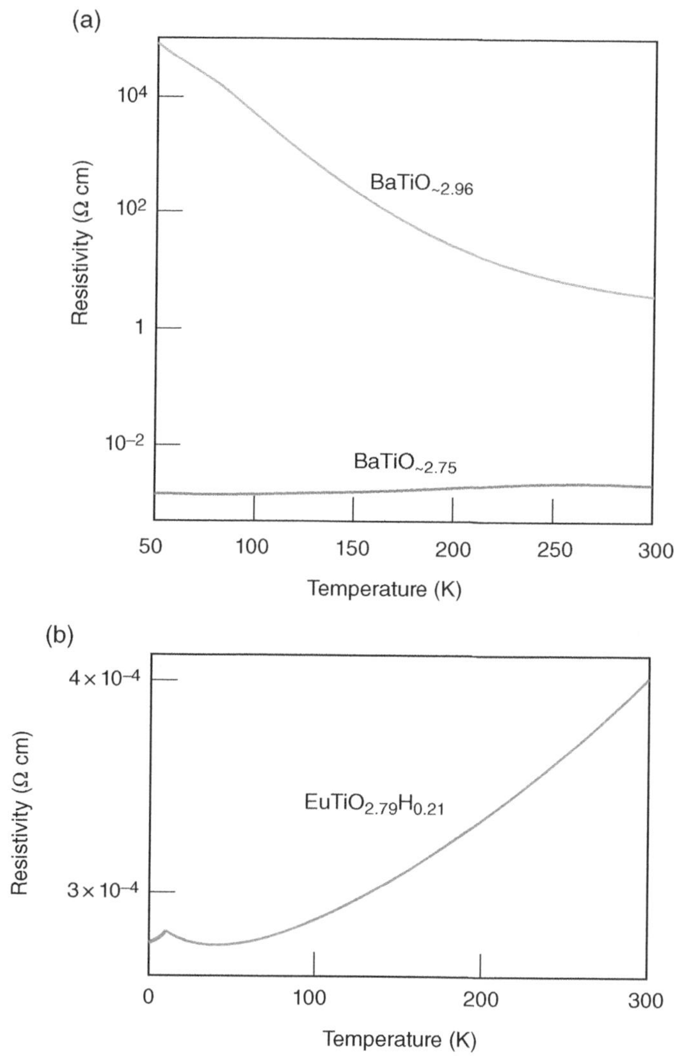

Figure 8.2 *(a) The resistivity of reduced* $BaTiO_{\sim 2.97}$ *(insulator) and* $BaTiO_{\sim 2.75}$ *(metal). (b) The metallic resistivity of a thin film of composition* $EuTiO_{2.79}H_{0.21}$

[1] The resistivity of a metal increases with increasing temperature, while that of a semiconductor decreases with increasing temperature.

equal and the Ti—O—Ti angle is 180°, thus allowing for a potentially metallic conduction band to form. In the fully oxidised phase, this conduction band is empty. However, the generation of oxygen vacancies is charge balanced by the formation of two Ti^{3+} (Ti'_{Ti}) cations:

$$2Ti_{Ti} + O_O \rightarrow V_O^{\bullet\bullet} + 2Ti'_{Ti} + \tfrac{1}{2}O_2(g)$$

The electrons on the Ti^{3+} (d^1) cations are allocated to the conduction band, giving rise to itinerant electron metallic conductivity.

A similar situation occurs in $SrTiO_3$ and $BaTiO_3$ when Nb^{5+} partially substitutes for Ti^{4+} in $SrNb_xTi_{1-x}O_3$ and $BaNb_xTi_{1-x}O_3$. The phase $SrNb_{0.01}Ti_{0.99}O_3$ is a metal. Each Nb^{5+} added to the structure is balanced by the presence of a Ti^{3+} ($3d^1$) cation, making electronic conductivity possible, as in the case of reduced $BaTiO_{3-\delta}$. Moreover, in these doped samples the Ti—O bond lengths are regular, and the conditions for metallic behaviour are fulfilled. In the case of $BaNb_{0.01}Ti_{0.99}O_3$, however, the material remains an insulator. Although the formation of counterbalancing Ti^{3+} ions occurs, structural studies reveal here that the Ti—O bond lengths remain irregular and the Ti—O—Ti bond angle departs from 180°, as in the undoped phase, so that the formation of a partly filled conduction band is inhibited.

The titanate $EuTiO_3$, containing Eu^{2+} and Ti^{4+}, is an AFM insulator with an empty 3d t_{2g} band. As with the alkali metal titanates it can be transformed into a metal by the introduction of electrons into the empty 3d band by A-site aliovalent substitution, as in $Eu_{1-x}La_xTiO_3$, or by reduction to form $EuTiO_{3-\delta}$ as expected. However, an alternative method to transform the insulator into a metal utilises the aliovalent anion substitution of hydride, H^-, for O^{2-}, to form an oxyhydride, $EuTiO_{3-\delta}H_\delta$. The replacement of an O^{2-} anion by H^- requires the replacement of a Ti^{4+} cation by a Ti^{3+} ($3d^1$) cation, thus populating the t_{2g} band. In addition, the change in X-site anions forces the phase to adopt a cubic structure with a random distribution of H and O over the X-sites. At H^- concentrations greater than approximately 0.2, $Eu^{2+}TiO_{2.8}H_{0.2}$, the material becomes a Pauli paramagnetic metal (Figure 8.2b). Similar metal–insulator transitions are found for $CaTiO_{3-\delta}H_\delta$, $SrTiO_{3-\delta}H_\delta$ and $BaTiO_{3-\delta}H_\delta$, which all form Pauli paramagnetic phases as the aliovalent H-doping increases.

8.2.2 $LnNiO_3$

The phase $LaNiO_3$, with Ni^{3+} B-site cations, has rhombohedral symmetry of the $LaAlO_3$ type and is a typical Pauli paramagnetic metal. The low-spin (LS) d^7 ion has a nominal electron configuration of $t_{2g}^6\,e_g^1$, and the partly filled e_g band, containing one d-electron, allows for metallic conduction. If the La cation is replaced by other smaller lanthanoids, on moving from La towards Lu, there is a change from metallic to semiconductor/insulator behaviour. This size drop leads to a change in crystal structure from the rhombohedral $LaAlO_3$ type to the orthorhombic $GdFeO_3$ type, which decreases the A-site geometry due to octahedral tilting (Chapter 1). As a result of this, the Ni—O—Ni bond angle changes from closer to 180° as the lanthanoid size decreases. In addition, the O—Ni—O bond lengths alternate from long to short, and a Jahn–Teller distortion occurs for LS Ni^{3+} in the larger of the octahedra. All of these combine to suppress d band formation and forces the d-electrons to remain localised on the Ni^{3+} cations. As the temperature increases, these local distortions decrease. For the larger lanthanoids, with the smallest degree of octahedral tilting, the reduction in the structural distortion is sufficient to transform the insulating room temperature form to a high-temperature metal. This metal–insulator transition occurs at approximately 460 K for $EuNiO_3$ and 400 K for $SmNiO_3$. In systems such $Eu_xLa_{1-x}NiO_3$ there is a linear dependence of the metal–insulator transition temperature, T; $x=0.4$, $T=125$ K, $x=0.5$, $T=190$ K, $x=0.6$, $T=260$ K and $x=0.8$, $T=380$ K. For the smaller lanthanoid ions, with greater degrees of octahedral tilting, the room temperature insulating state does not transform to a metallic state at higher temperatures.

8.2.3 Lanthanoid Manganites

The lanthanoid manganites, $LnMnO_3$, have a nominal formula $Ln^{3+}Mn^{3+}O_3$, containing a population of Mn^{3+} ($3d^4$, $t_{2g}^3\,e_g^1$, high-spin (HS)) cations. The Mn^{3+} cation is a well-known Jahn–Teller ion, and the oxygen coordination octahedra are elongated. This results in a distorted perovskite structure, and the phases are, in general, AFM insulators, attributed to superexchange interactions between the Mn^{3+} cations and the p orbitals of the linking oxygen atoms (Section 7.3).

A-site substitution with an alkaline earth cation, such as Ca^{2+}, Sr^{2+} or Ba^{2+}, or with Pb^{2+} necessitates the formation of one Mn^{4+} ($3d^3$ t_{2g}^3) cation for every substituent. Taking the well-studied $LaMnO_3$ as an example, the formula of the Ca-substituted phase can be written $La_{1-x}Ca_xMn^{3+}_{1-x}Mn^{4+}_xO_3$, assuming that no change in oxygen stoichiometry occurs. This has the effect of diluting the Jahn–Teller distortions, as Mn^{4+} is not Jahn–Teller sensitive. With the partial loss of this distortion, there is now the possibility that electrons can migrate from one Mn cation to another using the double exchange mechanism. At a critical composition, between the approximate limits of $x \approx 0.16$–0.5 for $La_{1-x}Sr_xMnO_3$ and $x \approx 0.2$–0.5 for $La_{1-x}Ca_xMnO_3$, there are sufficient Mn^{4+} cations present for this to lead to delocalisation of the electrons and metallic conductivity. Note that the conduction electrons, in terms of this model, all have parallel spins and are said to be spin polarised (Section 8.5).

At higher temperatures, the ordering forces involved in the double exchange mechanism are disrupted by thermal agitation. Eventually this overcomes the double exchange interaction, and the phase reverts to a paramagnetic state. In the case of $LaMnO_3$ the high temperature state is insulating. The metal–insulator transition temperature for $La_{0.7}A_{0.3}MnO_3$ is approximately 250 K for $A = Ca^{2+}$, 350 K for $A = Sr^{2+}$ and 320 K for $A = Ba^{2+}$.

The metal–insulator transition in related lanthanoid manganites depends upon the lanthanoid present. A critical factor is the degree of octahedral tilt, which changes the Mn—O—Mn bond angle. The double exchange mechanism operates best when the overlap between the orbitals is maximised, which occurs for a 180° bond angle, found in compounds of the larger lanthanoids. As lanthanoid radius falls, tilting increases and the Mn—O—Mn bond angle drops. This causes the resistivity of the metallic phase to increase and the metal–insulator transition temperature to fall.

8.2.4 Lanthanoid Cobaltites

The lanthanoid cobaltites $LnCoO_3$ show a strong spin correlation with a number of spin states (HS, IS or LS), available to the Co^{3+} cations. $LaCoO_3$ itself is reported as being diamagnetic at very low temperatures, transforming to paramagnetic at approximately 100 K and then transforming to a metallic phase at approximately 500 K. Although it is clear that the spin states on the Co^{3+} cations determine this transition, the spin states themselves are still open to question. At present it seems that the ground state of the system is LS (t_{2g}^6), and the transition to the metallic state corresponds to a change to an IS ($t_{2g}^5\,e_g^1$) state rather than an HS ($t_{2g}^4\,e_g^2$) state.

The transitions between these various possibilities have been explored extensively in mixed valence phases such as $Pr_{0.5}Ca_{0.5}CoO_3$ and $Pr_{0.5}Sr_{0.5}CoO_3$. These compounds have a nominal 1 : 1 ratio of Co^{3+} and Co^{4+} cations and are expected to engender a ferromagnetic phase due to double exchange. In agreement with this, $Pr_{0.5}Sr_{0.5}CoO_3$ is metallic to low temperatures and a ferromagnetic metal below a Curie temperature T_C of approximately 230 K. (Two other magnetic transitions also occur below this temperature.) In $Pr_{0.5}Ca_{0.5}CoO_3$ the room temperature form is also metallic, but there is sudden jump in the resistivity, signalling the formation of a low-temperature insulating phase, at approximately 75 K. The changes responsible for this transition are complex and still open to some uncertainty. What is clear is that there is a spin transition from a Co^{3+} LS (t_{2g}^6) state to an IS ($t_{2g}^5\,e_g^1$) or HS ($t_{2g}^4\,e_g^2$) state. However, this is complicated by the fact that there is also a valence change in the Pr^{3+} cations, some of which donate electrons to the Co^{3+} cations and take on a Pr^{4+} charge state. The Co^{4+} cations are not involved in the transition and remain in an LS (t_{2g}^5) state. Laser irradiation of the insulating material has been shown to induce small islands of metallic material within the insulating matrix, again mediated by a change in the Co^{3+} spin state.

8.2.5 (Sr, Ca)$_2$RuO$_4$ and Ca$_2$Ru$_{1-x}$Cr$_x$O$_4$

The Ruddlesden–Popper phase Sr_2RuO_4 is a metallic superconductor and is the only known non-cuprate superconductor isostructural with the cuprates (Sr, La)$_2$CuO$_4$ and (Ba, La)$_2$CuO$_4$ described later. In simple terms the d-electrons from Ru^{4+} (4d^4) cation would be expected to occupy the partly filled t_{2g} d band and show metallic behaviour (Figure 8.3a). The isostructural Ca_2RuO_4, however, is a Mott insulator. Compared to Sr_2RuO_4, the RuO_6 octahedral tilt is increased, and the Ru—O—Ru bond angle changes away from 180°. This narrows and splits the t_{2g} d band into a filled and empty pair due to Mott–Hubbard Coulomb interactions (Figure 8.3b). The Fermi level, which lies within the t_{2g} band in Sr_2RuO_4, is now situated in the band gap in Ca_2RuO_4 so that an insulating ground state results. These octahedral distortions are reduced at higher temperatures, and Ca_2RuO_4 becomes metallic at approximately 350 K. The system $Ca_{2-x}Sr_xRuO_4$ shows a continuum of behaviour from insulator to metal. In the phase range up to $Ca_{1.8}Sr_{0.2}RuO_4$ the phases show metal–insulator transitions, with the transition temperature decreasing as x increases. In the phase range $Ca_{1.8}Sr_{0.2}RuO_4$–Sr_2RuO_4 the material is a metal as the distortions have been sufficiently suppressed.

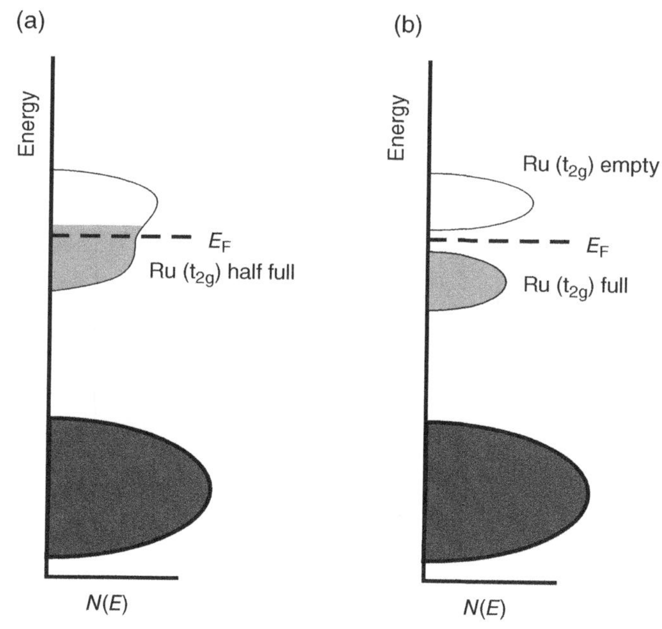

Figure 8.3 *Schematic density of states function for (a) metallic Sr_2RuO_4 and (b) the Mott insulator Ca_2RuO_4*

B-site doping, for example, in the phases $Ca_2Ru_{1-x}Cr_xO_4$, has a similar effect. Small amounts of Cr^{3+} inclusion lessen the distortions and the non-linear Ru—O—Ru bond angle, so that the metal–insulator transition temperature falls to 284 K for $Ca_2Ru_{0.968}Cr_{0.032}O_4$ and to 81 K in $Ca_2Ru_{0.908}Cr_{0.092}O_4$.

8.2.6 $NaOsO_3$

The perovskite $NaOsO_3$ contains Os^{5+} ($5d^3$) cations in the B-sites. At room temperature the phase is an AFM insulator which undergoes a transition to a metallic state at 410 K. The oxide has the $GdFeO_3$ structure, both above and below the metal–insulator transition, and so structural distortions cannot be the cause of the transformation. However, there is a major change in the magnetic behaviour on crossing the transition boundary. The transition occurs when the spins on the Os^{5+} t_{2g}^3 ($5d^3$) half-filled band is split into two, a filled and an empty band, thus creating a band gap in the high-temperature conduction band, as in the case of Sr_2RuO_4 and Ca_2RuO_4. Careful investigation has confirmed that this band splitting is driven by the AFM ordering of the spins on the Os^{5+} cations. The spins align along the **c**-axis (in space group *Pnma*) in a G-type AFM array in which every nearest neighbour has oppositely aligned spins, doubling the ideal perovskite unit cell (Figure 8.4). This band splitting is the first proven example of a Slater transition, a metal–insulator transition driven by magnetic spin ordering.

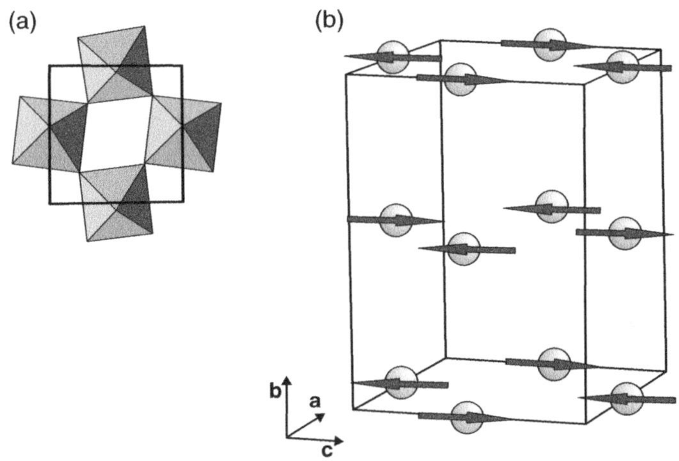

Figure 8.4 *$NaOsO_3$: (a) projection down [010] of the structure; (b) the G-type antiferromagnetic ordering below the metal–insulator transition (space group* Pnma*)*

8.3 Perovskite Superconductors

Although cuprate superconducting oxides have received most attention, the phenomenon of superconductivity in perovskite-related phases was observed well before the cuprate phases were investigated. Superconductivity in semiconducting $SrTiO_3$, albeit with a superconducting transition temperature T_c of only about 300 mK, was first reported in 1965. The insulating parent phase can be transformed into an electronically conducting state by A-site doping, typically with La^{3+}; by B-site doping, typically with Nb^{5+}; or by reduction, thus introducing oxygen vacancies (Section 8.2.1). It has also been shown that electron pairing in the form of Cooper pairs, necessary for superconductivity, occurs well *above* T_c in this material.

Superconductivity at higher temperatures, comparable to the alloy superconductors, was first observed in substituted $BaBiO_3$ in 1975. $BaBiO_3$ is a semiconductor with the Bi cations not in a 4+ state, which is appropriate to the perovskite composition, but half of the atoms have a charge close to Bi^{3+} and half a charge close to Bi^{5+}. The result of this is that the oxygen octahedra surrounding the Bi atoms are of two sizes, with the Bi^{3+} in the large octahedra and Bi^{5+} in the smaller octahedra. The unit cell is monoclinic as a result of the octahedral distortions. Moreover, because of the different geometries of the two types of octahedra, itinerant electron transfer is hindered, as described previously, and $BaBiO_3$ is a semiconductor. However, these distortions can be suppressed by doping, and this leads to metallic conductivity and superconductivity over some of the composition range.

The first superconductor was found in the $BaBiO_3$–$BaPbO_3$ system. The addition of $BaPbO_3$ allows a solid solution to form in which the Pb^{4+} ions substitute for Bi to form $BaPb_xBi_{1-x}O_3$. This solid solution remains semiconducting to a composition of about $BaPb_{0.6}Bi_{0.4}O_3$ at which point it becomes metallic and remains so down to pure $BaPbO_3$. Superconductivity with a superconducting transition temperature, T_c, of approximately 13 K occurs in a narrow composition range close to $BaPb_{0.75}Bi_{0.25}O_3$. Another $BaBiO_3$-derived superconductor is formed by substitution of Ba^{2+} by K^+ on the perovskite A-sites. The narrow range of compositions over which the phase is superconducting is close to $Ba_{0.6}K_{0.4}BiO_3$ and shows a superconducting transition temperature T_c of about 30 K. The perovskite $BaBiO_3$ appears to be unique to date in that it can be made superconducting by cation substitution on either the A or B sublattice.

The A-site ordered double perovskite $(Na_{0.25}K_{0.45})Ba_3Bi_4O_{12}$ has a superconducting transition temperature T_c of approximately 27 K. The structure is cubic, space group $Im\bar{3}m$ (229), $a = 0.8550$ nm, with ordering of the (Na, K) and Ba atoms in layers to give a doubled perovskite unit cell. As the formula suggests, one set of A-sites (A1) is only partially occupied and contains the Na^+ and K^+ cations and a population of alkali metal vacancies. The Ba atoms occupy the other set of A-sites (A2), while the B-sites are fully occupied by Bi. The value of the transition temperature is dependent upon the valence of the Bi ions, with the maximum T_c reported for a nominal valence of 4.35–4.40.

8.4 Cuprate High-Temperature Superconductors

8.4.1 Overview

Cuprate high-temperature superconductors are copper-containing perovskite-derived oxides. Many of these maintain the superconducting state to temperatures above that of liquid nitrogen (Table 8.1). The structures of the superconductors are built up of slices of perovskite type linked by slabs with structures (in the main) of the rock salt (NaCl) or fluorite (CaF_2) type (Section 4.6). The copper valence in most compounds lies between the formal values of Cu^{2+} and Cu^{3+}.

Table 8.1 *Some high-temperature cuprate superconducting oxides*

Compound[a]	T_c (K)
$La_{1.84}Sr_{0.16}CuO_4$	38
$Nd_{2-x}Ce_xCuO_4$	20
$YBa_2Cu_3O_{6.95}$	93
Bi_2O_2/Tl_2O_2 layers	
$Bi_2Sr_2CuO_6$	10
$Bi_2Sr_2CaCu_2O_8$	92
$Bi_2Sr_2Ca_2Cu_3O_{10}$	110
$Tl_2Ba_2CuO_6$	92
$Tl_2Ba_2CaCu_2O_8$	119
$Tl_2Ba_2Ca_2Cu_3O_{10}$	128
$Tl_2Ba_2Ca_3Cu_4O_{12}$	119
TlO/HgO layers	
$TlBa_2CuO_5$	—
$TlBa_2CaCu_2O_7$	103
$TlBa_2Ca_2Cu_3O_9$	110
$HgBa_2CuO_4$	94
$HgBa_2CaCu_2O_6$	127
$HgBa_2Ca_2Cu_3O_8$	133
$Hg_{0.8}Tl_{0.2}Ba_2Ca_2Cu_3O_{8.33}$	138
$HgBa_2Ca_3Cu_4O_{10}$	126

[a]The formulae are representative and do not always show the exact oxygen stoichiometry for the optimum T_c values given.

There are a number of metallic perovskite-related cuprates that do not show superconductivity, typified by $La_4BaCu_5O_{13}$, containing apex-linked CuO_5 square pyramids and CuO_6 octahedra. A comparison of these materials with the superconductors makes it apparent that superconductivity in the cuprates depends on two-dimensional sheets of CuO_2 planes. In these cuprates, there is a considerable spin–charge correlation which appears to be important in the appearance of a superconducting region.

The appearance of superconductivity and the magnitude of the superconducting transition temperature are also closely connected with the composition of these nonstoichiometric solids, which all exhibit considerable variation in oxygen content. The superconducting transition temperature is also affected by external factors such as pressure and crystal elastic strain.

8.4.2 Lanthanum Cuprate, La_2CuO_4

The structure of La_2CuO_4 is a Jahn–Teller distorted variant of the K_2NiF_4 structure (Figure 4.8). When prepared in air by heating the oxides CuO and La_2O_3, the compound is stoichiometric with oxygen content 4.00 and is an AFM insulator. The substance can be transformed into a superconductor by A-site substitution via replacement of some of the La^{3+} cations with alkaline earth cations A^{2+} (Ba^{2+}, Sr^{2+} or Ca^{2+}). In this process, small changes in the oxygen content may also occur, which will be ignored in the following discussion.

Charge neutrality can be maintained in one of two ways in this compound. If valence induction occurs, one Cu^{3+} forms for each A^{2+} substituent to give a formula $La_{2-x}A_xCu^{2+}_{1-x}Cu^{3+}_xO_4$. The Cu^{3+} cations are generally regarded as consisting of holes located on some of the Cu^{2+} ions to form ($Cu^{\bullet}_{Cu}$) defects within the CuO layers, although there is evidence that holes may also reside on O^{2-} ions to generate O^- species. Because Cu^{3+} can be looked on as Cu^{2+} together with a trapped hole, it is often convenient to refer to the substitution as hole doping or acceptor doping. The holes are regarded as forming in the LaO layers, also known, in general, as charge reservoir layers, before migrating into the CuO sheets, also known as superconducting layers (Section 8.4.5) to locate on some of the Cu^{2+} ions to form Cu^{3+} defects. (We can note that hole doping is the usual way in which the cuprate superconductors are transformed to the superconducting state. The only exception is Nd_2CuO_4, which is transformed by way of electron doping (Section 8.4.3).) As an alternative, it is possible to compensate for each two A^{2+} substituents added by the generation of an oxygen vacancy to give $La_{2-x}A_xCu^{2+}O_{4-x/2}$. The principal defects are now $V^{\bullet\bullet}_O$.

The balance between hole and vacancy creation is delicately poised. The process can be illustrated with respect to the most studied phase, $La_{1-x}Sr_xCuO_4$. Initially doping has little effect on the insulating properties of La_2CuO_4, but, as the Sr^{2+} concentration passes approximately 0.07, the hole concentration becomes significant and the material becomes a superconductor at low temperatures. The Cu^{3+} population rises, peaking when x is approximately 0.16, to give a maximum superconducting transition temperature of 38 K at a composition of approximately $La_{1.84}Sr_{0.16}CuO_4$.

Continued substitution forces the compensation mechanism to change to vacancy generation, and the oxygen content of the parent phase falls below 4.0. The number of Cu^{3+} ions decreases as oxygen vacancies form, and when the concentration of Sr^{2+} reaches approximately 0.27, all of the compensation is via vacancies and the material is no longer a superconductor. Superconductivity occurs within a temperature–composition region of the phase diagram known as the superconducting dome (Figure 8.5). High oxygen pressures modify the position of the dome and allow materials to remain superconducting up to a Sr^{2+} concentration of approximately 0.32.

8.4.3 Neodymium Cuprate, Nd_2CuO_4

The structure of Nd_2CuO_4 is very similar to La_2CuO_4, the principal difference lying in the disposition of the oxygen atoms, as the cations in the two structures are in almost identical positions (Figure 4.3). In the superconductor literature this structure is labelled T′. The structural and chemical similarity of La_2CuO_4 and Nd_2CuO_4 suggests that Nd_2CuO_4 should become metallic following acceptor doping, but hole superconductivity does not arise in this phase when Nd^{3+} is substituted by Ca^{2+}, Sr^{2+} or Ba^{2+}. However, Nd_2CuO_4 can be made metallic and superconducting by the substitution of the lanthanoid by a higher valence cation such as Ce^{4+} to form, for

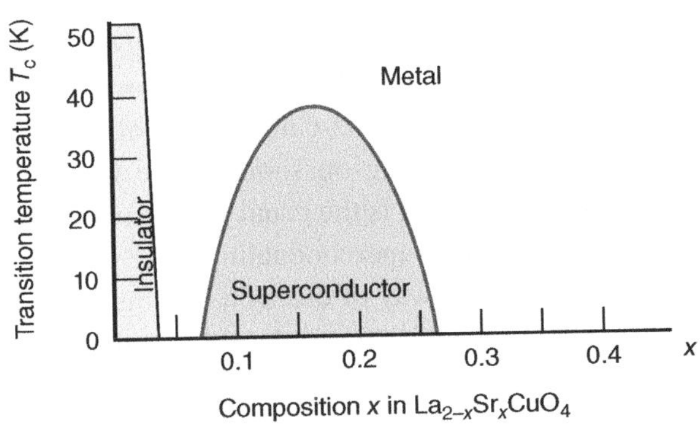

Figure 8.5 *The insulating antiferromagnetic and superconducting phase region for* $La_{2-x}Sr_xCuO_4$

example, $Nd_{1-x}Ce_xCuO_4$. As with La_2CuO_4, small changes in the oxygen content may also occur, which are ignored in the following discussion.

In order to maintain charge neutrality following Ce^{4+} substitution, the system can respond by the formation of one vacancy on the Nd^{3+} sublattice for every three Ce^{4+} substituents, the inclusion of one oxygen interstitial for every two Ce^{3+} substituents or the generation of one Cu^{+} in place of Cu^{2+} for every Ce^{4+} substituent to give a formula $Nd_{1-x}Ce_xCu^{2+}_{1-x}Cu^{+}_xO_4$. It is this latter process that generates the superconducting state. Because Cu^{+} can be looked on as Cu^{2+} together with a trapped electron, it is often convenient to refer to the substitution as electron doping or donor doping. As with the holes, the electrons are regarded as forming in the LaO layers (the charge reservoir layers) before migrating into the CuO sheets (the superconducting layers) to locate on some of the Cu^{2+} ions to form Cu^{+} defects. The compounds are *electron superconductors*. Initial doping turns the insulator into a metal and then creates a superconducting state which occurs under a superconducting dome with x taking values of approximately 0.12–0.18. The maximum value of the superconducting transition temperature, T_c, approximately 24 K, is reached in the compound $Nd_{1.85}Ce_{0.15}CuO_4$.

8.4.4 Yttrium Barium Copper Oxide, $YBa_2Cu_3O_7$

The compound $YBa_2Cu_3O_7$ has been widely studied because it was the first superconductor discovered with a T_c above the boiling point of liquid nitrogen. The crystal structure of $YBa_2Cu_3O_7$ consists of three perovskite-like unit cells stacked one on top of the other (Figure 4.9a). If the ions are allocated the formal charges of Y^{3+}, Ba^{2+} and O^{2-}, the Cu must take an average charge of 2.33, which can be considered to arise from the nominal presence of two Cu^{2+} ions and one Cu^{3+} ion per unit cell.

Stoichiometric $YBa_2Cu_3O_7$ is an insulator, but superconductivity appears when a small amount of oxygen is lost, resulting in a hole superconductor. In fact the oxygen content of the phase is almost always below 7.0 and depends upon the annealing temperature and the surrounding oxygen partial pressure (Figure 8.6a). The phase is able to accommodate a wide range of oxygen stoichiometry, and the curve shows no plateau at $YBa_2Cu_3O_{6.5}$, corresponding to a nominal Cu^{2+}-only phase. The maximum value of T_c, close to 93 K, is found at the composition $YBa_2Cu_3O_{6.95}$. As more oxygen is removed, the value of T_c falls to a plateau at approximately 60 K when the composition lies between the approximate limits of $YBa_2Cu_3O_{6.7}$ and $YBa_2Cu_3O_{6.5}$ (Figure 8.6b). The form of this curve suggests that there may be two superconducting structures forming, one centred near to $YBa_2Cu_3O_{6.95}$ and the other near to $YBa_2Cu_3O_{6.50}$. Continued oxygen removal rapidly leads to a loss of superconductivity, and the material again behaves as an insulating ceramic beyond $YBa_2Cu_3O_{6.35}$.

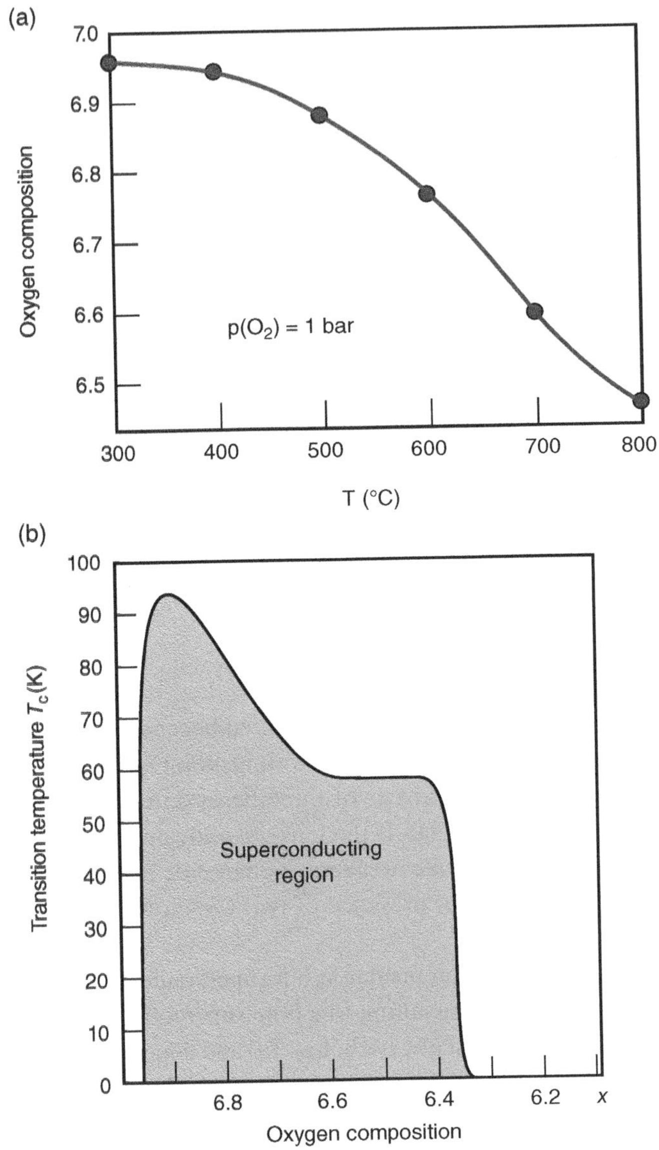

Figure 8.6 *Oxygen stoichiometry in $YBa_2Cu_3O_{6+\delta}$: (a) the composition of samples annealed at temperatures between 300 and 800°C under 1 bar of oxygen pressure; (b) the variation of the superconducting transition temperature with oxygen content*

The ordered way in which the oxygen atoms are lost following reduction appears to be of vital importance in allowing the superconducting transition to take place, and different ordering patterns may be responsible for the two superconducting transition temperatures observed.

8.4.5 Perovskite-Related Structures and Series

The highest transition temperatures found in the cuprate superconductors are shown by rather complicated homologous series of structures that are built from layers of a nominal $ACuO_3$ perovskite parent n octahedra in thickness (Section 4.6.2, Figure 4.10). In all of these phases there is a strong dependence of the appearance of superconductivity and the superconducting transition temperature T_c, upon the oxygen composition (and thus upon the defect content), of the phase. Typically the superconducting transition temperature rises to a maximum and then falls as the oxygen content changes, as described for La_2CuO_4 and $YBa_2Cu_3O_7$. For example, the maximum value of T_c for the superconductor $Bi_2Sr_2CaCu_2O_{8+\delta}$ reaches a maximum of 95 K at a composition close to $Bi_2Sr_2CaCu_2O_{8.19}$. The transition temperature also depends upon the number of CuO_2 layers present, rising to a maximum for three layers and falling off to either side of this value.

In all these compounds, the part of the structure which leads to superconductivity is the slab of CuO_2 sheets, derived from a CuO_6 octahedral layer by the removal of the apical oxygen ions. When more than one sheet is present, they are separated by cation layers, Q (usually Ca or Y), to give a sequence CuO_2-$(Q–CuO_2)_{n-1}$, which form the *superconducting layers* in the material. The index n is the total number of CuO_2 layers in the phase, which is equal to the formula number of Cu atoms present.

The introduction of holes is necessary to induce the change from the parent insulating phase to the superconducting state. In practice, this is achieved by oxygen composition change which mainly occurs within the *charge reservoir* layers which separate each CuO_2-$(Q–CuO_2)_{n-1}$ superconducting slab. The general structure of a charge reservoir sheet is AO–$[MO_x]_m$–AO, where A is a lanthanoid, such as La, or alkaline earth, typically Sr or Ba; M is a metal, typically Bi, Pb, Tl or Hg, present as a non-stoichiometric oxide layer with x usually close to 0 or 1.0; and m takes values of 0, 1, 2,.... Addition of oxygen to the charge reservoir layers favours the creation of holes, which are subsequently transferred to the CuO_2 slab to induce metallic and superconducting properties (Figure 8.7). The reaction can be written schematically in terms of defects as

$$\tfrac{1}{2}O_2 \rightarrow O_i'' + 2h^{\bullet}$$
$$2Cu_{Cu} + 2h^{\bullet} \rightarrow 2Cu^{\bullet}$$

8.4.6 The Generic Superconductivity Phase Diagram

The behaviour of the majority of cuprate superconductors is similar to that of La_2CuO_4, and a 'generic phase diagram' has been proposed that applies to most materials in outline (Figure 8.8). The stoichiometric parent phase is generally an AFM Mott insulator, such as La_2CuO_4. Doping with holes over a modest composition range gives rise to a superconducting phase, situated under the superconducting dome. The superconducting range is rather narrow, and the value of the

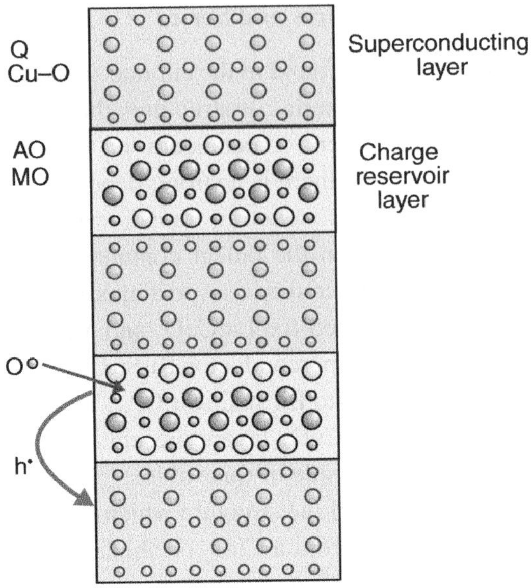

Figure 8.7 Doping of oxygen into charge reservoir layers (schematic) results in the transfer of holes into the superconducting CuO_2-containing layers

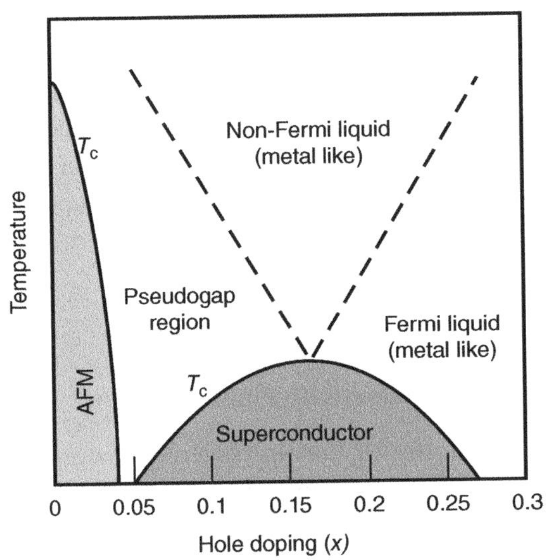

Figure 8.8 Generic phase diagram for the perovskite-related superconducting cuprate phases

superconducting transition temperature, T_c, peaks at a doping level of about 0.15–0.20 holes per CuO_2 sheet.

These insulating and superconducting regions are well delineated, but the region above the dome is still open to controversy and subject to continuing study. At present, the region to the higher dopant side of the AFM insulator is believed to be a type of metallic material known as a pseudogap phase. In this region the density of states near to the Fermi energy is very sparsely populated. The origin of the pseudogap is still uncertain, and it is not yet clear whether this physical state is relevant to the appearance of superconductivity or not. Adjacent to the pseudogap phase field is a metallic phase field. This again is poorly understood at present and is tentatively divided into a region labelled as a Fermi liquid and a region labelled a non-Fermi liquid or strange metal. A Fermi liquid is a metallic phase in which the electrons in the conduction band interact with each other at low temperatures, and is considered to accurately describe most metals at low temperatures. (This is to be distinguished from high temperatures where the electrons can be considered to behave as a non-interacting Fermi gas.) The non-Fermi liquid portion of the diagram has properties that do not concur with those expected of a normal Fermi liquid.

8.4.7 Defects and Conductivity

Oxygen gain and loss introduce a wide range of defects into these structures, all of which may have important consequences for the conductivity and hence device operation. Stacking defects, consisting of different thicknesses of the CuO_2-containing slabs, commonly occur in all members of the superconducting cuprate series (Section 4.6.2). As the superconducting transition temperature appears to peak when three CuO_2 layers are present, wider or narrower slabs will have a degrading effect.

Twin boundaries break up the crystal structure and invariably increase the resistivity of these phases. Such faults are commonly observed in solids which undergo a symmetry change from a high-temperature to a low-temperature structure and are known as transformation twins. For example, La_2CuO_4 adopts tetragonal symmetry, in which $a=b$, at 260°C. At room temperature the structure is orthorhombic with lattice parameters of $a=0.5363$ nm, $b=0.5409$ nm, $c=1.317$ nm. Cooling produces a heavily twinned structure, as either of the high-temperature **a**- = **b**-axes can become the low-temperature orthorhombic **a**- or **b**-axes. These defects, have {110} twin planes (Figure 8.9).

Transformation twinning also occurs in the $YBa_2Cu_3O_7$-related phases with compositions between approximately $YBa_2Cu_3O_{6.25}$ and $YBa_2Cu_3O_{6.6}$. Within this composition range the high-temperature structure is tetragonal. Cooling to room temperature induces a tetragonal to orthorhombic change which is opposite to that encountered in La_2CuO_4. The result, however, is the same, as either of the

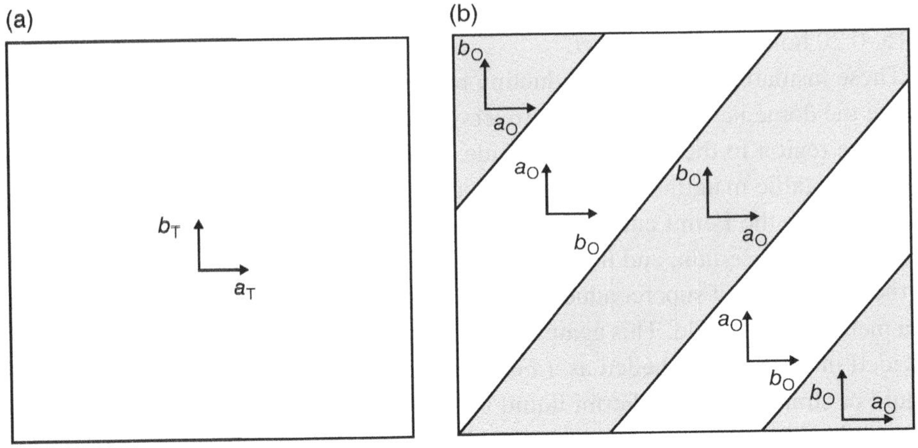

Figure 8.9 Transformation of a high-temperature tetragonal crystal with lattice parameters $a_T = b_T$ into a multiply twinned orthorhombic phase with lattice parameters a_o, b_o and c_o, leading to {110} twin planes

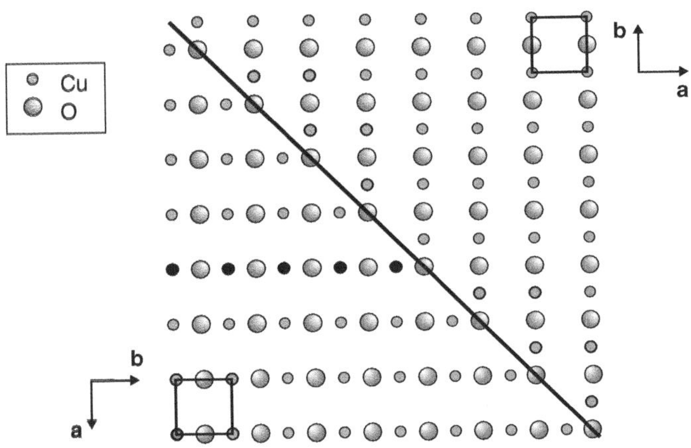

Figure 8.10 Basal (001) surface of $YBa_2Cu_3O_7$ containing a {110} twin plane

high-temperature orthorhombic **a**- or **b**-axes can transform into low-temperature **a**- = **b**-axes and subsequently the low-temperature crystals are heavily twinned, with twin boundaries on {110}.

At more oxygen-rich compositions, including $YBa_2Cu_3O_7$, there is no symmetry change on cooling, and transformation twinning cannot occur. Nevertheless most crystals are also heavily twinned on {110} planes at room temperature (Figure 8.10). In this case it appears that these faults are growth twins. Here, the chains of dopant oxygen atoms may line up along either of two possible directions as the crystallites form. This alignment then defines the **b**-axis in the crystal. These are selected at random, so that different crystal nuclei will have different orientations of the **b**-axes, within the symmetry imposed by the crystallography of the system. As the nuclei grow and abut on one another, twinned crystallites form.

8.5 Spin Polarisation and Half-Metals

The relatively simple band structure of a cubic perovskite becomes more complex when the electron spin is taken into account. Calculations show that the mainly t_{2g} and e_g bands split into two components, one for 'spin-up' electrons and one for 'spin-down' species. The original two bands become four. An example is provided by the perovskite $CaMnO_3$, containing Mn^{4+} (d^3 t_{2g}^3), which is an AFM semiconductor. The lower-energy t_{2g} band should contain three electrons with parallel spins. As the t_{2g} state can hold up to six electrons, it is only half full in $CaMnO_3$, thus predicting metallic conductivity. However, because of electron correlation (Hund's coupling between electrons, which results in Hund's rules for electron occupation of similar orbitals), both the t_{2g} and e_g orbital sets are split into two sub-bands, one for 'spin-up' (↑) electrons and one for 'spin-down' (↓) electrons. The simplified density of states diagram (Figure 8.11) shows that the valence band, derived mainly from O 2p states,

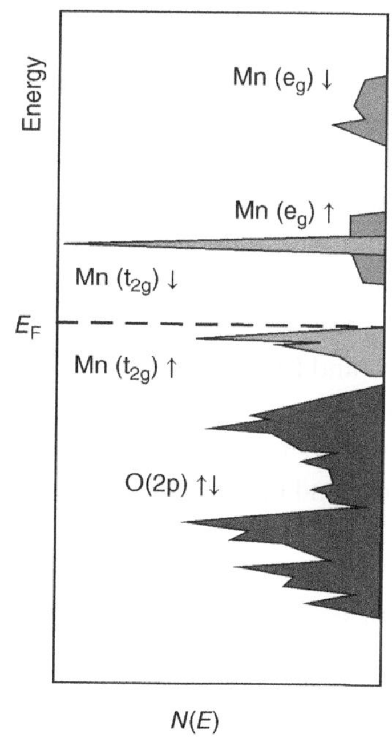

Figure 8.11 *The simplified density of states diagram for $CaMnO_3$*

contains both spin-up and spin-down (↑↓) electrons. The t_{2g} band is split into a lower-energy spin-up (↑) and a higher-energy spin-down (↓) sub-band, as is the empty e_g band. The lowest-energy t_{2g} spin-up band lies just above the mainly oxygen 2p band and is filled by the Mn^{4+} t_{2g}^{3} electrons, while the upper t_{2g} spin-down band remains mainly empty. The material is therefore expected to be an insulator/semiconductor.

This situation raises the possibility that only one spin state might contribute to the conductivity of a metallic material. If the Fermi level of the metallic phase lies in a band containing electrons in a t_{2g} single spin state (or polarisation), it implies that the electric current will be carried by electrons with only one spin. Such materials are called half-metals. That is, a half-metallic phase is a metallic conductor for electrons of one spin orientation, say, spin-up, and an insulator for electrons of the other spin orientation, say, spin-down. Half-metallic conductivity has been found in the metallic A-site substituted materials $La_{1-x}Ca_xMnO_3$ which exist over the composition range $La_{0.8}Ca_{0.2}MnO_3$–$La_{0.6}Ca_{0.4}MnO_3$ and in $La_{0.7}Sr_{0.3}MnO_3$. The strict partitioning of the electron spins is not inflexible, but depends upon temperature, among other factors. The perovskite $La_{0.7}Sr_{0.3}MnO_3$, a half-metallic ferromagnetic phase, changes above the Curie temperature, T_C, at which point the electron spins disorder and the single spin current is lost.

More recent studies of half-metallic behaviour have involved the double perovskites Ca_2FeMoO_6, Ba_2FeMoO_6 and especially Sr_2FeMoO_6. The band structure is made up from hybridisation of Fe^{3+} ($3d^5$ t_{2g}^{5}), Mo^{5+} ($4d^1$ t_{2g}^{1}) and O 2p orbitals (Figure 8.12). None of the spin-up bands intersect the Fermi level, and for these charge carriers the material is an insulator. For the spin-down carriers a band mainly made up of contributions from the Fe t_{2g} and Mo t_{2g} orbitals is intersected by the Fermi level, and for spin-down electrons the material is a metallic conductor. Broadly speaking the conductivity arises mainly from the d-electrons associated with the Mo^{5+} cations, and in the ground state the current is totally spin polarised and carried by 'spin-down' electrons only. To some extent, in a similar way to that described for $La_{0.7}Sr_{0.3}MnO_3$, the degree of spin polarisation is constrained by the amount of disorder on the Fe and Mo sites. In particular, antisite disorder, in which Fe occupies some of the Mo sites and vice versa, has a significant effect on the half-metallic behaviour of the system. Careful preparation methods are needed to minimise this antisite disorder and optimise properties.

8.6 Charge Ordering and Orbital Ordering

Cation charge ordering may also occur in a perovskite containing cations with two different charge states, and this can be instrumental in preventing the formation of a metallic phase where one might be expected. This has been well documented in the

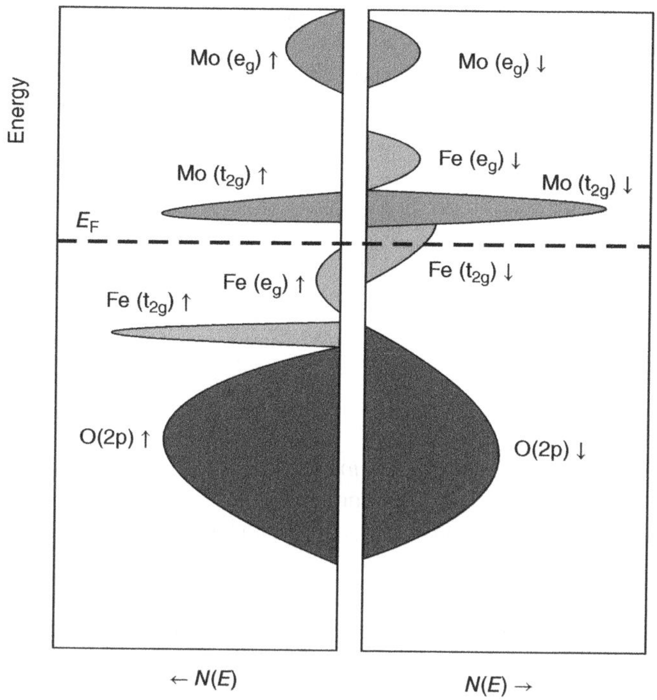

Figure 8.12 *Schematic density of states diagram for spin-up and spin-down electrons in the double perovskite* Sr_2MoFeO_6

A-site substituted lanthanoid manganites. The parent phase, $LnMnO_3$, has a population of Mn^{3+} ($3d^4$, $t_{2g}^3e_g^1$, HS) cations. The Mn^{3+} cation is a well-known Jahn–Teller ion, and the oxygen coordination octahedra are elongated, resulting in a distorted perovskite structure. A-site substitution with Ca^{2+} necessitates the formation of one Mn^{4+} ($3d^3$ t_{2g}^3 HS) cation for every Ca^{2+} cation to maintain charge balance. The composition $La_{0.5}Ca_{0.5}MnO_3$ has equal numbers of Mn^{3+} and Mn^{4+} cations present. Of these, the octahedra containing Mn^{4+} are regular, while those containing Mn^{3+} are Jahn–Teller distorted. Ordering of these two different octahedra leads to a lower-energy structure, resulting in a charge-ordered (CO) state. Electron movement between Mn^{3+} and Mn^{4+} is hindered by the distortion, and the t_{2g} electrons stay localised, so that the material remains an insulator.

The Jahn–Teller distortion has another consequence. The d states that split in energy in an octahedral crystal field into the t_{2g} and e_g pair are further spilt in a Jahn–Teller environment (Figure 1.7). In the case of an elongated octahedron, the most common situation is that the t_{2g} set are split slightly, but there is a bigger splitting for the e_g pair, with the d_{z^2} orbital at an appreciably lower energy than the $d_{z^2-y^2}$ orbital. This means that the orbital that is preferentially occupied will be d_{z^2} and orbital

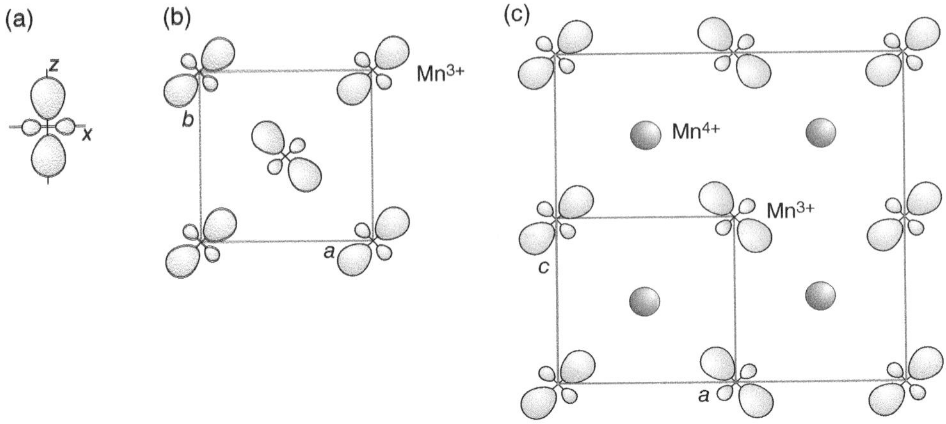

Figure 8.13 *Charge and orbital ordering in manganites: (a) the* d_{z^2} *orbital; (b) the orbital ordering in* $LaMnO_3$ *(axes refer to the Pbnm setting); (c) charge and orbital ordering in* $La_{0.5}Ca_{0.5}MnO_3$ *(axes refer to Pnma setting). In (b) and (c) only the Mn cations are shown*

ordering will now be present. These are best documented for the mixed valence manganites such as $La_{1-x}Ca_xMnO_3$ (Figure 8.13), but other charge and orbital ordered arrangements have been characterised in a variety of mixed valence perovskites.

8.7 Magnetoresistance

8.7.1 Collosal Magnetoresistance (CMR) in Manganites

The magnetoresistance of a solid is the change of resistance that occurs when a magnetic field is applied. The magnetoresistive (MR) ratio is defined as the change in resistance when a magnetic field is applied to that in zero field:

$$\text{MR ratio} = \frac{(R_H - R_0)}{R_0} = \frac{\Delta R}{R_0}$$

where R_H is the resistance in a magnetic field and R_0 is the resistance in the absence of the field. Magnetoresistance can be positive or negative and can be an intrinsic property of the material or arise within extrinsic factors such as grain boundaries, dislocations and domain walls. Interest in magnetoresistance intensified when it was discovered that manganite perovskites show very high negative values of the MR ratio such that the resistance of the material falls from a high value to almost zero (generally at low temperatures) in a high enough magnetic field. This extreme behaviour is called colossal magnetoresistance (CMR). In general, rather large applied magnetic fields, usually several tesla, are required to promote large changes in the electrical resistance.

Intrinsic CMR was initially encountered in $LaMnO_3$ in which the A-sites have been donor doped with an alkaline earth element, such as Ca, Sr or Ba, or else with Pb^{2+}, to form $La_{1-x}A_xMnO_3$ phases, so that the crystals contain both Mn^{3+} and Mn^{4+}. The important composition for CMR is close to $La_{0.7}Ca_{0.3}MnO_3$ or $La_{0.7}Sr_{0.7}MnO_3$. At this composition the materials are ferromagnetic metallic phases as a result of double exchange. At the Curie temperature, T_C, of the order of 375 K for the manganites, this metallic phase is transformed into a paramagnetic insulator. Near to, but above this transition temperature, although the electron spins are disordered, little energy is required to reverse this state of affairs. The application of a sufficiently high magnetic field will force realignment, reproducing the ferromagnetic metallic state. The resistivity falls sharply as a consequence, and the MR change corresponds to the metal–insulator transition (Figure 8.14a). The largest magnetoresistance values are found close to T_C, and rather large magnetic fields (above 1 T) are required to bring appreciable change about.

Substitution of some of the Mn with both magnetic or non-magnetic ions, including Ti^{4+}, Sn^{4+}, Fe^{3+}, Cr^{3+}, Al^{3+}, Ga^{3+}, In^{3+}, Mg^{2+} and Ni^{2+}, increases the CMR effect considerably. For $Pr_{0.7}Ca_{0.2}Sr_{0.1}MnO_3$ the ratio R_0/R_H is approximately 230. This jumps to 4×10^5 for the modestly doped $Pr_{0.7}Ca_{0.2}Sr_{0.1}Mn_{0.98}Mg_{0.02}O_3$.

The physical causes of intrinsic CMR are not yet clear. Originally believed to be associated with superexchange or double exchange, it is now thought that the important factor is the presence of ferromagnetic half-metallicity with majority spin

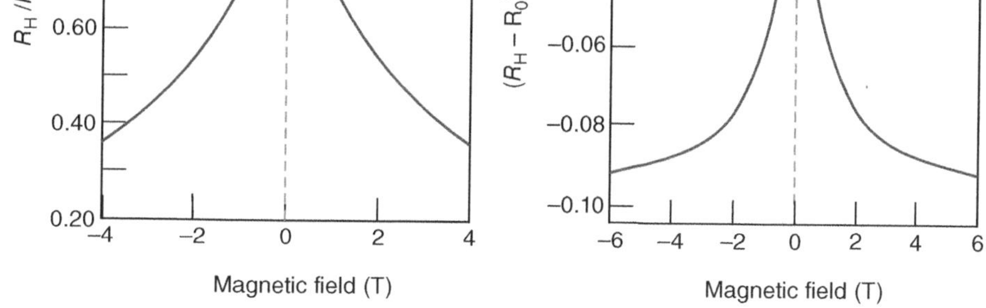

Figure 8.14 *(a) Typical magnetoresistance curve for an approximately $La_{0.7}A_{0.3}MnO_3$ ($A = Ca^{2+}$, Sr^{2+}) crystal close to the magnetic Curie temperature. Details of the curve depend on temperature and A^{2+} concentration. (b) The low-field magnetoresistance of Sr_2FeMoO_6*

conduction. The phenomenon has also been treated in a semi-empirical way by treating these materials as a mixture of two phases, a ferromagnetic metal and a paramagnetic insulator. The resistance, *R*(total), is then given by the sum of the resistance of the two components:

$$R(\text{total}) = vR(\text{ferro}) + (1 - v)R(\text{para})$$

where *v* represents the volume fraction of the ferromagnetic component, *R* (ferro) is the resistance of the ferromagnetic component and *R* (para) is the resistance of the paramagnetic insulating phase.

8.7.2 Low-Field Magnetoresistance

Low-field magnetoresistance (LFMR) is an extrinsic property of a number of ceramic perovskites. Whereas high magnetic fields of the order of 5–6 T are required to produce the effects of CMR in crystals and single crystal films, here the fields are generally less than 0.5 T. A typical material showing this LFMR is the ordered double perovskite Sr_2FeMoO_6 (Figure 8.14b). This has a high Curie temperature of approximately 415 K and is a half-metal, showing magnetoresistance to temperatures of approximately 400 K. The magnetoresistance is attributed to the tunnelling of the spin-down electrons that originate with the Mo cations, as the Fe cations contribute electrons to the filled spin-up band only (Section 8.5).

The mechanism appears to be related to tunnelling across the grain boundaries or magnetic domain walls in the ceramic by spin-polarised electrons. Because of this, much effort has been expended on the modification of grain boundaries, including decreasing grain size and the use of nanoparticles. The microstructure of the ceramic is also important and the introduction of insulating secondary phases such as TiO_2 or $SrTiO_3$, both of which preferentially segregate at the grain boundaries, can also have a beneficial effect. The magnitude of the resistance change can be varied by chemical disorder, including, in double or ordered perovskites, the presence of antisite defects. As with CMR, the LFMR effect is most pronounced at low temperatures.

8.8 Semiconductivity in Perovskites

Many perovskites in the 'insulating' state are, in fact, semiconductors. There are several models for the mechanism of charge transport, which can generally be differentiated by considering the temperature variation of the conductivity or resistivity of the sample. If the conductivity can be considered to be due to itinerant electrons promoted to an empty conduction band by thermal energy and holes left in the valence band, so that conductivity is similar to that of an elemental semiconductor, the resistivity follows a simple activated energy law:

$$\rho = A\exp\left(\frac{E}{2k_{\mathrm{B}}T}\right) \tag{8.1}$$

where A is an experimentally determined constant and E is the thermal activation energy involved in the conduction process (equal to the band gap for elemental semiconductors). In many perovskites, only one charge carrier is important, in which case the activated resistivity is given by Equation (8.1). The insulating double perovskite Pb_2MnWO_6 has a low conductivity at higher temperatures which follows this type of law with E equal to 0.53 V. The cubic organic–inorganic perovskite $CH_3NH_3SnI_3$ behaves similarly, but is a hole conductor. The band structure calculations show that a valence band, mainly derived from I 5p orbitals with some Sn^{2+} 5s contribution, is separated by about 0.4 eV from an empty valence band mainly derived from the Sn^{2+} 5p orbitals, and although experimentally determined band gap values are considerably higher, it seems certain that the material behaves in an analogous fashion to a hole-doped elemental semiconductor with a hole concentration of the order of $10^{18}\,\mathrm{cm}^{-1}$ in as-grown crystals.

In practice, the results of many conductivity measurements made on semiconducting perovskites are most easily explained if electronic conduction takes place by a diffusion type of process in which electrons hop from one cation to another. Thus in appropriately A-site substituted manganites containing Mn^{3+} and Mn^{4+} or cobaltites containing Co^{2+} and Co^{3+}, conductivity can be imagined to take place in this way. The electrons or holes are now thought of as being localised at cations. In many cases electron transfer may be accompanied by a lattice distortion. The migrating electron plus distortion unit can be considered as a quasiparticle called a polaron. For many perovskites it is believed that conductivity is via polaron hopping, which can be described by a general equation of the form

$$\rho(T) = AT^{m}\exp\left[\left(\frac{T_0}{T}\right)^{p}\right]$$

where $\rho(T)$ is the resistivity of the phase, A is a constant, T is the temperature, T_0 is called the characteristic temperature of the material and m and p are interrelated and depend upon the exact mechanism of the hopping process. The value of T_0 can be calculated assuming specific models for the charge migration. In general if hops are always to nearest-neighbour cations, $m = p = 1$ and the equation can be written as

$$\rho(T) = AT\exp\left(\frac{T_0}{T}\right)$$

This form of the resistivity is often found at higher temperatures. If the value of T_0 is small so that $T_0/T \ll 1$, the exponential term can be expanded to give the approximation

$$\rho(T) = AT + \rho_0$$

where ρ_0 can be considered to be a temperature independent term. At lower temperatures, perovskites often behave as if the hopping jumps are variable. There are two cases that have been explored. If there are no long-range Coulombic interactions between the migrating charge carriers, the Mott variable-range hopping law is followed:

$$\rho(T) = AT \exp\left[\left(\frac{T_0}{T}\right)^{0.25}\right]$$

If it is assumed that a migrating electron suffers a Coulombic attraction with the hole remaining on the parent cation, then the Efros–Shklovskii variable-range hopping law is appropriate:

$$\rho(T) = AT \exp\left[\left(\frac{T_0}{T}\right)^{0.5}\right]$$

It is found that the Mott variable-range hopping law often holds at intermediate temperatures and the Efros–Shklovskii variable-range hopping law at lower temperatures.

It is not always easy to decide which mechanism holds in these phases, and the mechanism may change as the concentration of mobile charge carriers changes. In general the mechanism may be clarified by a plot of $\ln(\rho/T)$ versus $(1/T)$ or similar. For example, the perovskite bronze Ca_xWO_3 shows both activated and variable-range hopping as the composition changes. The structure of WO_3 is closely related to perovskite (Section 2.3) and has a band structure similar to $SrTiO_3$. As W^{6+} is a $5d^0$ ion, the d band is empty. As Ca is interpolated into WO_3 to form a perovskite bronze, the Ca^{2+} 4s electrons are introduced into the empty d band. At Ca concentrations x below 0.08, the resistivity versus temperature plots suggest that a Mott variable-range hopping mechanism holds, whereas for concentrations above 0.12 it now appears that nearest-neighbour hopping operates.

In many perovskites, the distortion may be considerable, as with the Jahn–Teller effect or the influence of lone pair electrons. In such cases the distortion may prevent any charge carrier migration at all, and semiconducting behaviour is suppressed. Changing the degree of distortion by substitution will often reverse this.

8.9 Thin Films and Surface Conductivity

The surfaces of perovskite crystals may show many curious features. One of the most unexpected is the presence, on the surface of pure (undoped) $SrTiO_3$ crystals, of a metallic conducting layer. $SrTiO_3$ is, as described earlier, an archetypal transparent non-magnetic wide band gap insulator. It is found that (001) surfaces made up of TiO_2 layers supports a two-dimensional electron gas, in which the electrons, although confined in the third dimension, can move freely over the surface. As in some of the three-dimensional materials described earlier, the electrons can be thought of as occupying two spin distinguished sub-bands. In these bands the electron spins are aligned parallel to the surface, but they have a rotational dissymmetry and are chirally opposite. In one band the spins rotate in a clockwise direction, while in the other the rotation direction is anticlockwise. These sub-bands are formed from the electrons contained in the t_{2g} conduction band (Figure 8.1) and are separated by a band gap of approximately 90 meV.

An equally counterintuitive finding is that the interface between two insulating dielectric perovskite oxides, $LaAlO_3$ and $SrTiO_3$, can be superconducting. The interface supports a metallic electron gas, and it is this that forms the superconducting state, with a superconducting transition temperature of approximately 200 mK for interfaces of thickness of the order of 10 nm.

References

J. B. Goodenough, *Chem. Mater.*, **26**, 820–829 (2014).

S. Piskunov, E. Heifets, R. Eglitis, G. Borstel, *Comput. Mater. Sci.*, **29**, 165–178 (2004).

Further Reading

For a perspective on electronic conductivity in transition metal oxides, with relevance to perovskites see:

J. B. Goodenough, *Chem. Mater.*, **26**, 820–829 (2014).

J.-G. Zhou, J. B. Goodenough, B. Dabrowsky, *Phys. Rev.*, **B67**, 020404(R) (2003).

For comprehensive reviews covering early studies of manganites see:

J. M. D. Coey, M. Viret, S. von Molnar, *Adv. Phys.*, **48**, 167–293 (1999).

M. B. Salamon, M. Jaime, *Rev. Mod. Phys.*, **73**, 583–628 (2001).

For details of the Slater transition in $NaOsO_3$ see:

S. Calder *et al.*, *Phys. Rev. Lett.*, **108**, 257209 (2012).

For reviews of superconductivity of relevance to perovskite phases see:

R. J. Cava, *J. Am. Ceram. Soc.*, **83**, 5–28 (2000).

P. M. Grant, *Nature*, **476**, 37–39 (2011).

W. Grochala, *J. Mater. Chem.*, **19**, 6949–6968 (2009).
B. Keimer *et al.*, *Nature*, **518**, 179–186 (2015).
A. Mann, *Nature*, **475**, 280–282 (2011).

Half-metallic materials, including perovskites, are discussed in:
C. Y. Fong, J. E. Pask, L. H. Yang, Half-Metallic Materials and Their Properties, World Scientific, Singapore (2013).

For a review of correlated electron systems, including perovskite based superconductors, colossal magnetoresistance, manganites and cobaltites see:
B. Raveau, *Angew. Chem. Int. Ed.*, **52**, 167–175 (2013).

The conductivity/superconductivity at $SrTiO_3$ interfaces is described by:
N. C. Plumb *et al.*, *Phys. Rev. Lett.*, **113**, 086801 (2014).
N. Reynan *et al.*, *Science*, **317**, 1196–1199 (2007).
A. F. Santander-Syro *et al.*, *Nat. Mater.*, **13**, 1085–1090 (2014).

9

Thermal and Optical Properties

In this chapter, only those thermal and optical properties that can be considered in some way to be unique, or at least emphasised, in perovskites are described. Thus general properties of solids (such as heat capacity, where the focus is upon lattice dynamics, or photoluminescence, where the focus is upon dopant ions), in which the perovskite serves only as an inert matrix, are neglected.

9.1 Thermal Expansion

9.1.1 Normal Thermal Expansion

For perovskites that are used in high-temperature applications, such as solid oxide fuel cells that operate above 800°C or so, thermal expansion becomes an important physical property, as mismatch of the thermal expansion of the cell components, cathode, electrolyte and anode, will cause early cell failure. The magnitude of the thermal expansion of many perovskites is rooted in the thermal behaviour of the BX_6 octahedra and is associated with octahedral tilt, distortion and the bonding between the B-cation and the surrounding anions. These are all susceptible to modification as the temperature rises and can contribute to anomalies in thermal expansion characteristics.

Perovskites: Structure–Property Relationships, First Edition. Richard J. D. Tilley.

The effect of unequal bonding can be illustrated with respect to the thermal expansion of a number of K_2NiF_4 structure phases. Both Sr_2TiO_4 and Sr_2SnO_4 have isotropic thermal expansion while many others including La_2NiO_4 and the mixed lanthanoid aluminates $SrLaAlO_4$, $CaErAlO_4$ and $CaYAlO_4$ have anisotropic thermal expansion. The difference between Sr_2TiO_4 and $SrLaAlO_4$ is illustrative. In Sr_2TiO_4, the bonding around the TiO_6 octahedra is regular, and the bond lengths between Ti and the O1 equatorial oxygen are almost equal to the bond lengths between the Ti and the apical O2 neighbours (Figure 9.1a). Bond valence sums are equal and the

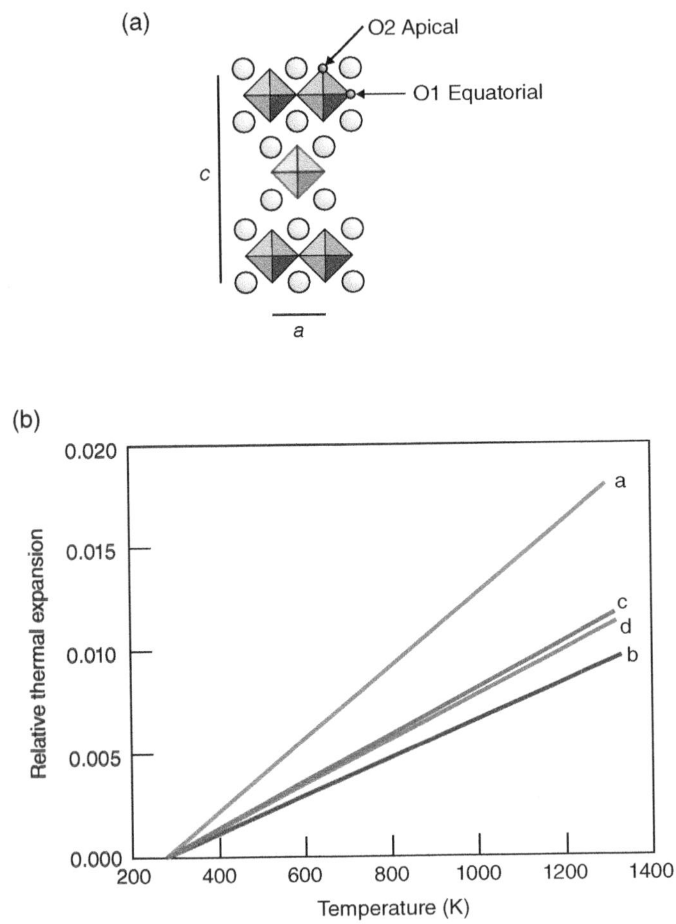

Figure 9.1 *Thermal expansion of Sr_2TiO_4 and $SrLaAlO_4$: (a) idealised K_2NiF_4 structure of both phases; (b) relative thermal expansion; $SrLaAlO_4$, plot a, $\Delta c/c_0$, plot b, $\Delta a/a_0$; Sr_2TiO_4, plot c, $\Delta c/c_0$, plot d, $\Delta a/a_0$*

bonds are equally strong. The relative expansion and the thermal expansion of the **a**- and **c**-axes, αa, αc, over a temperature range T_0 to T, defined by

$$\Delta a/a_0 = \left[a(T) - a(T_0)\right]/a(T_0);\ \Delta a/a_0 = \Delta b/b_0,\ \text{defined similarly}$$
$$\alpha a = (\Delta a/a_0)/(T - T_0) = 1.225\times10^{-5}\,\text{K}^{-1}\,(298 - 1273\,\text{K});\ \alpha b = \alpha a$$
$$\Delta c/c_0 = \left[c(T) - c(T_0)\right]/c(T_0)$$
$$\alpha c = (\Delta c/c_0)/(T - T_0) = 1.270\times10^{-5}\,\text{K}^{-1}\,(298 - 1273\,\text{K})$$

are found to be almost identical (Figure 9.1b). Moreover, the values closely mimic the relative thermal expansion of the equatorial and apical Ti–O distances.

In the case of $SrLaAlO_4$, the bonds between Al and the apical oxygen O2 are longer than those between Al and the equatorial O1 atoms. Bond valence sums indicate that the apical bonds are weaker than the equatorial bonds, and this is reflected in the resultant axial relative thermal expansion values (as defined earlier) and the axial thermal expansion, which is found to be greater along the **c**-axis than along the **a**- (or **b**-)axes (Figure 9.1b):

$$\alpha a = 0.943\times10^{-5}\,\text{K}^{-1}\,(298 - 1273\,\text{K})$$
$$\alpha c = 1.775\times10^{-5}\,\text{K}^{-1}\,(298 - 1273\,\text{K})$$

Again, this relative thermal expansion is identical to the change in the apical and equatorial bond lengths over the same temperature range.

The thermal expansion of perovskites may also depend upon features such as magnetic and spin ordering, and if a perovskite passes from a magnetically ordered phase such as antiferromagnetic to paramagnetic, the thermal expansion will invariably show an anomaly. This has been reported, for example, in the well-studied lanthanoid titanates, $LnTiO_3$, which show both antiferromagnetic and ferromagnetic regimes.

A similar situation occurs in $LaCoO_3$. Here the thermal expansion passes through an anomalous peak at about 250°C (Figure 9.2). The material nominally contains Co^{3+} cations, but at lower temperatures, disproportionation occurs so that

$$2Co^{3+} \rightarrow Co^{4+} + Co^{2+}$$

As the temperature increases, this reaction is reversed:

$$Co^{4+} + Co^{2+} \rightarrow 2Co^{3+}\left(\text{IS}, t_{2g}^5 e_g^1\right)$$

The lowest energy state for the Co^{3+} cations is the low-spin (LS, t_{2g}^6) configuration. However, the intermediate-spin (IS, $t_{2g}^5 e_g^1$) and high-spin (HS, $t_{2g}^4 e_g^2$) states are

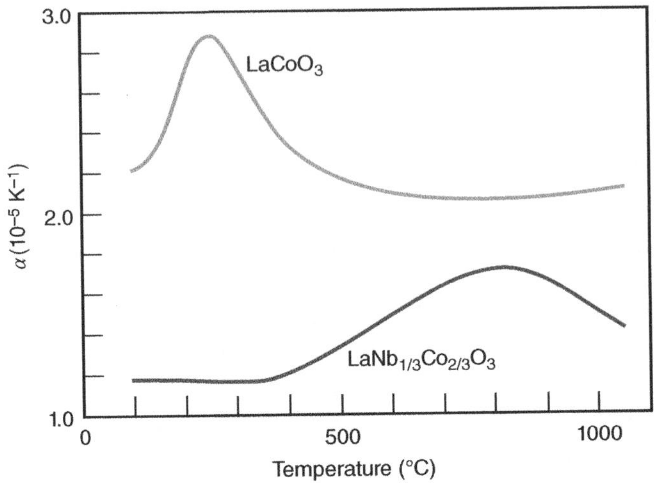

Figure 9.2 *The anomalous thermal expansion of* $LaCoO_3$ *and* $LaNb_{1/3}Co_{2/3}O_3$

also accessible. Thus, as the temperature increases from room temperature, there are two reactions occurring, the generation of Co^{3+} cations and their conversion from the LS state to the IS state. The peak in the thermal expansion is due to cumulative effect of these two changes. At higher temperatures all cations can be considered to be in the IS state and the thermal expansion increases smoothly. As would be anticipated, a change in the Co^{3+} population will modify this effect. B-site substitution with Nb^{5+} in $LaNb_{0.333}Co_{0.667}O_3$ alters the number of Co^{2+} cations present, as each Nb^{5+} must be balanced by the transformation of two Co^{3+} cations to Co^{2+}. This causes a displacement of the disproportionation reaction which forms Co^{3+} to a higher temperature, thus moving the peak in the thermal expansion curve in the same direction (Figure 9.2).

The thermal expansion of many perovskites can often be tailored by changing the composition of the material via A-, B- and X-site substitution, a feature of considerable importance in ceramic components. For example, the phase $LaCo_{1/3}Nb_{2/3}O_3$ described earlier shows reduced thermal expansion due to Nb^{5+} substitution compared to the parent $LaCoO_3$ perovskite (apart from the transposition of the thermal expansion anomaly) (Figure 9.2).

9.1.2 Thermal Contraction

There are an increasing number of perovskite phases known that show a contraction as the temperature increases, at least over a certain temperature range. This effect is usually known, somewhat circuitously, as *negative thermal expansion* (NTE). Thermal contraction is not the result of a single mechanism. Here two examples are given.

In many perovskite structures the contraction is associated with the distortions that lower the symmetry of phases from the ideal cubic structure. As the temperature rises, these distortions tend to decrease. If the decrease outweighs the normal expansion that might also take place simultaneously, then an overall contraction may be observed. A well-documented example is provided by $PbTiO_3$. At lower temperatures this phase has a tetragonal unit cell, due to a displacement of the Ti^{4+} cations away from the centres of the TiO_6 octahedra, along the **c**-axis (Figure 9.3a). As the temperature increases the distortion tends to decrease, because of changes in vibrational energy and a decrease in anion–anion repulsion, so that the long diagonal shortens and the **c**-axis decreases. At the same time the undistorted diagonals expand, causing the **a**- and **b**-axes to increase normally (Figure 9.3b). The contraction outweighs the expansion so that the cell volume shows overall thermal contraction up to the Curie temperature of 490°C. The mean thermal expansion coefficient, α, for ceramic samples, is approximately $-1.99\times10^{-5}\,K^{-1}$. Above the transition temperature the phase transforms to the ideal cubic structure, and thereafter normal thermal expansion is observed as the temperature increases, with a thermal expansion coefficient of approximately $3.72\times10^{-5}\,K^{-1}$.

Doping the A-sites of $PbTiO_3$ with La^{3+} causes the transition temperature from tetragonal to cubic to fall, and the thermal contraction to diminish steadily, so that at an approximate composition of $Pb_{0.8}La_{0.2}TiO_3$, ceramic samples show a mean thermal expansion coefficient of $\alpha=-0.11\times10^{-5}\,K^{-1}$ from room temperature to 130°C. Above this temperature the cubic polymorph has a thermal expansion coefficient of approximately the same value as the undoped material, of $3.82\times10^{-5}\,K^{-1}$, indicating that the substituent influences the distortion of the low-temperature phase rather than other factors. A-site substitution with even modest amounts of Cd^{2+} has the opposite effect, as $\alpha=-2.40\times10^{-5}\,K^{-1}$ for ceramic $Pb_{0.94}Cd_{0.06}TiO_3$.

Lanthanoid manganites, such as $LaMnO_3$, $NdMnO_3$ and $GdMnO_3$, are of potential value in solid oxide fuel cell cathodes. However, many of these phase show thermal contraction because of the diminishing Jahn–Teller distortion of the Mn^{3+} cations as the temperature is increased. Such effects tend to rule out these materials for real cell applications, although A- and B-site substitution, as demonstrated for $PbTiO_3$ earlier, can ameliorate the problem.

A perovskite phase with large thermal contraction is the $AA'_3B_4O_{12}$ cubic perovskite $SrCu_3Fe_4O_{12}$ (Section 2.1.3), which contracts between the temperatures of 170 and 270 K (Figure 9.4). The coefficient of linear thermal expansion in this range is $\alpha=-2.26\times10^{-5}\,K^{-1}$. Outside of this range, both above and below it, the thermal expansion is normal. The cause of the contraction is a continuous charge transfer between the Cu on A′-sites and Fe on B-sites. At the lower temperature, the charge distribution is $Sr^{2+}Cu_3^{\sim2.8+}Fe_4^{\sim3.4+}O_{12}$. The Cu charges are $(Cu_{0.6}^{2+}Cu_{2.4}^{3+})$ and the Fe charge distribution is nominally $(Fe_{2.4}^{3+}Fe_{1.6}^{4+})$. At the high-temperature

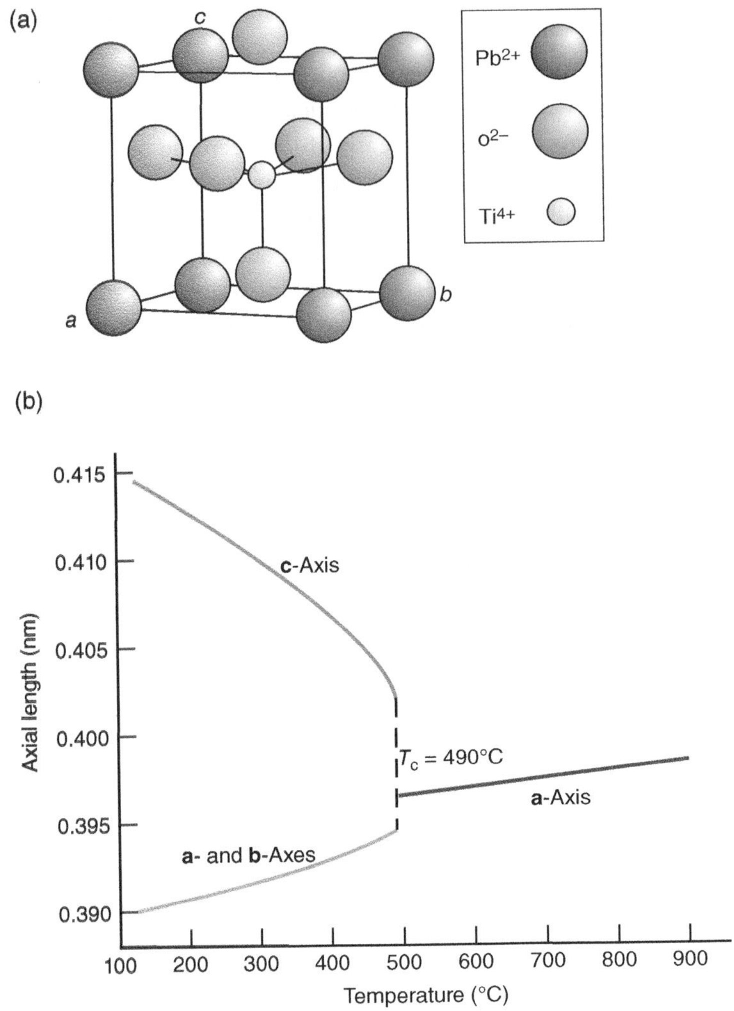

Figure 9.3 *Thermal expansion of $PbTiO_3$: (a) the TiO_6 octahedral unit showing off-centre Ti^{4+}; (b) thermal expansion/contraction of the unit cell axes as a function of temperature*

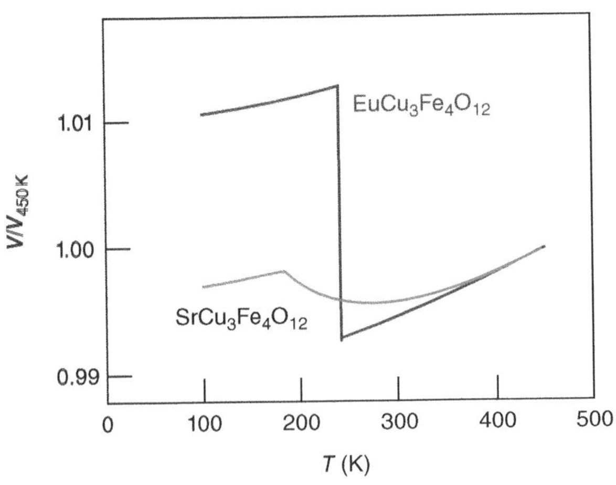

Figure 9.4 *Thermal expansion/contraction of $SrCu_3Fe_4O_{12}$ and $EuCu_3Fe_4O_{12}$*

extreme, this phase has a charge distribution of $Sr^{2+}Cu_3^{\sim 2.4+}Fe_4^{\sim 3.7+}O_{12}$, equivalent to $(Cu_{1.8}^{2+}Cu_{1.2}^{3+})$, and the Fe charge distribution is nominally $(Fe_{1.2}^{3+}Fe_{2.8}^{4+})$. (Note that at very much lower temperatures, the Fe distribution changes abruptly by charge disproportionation to $(Fe_{3.2}^{3+}Fe_{0.8}^{5+})$.) The charge transfer effectively changes the coordination polyhedra around the cations in a continuous fashion. As the temperature increases from 170 to 270 K, the Fe—O bond distance and Sr—O bonds gradually shrink while the Cu—O bond distance gradually increases and the Fe—O—Fe bond angle in the tilted octahedra, ψ, increases. The unit cell dimension of the phase is simply related to the Fe—O bond length d and the Fe—O—Fe bond angle ψ as

$$a = 4d \sin\left(\frac{\psi}{2}\right)$$

The contraction or expansion of the perovskite is thus determined by the shrinkage of the Fe—O bond length compared to the increase in the Fe—O—Fe bond angle. Over the anomalous thermal contraction range, the Fe—O bond length is of most importance. Above and below the thermal contraction range, no charge disproportionation occurs and the phase behaves as a normal expanding solid as the temperature increases.

We can note that this is in distinction to the behaviour of similar materials such as the lanthanoid-containing phases $LaCu_3Fe_4O_{12}$ and $EuCu_3Fe_4O_{12}$. At room temperature, $LaCu_3Fe_4O_{12}$ has the ideal charge distribution $La^{3+}Cu_3^{3+}Fe_4^{3+}O_{12}$, and as the temperature increases, the sample expands normally. At 393 K (120°C), there is an abrupt charge transfer between the Cu on A′-sites and Fe on B-sites to give $LaCu_3^{2+}Fe_4^{3.75}O_{12}$, where the Fe state is $(Fe^{3+}Fe_3^{4+})$. This charge transfer causes an abrupt volume decrease, after which normal thermal expansion is found. The same is true of $EuCu_3Fe_4O_{12}$, with a transition at approximately 245 K (Figure 9.4).

9.1.3 Zero Thermal Expansion Materials

Materials that do not expand at all as the temperature rises, *zero thermal expansion* (ZTE) materials, would be of value for many purposes, especially in microelectronic devices that may become warm due to power consumption. For this reason, materials that show ZTE are being actively sought. In the past, a ZTE solid could be constructed by using a composite of two materials, one showing thermal contraction and one showing thermal expansion. However, composite materials often have accompanying drawbacks in use, and a number of solids have now been fabricated that show virtually no expansion at all as the temperature rises.

The previous two sections have shown that some oxide perovskites have a marked thermal contraction and that this property can be modified by the traditional ceramic route of substitution, for which the perovskite structure is well adapted. This is well illustrated by $PbTiO_3$, as described earlier. Complex ceramics with compositions close to $0.7PbTiO_3$–$0.3Bi(Zn_{0.5}Ti_{0.5})O_3$ show virtually ZTE over the temperature range from room temperature to 400°C, while for $0.6PbTiO_3$–$0.3Bi(Zn_{0.5}Ti_{0.5})O_3$–$0.1BiFeO_3$, this range extends to approximately 700°C.

There has also been considerable effort in developing ZTE metallic phases. Notable among these are the nitride phases $(Cu,Sn)NMn_3$, $(Zn,Sn)NMn_3$ and $(Cu,Ge)NMn_3$, such as $Ge_{0.5}Cu_{0.5}NMn_3$. These latter solids have the antiperovskite structure (Section 1.10) in which the Cu and Ge atoms (or their equivalents) occupy the A positions, the N atoms take the octahedrally coordinated B positions and the Mn atoms occupy the positions of oxygen in the ABO_3 perovskite structure. In them, the ZTE is due to a combination of a normal thermal expansion of the structure on the one hand and a shrinkage of the unit cell due to *magnetic ordering* between the Mn atoms on the other. This means that the degree of expansion or contraction can be controlled by manipulating the Mn contribution via the introduction of defects on the Mn sites in the structure.

9.2 Thermoelectric Properties

Electrical conduction and heat transport are closely linked, the connection being described by three thermoelectric coefficients, the Seebeck coefficient (or thermopower), the Peltier coefficient and the Thomson coefficient, all of which have relevance to thermoelectric power generation and refrigeration. In perovskites, the most reported values are for the Seebeck coefficient. The magnitude and sign (+ or −) of the Seebeck coefficient are related to the concentration and type of mobile charge carriers present. For band-like perovskites, the magnitude of the Seebeck coefficient is proportional to the density of states, either in the conduction band, for electron transport, or the valence band for hole transport.

The inorganic–organic perovskite $CH_3NH_3SnI_3$ is an example of this type of material, as it behaves as a band-like p-type semiconductor across the three crystallographic structural modifications, orthorhombic, tetragonal and cubic, that exist below room temperature. The value of the Seebeck coefficient, S, is positive (Figure 9.5), indicating that the charge carriers are holes. The almost linear dependence of the Seebeck coefficient, except for small discontinuities at phase changes, increasing steadily over a broad temperature range, is considered to be indicative of band-like conduction. The p-type hole conduction is believed to be due to the incorporation of small amounts of oxidised Sn^{4+} in place of Sn^{2+} within the crystal structure as each of these inclusions contributes two holes to the top of the valence band.

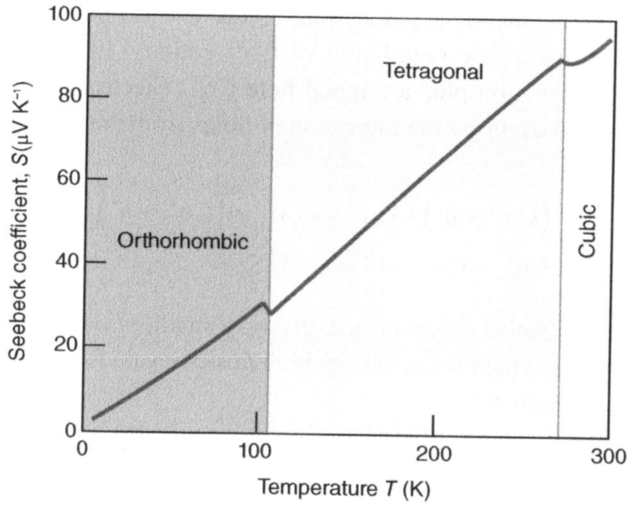

Figure 9.5 *The Seebeck coefficient of the perovskite $CH_3NH_3SnI_3$ as a function of temperature*

For perovskites that can be treated as ionic with charge carriers jumping from one cation to another, the Seebeck coefficient can be discussed in terms of the defect chemistry of the phase. An approximate relationship between the number of defects and the Seebeck coefficient is frequently given by the simple relationship:

$$\alpha = \pm\left(\frac{k}{e}\right)\left[\ln\left(\frac{n_0}{n_d}\right)\right]$$

where n_0 is the number of sites in the sublattice containing defects and n_d is the number of defects giving rise to mobile electrons or holes and $k/e = 86.17\,\mu V\,K^{-1}$. The positive version applies to p-type materials and the negative expression to n-type materials. Note that n_0/n_d increases as the number of defects falls and so the value of α is expected to be greatest for lowest defect populations.

This equation is formally equivalent to the Heikes equation:

$$\alpha = \pm\left(\frac{k}{e}\right)\ln\left[\frac{(1-c)}{c}\right]$$

where c is the fraction of defects (or mobile charge carriers) present. This form is useful because the value of c is directly related to the composition of the sample.

This type of behaviour is shown by many cobaltites. At lower temperatures, where complexities due to magnetic transitions between high, low or intermediate spin states can be ignored, the Seebeck coefficient is well described by the aforementioned equations. The parent phase, $La^{3+}Co^{3+}O_3$, is frequently slightly oxidised and contains a low population of Co^{4+} cations. Each Co^{4+} ion can be considered to be a Co^{3+} ion plus a trapped hole $Co^{\bullet}_{Co}$. Electronic conductivity can then be considered to occur by the migration of holes from one Co^{4+} ion to a neighbouring Co^{3+} ion:

$$\left(Co^{3+} + h^{\bullet}\right) + Co^{3+} \rightarrow Co^{3+} + \left(Co^{3+} + h^{\bullet}\right)$$

$$Co^{\bullet}_{Co} + Co_{Co} \rightarrow Co_{Co} + Co^{\bullet}_{Co}$$

The number of these native defects is usually very small, of the order of 10^{-4} of the total Co^{3+} population, which leads to very high positive values of S that fall slightly with temperature (Figure 9.6).

The Seebeck coefficient can be altered significantly by both A-site and B-site substitutions. For example, A-site doping with Sr^{2+} to form $La_{1-x}Sr_xCoO_3$ forces one Co^{3+} ion to transform to Co^{4+} to maintain charge neutrality. The value of the Seebeck coefficient remains positive, as the Seebeck current is still due to hole migration, but the number of $Co^{\bullet}_{Co}$ defects will increase in proportion to the amount of substituent. The number of defects is equal to the number of impurity Sr^{2+} ions (ignoring the small concentration of $Co^{\bullet}_{Co}$ in the parent compound), and so the fraction c in the Heikes equation is equal to x in $La_{1-x}Sr_xCoO_3$ and as a consequence the value of the Seebeck coefficient will fall (Figure 9.6).

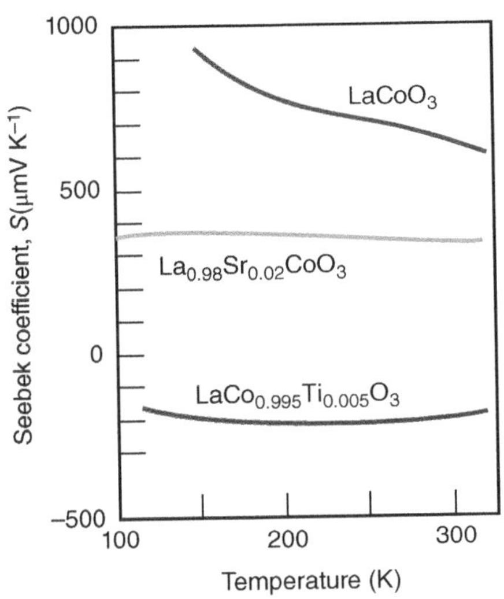

Figure 9.6 *The Seebeck coefficient as a function of temperature for slightly non-stoichiometric $LaCoO_3$, A-site substituted $La_{0.98}Sr_{0.02}CoO_3$ and B-site substituted $LaTi_{0.005}Co_{0.995}O_3$*

An analogous situation occurs for appropriate B-site substitution. For example, the B-site substitution of Ti^{4+} for Co^{3+} in the parent $LaCoO_3$ structure forces the transformation of two Co^{3+} cations to Co^{2+} to maintain charge balance in the substituted perovskite $LaTi_xCo_{1-x}O_3$. Each Co^{2+} cation can be considered to be a Co^{3+} cation ion plus a trapped electron Co'_{Co}. Electronic conductivity can then be considered to occur by the migration of electrons from one Co^{2+} ion to a neighbouring Co^{3+} ion:

$$Co^{2+} + Co^{3+} \rightarrow Co^{3+} + Co^{2+}$$

$$Co'_{Co} + Co_{Co} \rightarrow Co_{Co} + Co'_{Co}$$

The value of c in the Heikes equation is equal to x in $LaTi_xCo_{1-x}O_3$, and as a consequence, the numerical value of the Seebeck coefficient will fall compared to the parent phase, and as electron transport is involved, the Seebeck coefficient is accordingly negative (Figure 9.6).

9.3 The Magnetocaloric Effect

Magnetic and thermal properties are linked via the magnetocaloric effect (MCE), which describes the fact that ordinary magnetic materials heat up when placed in a magnetic field and cool down when the field is removed. The inverse MCE applies to materials that cool when placed in a magnetic field and warm up when the field is removed. The MCE at room temperature is of interest as a benign refrigeration method. The cause of the MCE is an order–disorder transformation in the magnetic structure of the solid brought about by the application of a magnetic field. In essence, a material in a disordered magnetic state is changed to a material in a magnetically ordered form by the application of a magnetic field. The magnetically aligned material (in a magnetic field) will disorder again when the magnetic field is removed. The energy for this transformation will come from the lattice energy of the solid, which, in adiabatic conditions, will cause the bulk phase to cool. It is quantified by the isothermal magnetic entropy change, often written as $|\Delta S_M|$. A large value of $|\Delta S_M|$ will arise from a large spin magnetisation (typically ferromagnetic), which is accompanied by a sharp drop in magnetisation associated with the loss of magnetic alignment (typically paramagnetic). Adiabatic cooling, used to reach temperatures well below 1 K in the laboratory, is an important application of the MCE.

The MCE is exhibited by a number of perovskite oxides, especially $La_{0.7}Ca_{0.3}MnO_3$ and $La_{0.7}Sr_{0.3}MnO_3$. The phase $La_{0.7}Ca_{0.3}MnO_3$ is ferromagnetic below a Curie temperature T_C of approximately 220 K and $La_{0.7}Sr_{0.3}MnO_3$ below a T_C of approximately 370 K, making them potential magnetocaloric materials, but unfortunately

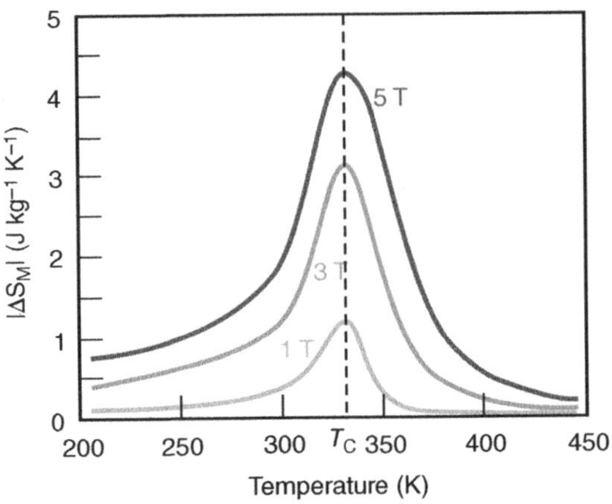

Figure 9.7 *The magnetic entropy change, $|\Delta S_M|$, as a function of temperature and magnetic field for $La_{0.8}Ca_{0.18}Na_{0.2}MnO_3$*

the magnetocaloric performance of these pure materials is poor. However, the flexibility of the perovskite structure to A-, B- and X-site substitution may be exploited to develop materials with improved performance in this respect. Perovskites studied with this in mind include complex A-site substituted phases such as $La_{0.8}Ca_{0.2-x}Na_xMnO_3$, similar B-site substituted phases such as $La_{0.7}Sr_{0.3}Mn_{0.9}Cr_{0.1}O_3$ and $La_{0.7}Sr_{0.3}Mn_{1-x}Co_xO_3$ and the double A- and B-site substituted phases $La_{0.6}Pr_{0.1}Sr_{0.3}Mn_{0.9}Fe_{0.1}O_3$, $La_{0.67}Ba_{0.33}Mn_{1-x}Fe_xO_3$ and $La_{0.67}Sr_{0.33}Mn_{1-x}V_xO_3$.

For all of these materials, the value of the isothermal magnetic entropy change, $|\Delta S_M|$, rises to a peak at the Curie temperature, T_C, of the phase. The peak height is dependent upon the applied magnetic field. The measure of the potential heat transfer between the hot volume and the cold sink during a single ideal refrigeration cycle is given by the relative cooling power, RCP, or refrigerant capacity, defined by

$$\mathrm{RCP} = |\Delta S_M| \times \delta T$$

where δT is the width of the peak at half-maximum. The relative cooling power for $La_{0.8}Ca_{0.6}Na_{0.2}MnO_3$ (Figure 9.7) is 89 J kg^{-1} K^{-1} in a magnetic field of 2 T.

9.4 The Pyroelectric and Electrocaloric Effect

The pyroelectric effect is used in pyroelectric energy harvesting, and the electrocaloric effect is currently being explored for refrigeration. Perovskite crystals that show the pyroelectric effect contain a unique polar axis and a spontaneous electric polarisation, P_s, directed along this axis. As all ferroelectrics are also pyroelectrics (Chapter 6), most studies refer to perovskite ferroelectric phases. In the pyroelectric effect, a change in temperature induces a change in the observed spontaneous polarisation, resulting in a change in the surface electric charges and so generating an electric field (i.e. voltage) across the crystal. (The pyroelectric effect is made up of two components, the primary pyroelectric effect, which is measured on a crystal with a constrained (constant) size, and the secondary pyroelectric effect, measured on an unconstrained crystal, so that it is able to expand or contract freely in the thermal gradient. For practical purposes, these effects are usually considered as one.) The electrocaloric effect is the converse of the pyroelectric effect. Here an electric field applied with a component along the polar axis results in a change in temperature of the crystal.

The principal ferroelectric perovskites that have been investigated for potential pyroelectric and electrocaloric applications include complex formulations involving $PbTiO_3$ (Curie temperature $T_C \sim 490°C$), $LiTaO_3$ ($T_C \sim 618°C$) and the relaxor ferroelectrics $Pb(Mg_{1/3}Nb_{2/3})O_3$, $Pb(Sc_{0.5}Ti_{0.5})O_3$ and $PbSc_{0.5}Sb_{0.5}O_3$. Although the theoretical framework for these physical phenomena is still being worked out, one suggested correlation between the expected temperature change in an electrocaloric material, ΔT, and fundamental properties is given by

$$\Delta T = (T \ln \varphi) P_s^2 / (3\varepsilon_0 \Theta C)$$

where T is the temperature, φ is the number of polar states accessible in the system, P_s is the saturation polarisation, ε_0 is the dielectric permittivity of free space, Θ is the Curie constant of the phase and C is the specific heat capacity. The value of the number of polar states accessible, φ, is most diverse in the relaxor ferroelectrics, where the complex micro- and nanostructure can be altered by thermal and chemical methods. It is for this reason that relaxors are now actively investigated for potential device applications.

As with many phenomena involving electric or magnetic transitions, the effect is most pronounced near the relevant transition point in the solid. As an example, a 350 nm film of $PbZr_{0.95}Ti_{0.05}O_3$, which is a rhombohedral ferroelectric between approximately 120°C and the Curie temperature of approximately 225°C, shows a peak value in the induced temperature change, ΔT, of 12 K at a temperature of approximately 226°C when the electric field change, ΔE, is 480 V cm^{-1}.

9.5 Transparency

There are many applications of perovskites where single crystals are used. However, polycrystalline ceramic materials are often preferable if they can be fabricated in a suitable form, especially in a transparent state. The problem of obtaining transparent ceramics, perovskites or other oxides, is well understood. It is necessary to sinter the starting materials to obtain a fully dense body with no pores or impurity phases present, especially at grain boundaries. This is because light scattering, mainly at internal boundaries, is the cause of translucency or opacity.

Ceramic perovskites are unique, to date, in that the transparency of a sample can be manipulated by an external electric field, employing electric-field-induced scattering. The materials that show this effect are the complex perovskites $Pb_{1-x}La_x(Zr_{1-y}Ti_y)_{1-x/4}O_3$, usually written as PLZT($x/(1-y)/y$), for example, PLZT (8/70/30). This means that they have potential applications as light modulators and attenuators without the need for polarisers as an adjunct. The transparent ceramics are made by conventional ceramic processing techniques. They have a high transparency that is composition dependent and a high surface reflectivity. For example, taking the refractive index as approximately 2.5, the reflectivity, R, of a surface in air is given by

$$R = \left[\frac{(n-1)}{(n+1)}\right]^2 = \left[\frac{1.5}{3.5}\right]^2 = 0.184$$

The loss due to reflection from both surfaces is then, approximately 0.37.

On the application of an electric field, the material shows a hysteresis loop, the exact shape of which is sensitive to temperature and composition. At the same time, the sample loses or gains transparency, the upper limit of which is determined by the surface reflectivity (Figure 9.8). The optical changes are believed to be due to the

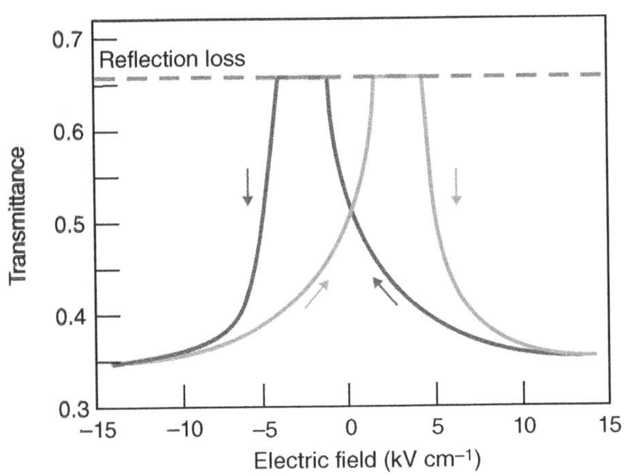

Figure 9.8 *Transparency versus electric field for a PLZT (7.45/71/29) ceramic at 40°C. The electric field is applied parallel to the light beam, across a disc of thickness 0.35 mm (Original data in Zhang et al. (2014))*

formation and destruction of macroscopic domains and polar nanodomains in the sample. As the domains have different polarisation directions, considerable refractive index changes are encountered as the light beam crosses domain boundaries, which gives rise to considerable light scattering. The behaviour is very dependent upon composition and temperature, which alter domain formation and the ease of domain wall movement or elimination.

9.6 Electrochromic Films

Electrochromic materials, which change colour when subjected to an electric field, are widely explored for application as 'smart windows' that control the amount of light reflected or transmitted. Colour production is linked to defect formation in an otherwise colourless matrix. The material most widely explored for such devices is tungsten trioxide, WO_3, which, ordinarily, is pale yellow and an insulator. Thin films are transparent. Electrochromic films rely on the formation of perovskite structure tungsten bronzes, M_xWO_3 (Section 2.3) in which cations such as H^+, Li^+, Na^+ or K^+ occupy the large A-type cage sites between the corner-linked WO_6 octahedra of the parent phase (Figure 9.9a). At low concentrations of the interpolated ion, the colour is believed to be due to charge transfer between two valence states of tungsten, W^{5+}–W^{6+} or W^{4+}–W^{6+} couples.

In an electrochromic device, a voltage is used to drive the reducing metal such as Li into a WO_3 film using an applied voltage. This results in the loss of an electron from the Li atom to form Li^+. The released electrons are associated with W^{6+} cations to form W^{5+} or W^{4+} species, allowing for charge transfer to occur (Figure 9.9b). As a result the transparent tungsten trioxide film turns into a blue-black perovskite tungsten bronze. Reversal of the voltage removes the interpolated metal and regenerates the colourless state, a process referred to as bleaching. The reaction can be schematically written:

$$WO_3\left(\text{oxidised, transparent}\right) + x\text{M} \rightarrow \text{M}_x WO_3\left(\text{reduced, blue-black}\right)$$

In principle, devices are constructed as a series of thin films on glass. Transparent conducting electrodes, usually indium tin oxide (ITO), serve as electrodes, sandwiching a film of WO_3, an ion-conducting electrolyte and a source/sink of metal ions (Figure 9.9c). In practice, many designs have been explored.

A notable application of electrochromic cells is in car mirrors, which can be electrically dimmed so as to cut down dazzling reflections from bright lights. These do not use Li^+ but instead rely upon the decomposition of atmospheric water vapour to

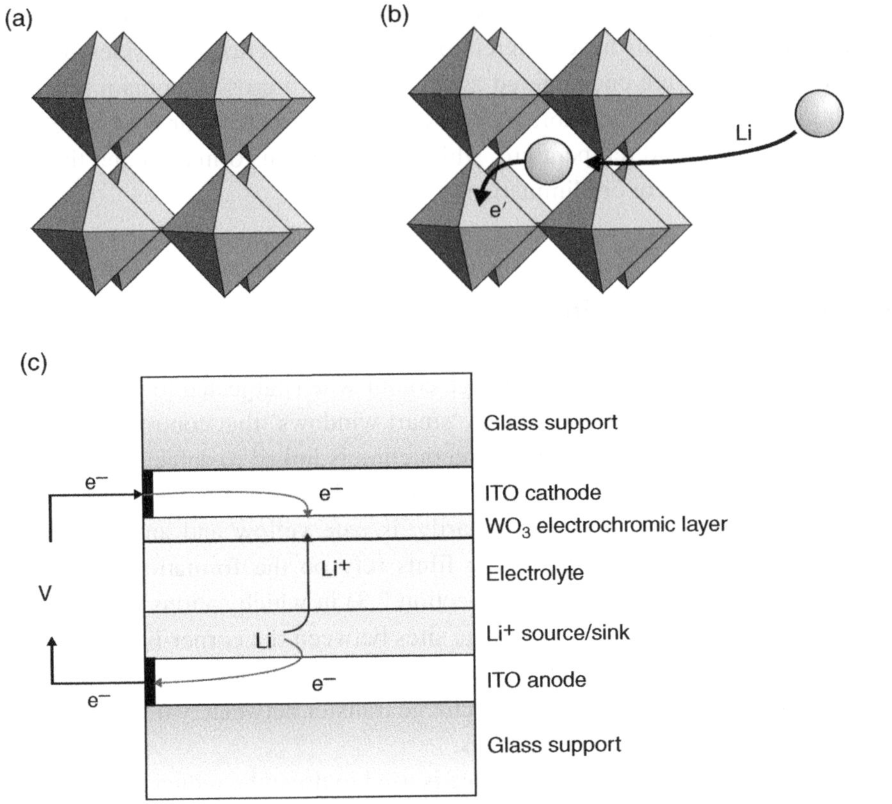

Figure 9.9 *Electrochromic cells: (a) ideal WO_3 structure; (b) incorporation of Li into the A-sites under an electric field; (c) schematic of an electrochromic cell using Li^+*

generate H^+ ions, which in turn are used to produce a hydrogen tungsten bronze H_xWO_3. Decomposition takes place on the outer indium tin oxide electrode:

$$2H_2O \rightarrow O_2(g) + 4H^+ + 4e^-$$

This electrochemical decomposition requires about 1 V at the electrode surface. To drive the protons into the WO_3 film, a proton-conducting electrolyte, typically hydrogen uranyl phosphate, $HUO_2PO_4.4H_2O$ (HUP), is utilised. The H^+ produced can pass through the proton-conducting electrolyte to form a hydrogen tungsten bronze, using electrons from the other electrode. If troublesome reflections occur, the unit is switched on and the WO_3 layer is darkened by the formation of H_xWO_3, which cuts down the dazzling reflection. When the problem no longer occurs, the voltage is reversed. The H^+ ions are pulled out from the bronze and the film becomes colourless once more. If a photocell is incorporated into the circuit, the whole device can be automated. In this design, the storage reservoir layer for H^+ is the surrounding atmosphere.

9.7 Electro-optic Properties

9.7.1 Refractive Index Changes

Electro-optic effects describe a change in optical properties arising from an applied electric field. This can include changes in colour or optical absorption, characteristic of electrochromic materials (Section 9.6), or a change in refractive index. In perovskites, it is this latter effect that is important. These crystals have been used in electro-optic devices to modulate the phase, the amplitude or the polarisation of a light beam traversing the medium and so function as shutters and other components in optical/electronic circuits.

The electro-optic effect comes about in the following way. When a static or slowly varying electric field is applied to a solid, it causes a displacement of the atom cores and electrons and hence alters the refractive index of the solid. The change is given by

$$n(E) = n - \tfrac{1}{2} r n^3 E - \tfrac{1}{2} R n^3 E^2 \ldots$$

or

$$\frac{1}{n(E)^2} = \frac{1}{n^2} + rE + RE^2 + \cdots$$

where $n(E)$ is the refractive index in the electric field, n is the normal refractive index of the material, and r and R are electro-optic coefficients. The effect is usually small and the higher terms can be safely neglected. In centrosymmetric crystals, including all cubic perovskites, the first term, rE, is zero. The change in refractive index is then seen to be proportional to the square of the electric field. This is the quadratic or Kerr effect, and the coefficient R is called the Kerr constant or coefficient. Devices that use the Kerr effect are normally called Kerr cells. In non-centrosymmetric crystals, the first term has to be retained. This results in a linear variation of the refractive index with electric field, called the Pockels effect. The coefficient r is the Pockels constant or coefficient. Devices that use the Pockels effect are normally called Pockels cells. In crystals the Kerr and Pockels coefficients are dependent upon the crystal symmetry and are represented by coefficients r_{ij} or R_{ij}, usually displayed in matrix form. The magnitude of both effects then depends upon the magnitude and direction of the electric field with respect to the crystal axes.

For many applications of perovskites, it is possible to focus attention on the linear term and neglect the smaller quadratic term. In this respect, the most important electro-optic perovskite crystal at present is $LiNbO_3$. The unit cell is trigonal although here it is convenient to refer the structure to a hexagonal unit cell.

In terms of crystal optics, the crystal is uniaxial negative, with the hexagonal **c**-axis being the optic axis and with refractive indices $n_o = 2.3149$ and $n_e = 2.2265$ at 532 nm. A beam of unpolarised light impinging upon a crystal of $LiNbO_3$ at an arbitrary angle will be split into two linearly polarised beams, with their polarisation directions lying perpendicular to each other. One beam will experience a refractive index of n_o and the other a refractive index somewhere between n_o and n_e dependent upon direction. Upon applying an electric field to the crystal, these refractive indices will be modified. A relatively simple case is when the electric field, E_3, is parallel to the **c**-axis. In this case the change in refractive indices is given by

$$n_o(E) = n_o - \frac{1}{2} n_o^3 r_{13} E_3$$
$$n_e(E) = n_e - \frac{1}{2} n_e^3 r_{33} E_3$$

where the appropriate Pockels coefficient is r_{13} for n_o, with a value of $9.6 \times 10^{-12}\,\text{m V}^{-1}$ and r_{33} for n_e, which has a value of $30.9 \times 10^{-12}\,\text{m V}^{-1}$. The induced changes in refractive index are very small. For example, when a field of 10 kV is applied across a crystal of 1 cm dimension, the change in n_e is given by

$$\Delta n_e = \frac{1}{2} \times (2.2265)^3 \times 30.9 \times 10^{-12} \times 10^6 = 1.71 \times 10^{-4}$$

Despite the small magnitude of the effect, it is widely used for a variety of electro-optical components.

There are a number of considerations that are important for a working device. The crystal can be cut in a variety of ways relative to the crystal axes; the field can be applied parallel, perpendicular or at an angle to the optic axis; and the light beam can also be directed parallel, perpendicular or at an angle to the optic axis (Section 5.1 and Figure 5.1). Working devices optimise all of these possibilities so as to produce a robust and easy-to-use component.

9.7.2 Electro-optic Phase Modulators

Perovskite $LiNbO_3$ crystals are widely used as phase modulators. Phase modulators make use of the fact that a beam of light entering a crystal will slow down compared to that of the beam in air (or vacuum) and on emerging from the crystal will have a phase shift compared to a beam that has bypassed the crystal. For simplicity, suppose that the beam is transmitted down the optic axis, so that it only experiences one refractive index, n_o. The phase shift is

$$\theta_0 = \frac{2\pi L n_o}{\lambda_0}$$

where L is the length of the crystal and λ_0 is the vacuum wavelength of the light. In an electric field, n_o is replaced by

$$n_o(E) = n_o - \tfrac{1}{2} n_o^3 r_{13} E_3$$

The phase shift due to the applied electric field is then

$$\begin{aligned}\theta &= \left(\frac{2\pi L}{\lambda_0}\right)\left[n_o - \tfrac{1}{2} n_o^3 r_{13} E_3\right] \\ &= 2\pi L n_o / \lambda_0 - \pi L n_o^3 r_{13} E_3 / \lambda_0 \\ &= \theta_0 - \pi L n_o^3 r_{13} E_3 / \lambda_0 = \theta_0 - \Delta\theta\end{aligned}$$

It is convenient to define the phase shift in terms of the voltage required to produce a phase shift of π in the exit beam, $V\pi$. Writing the field, E_3, as V/d, where V is voltage applied across a crystal of thickness d,

$$\theta = \theta_0 - \pi V / V_\pi$$

where

$$V_\pi = d\lambda_0 / L n_o^3 r_{13}$$

A typical $LiNbO_3$ modulator consists of an oriented slab of crystal such that the applied electric field is parallel to the optical axis (**c**-axis or z direction) and the light beam runs normal to it, parallel to the **x**- or **y**-axis (Figure 9.10a). Under normal circumstances, an unpolarised beam of light will split into the ordinary and extraordinary rays. This is prevented by arranging a polariser parallel to the optic axis so that the only ray propagating in the crystal is the extraordinary ray. The appropriate value of the refractive index is that for n_e, and the appropriate Pockels coefficient is r_{33}. For light of wavelength 550 nm, the value of $V\pi$ is then given by

$$V_\pi = \frac{d \times 550 \times 10^{-9}}{\left(L \times 2.2265^3 \times 30.9 \times 10^{-12}\right)} = \frac{1613d}{L}$$

Taking a crystal dimension of $L = 1$ cm and $d = 1$ mm,

$$V_\pi = 161\text{V}$$

Phase modulation is often used to create additional frequencies to a monochromatic laser beam of frequency ω. To do this, the fixed voltage V applied to the crystal is replaced by a sinusoidally varying voltage of frequency f. The phase-modulated output signal will now contain sidebands to the primary output with frequency ω, $\omega \pm f$, $\omega \pm 2f$ ... (Figure 9.10b).

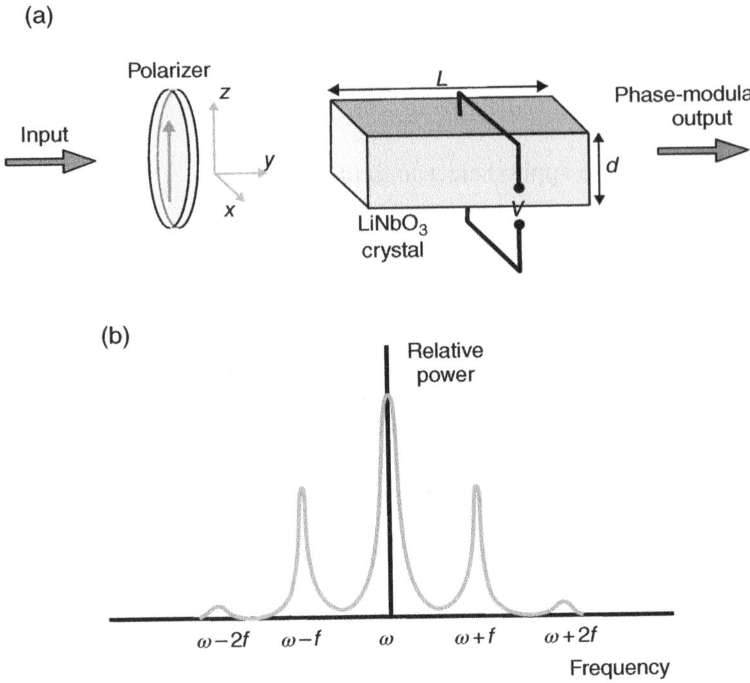

Figure 9.10 *An electro-optic phase modulator: (a) schematic of experimental arrangement; (b) output power for a sinusoidally varying voltage*

9.7.3 Electro-optic Intensity Modulators

There are a number of ways in which the intensity of a light beam can be modulated using electro-optic crystals such as $LiNbO_3$. Conceptually the simplest intensity modulator makes use of the same type of arrangement as in a Mach–Zehnder interferometer. The source light beam is split so that one part goes through the electro-optic crystal and one part does not (Figure 9.11a). The set-up of the crystal must take into account the fact that in general both an ordinary and an extraordinary ray will be generated, but this can be overcome in several ways. For example, using the same geometry as in the phase modulator (Figure 9.10a) will allow only the ordinary ray to be transmitted. This will be slowed compared to the beam passing through air by an amount dependent upon the voltage applied across the crystal. The beams are then recombined, and constructive or destructive interference will occur, dependent upon the phase shift imposed on the beam traversing the crystal. This arrangement is similar to that employed in a Mach–Zehnder interferometer except that the relative phases of the two beams are manipulated via an applied voltage rather than adjusting the length between reflecting mirrors. The output

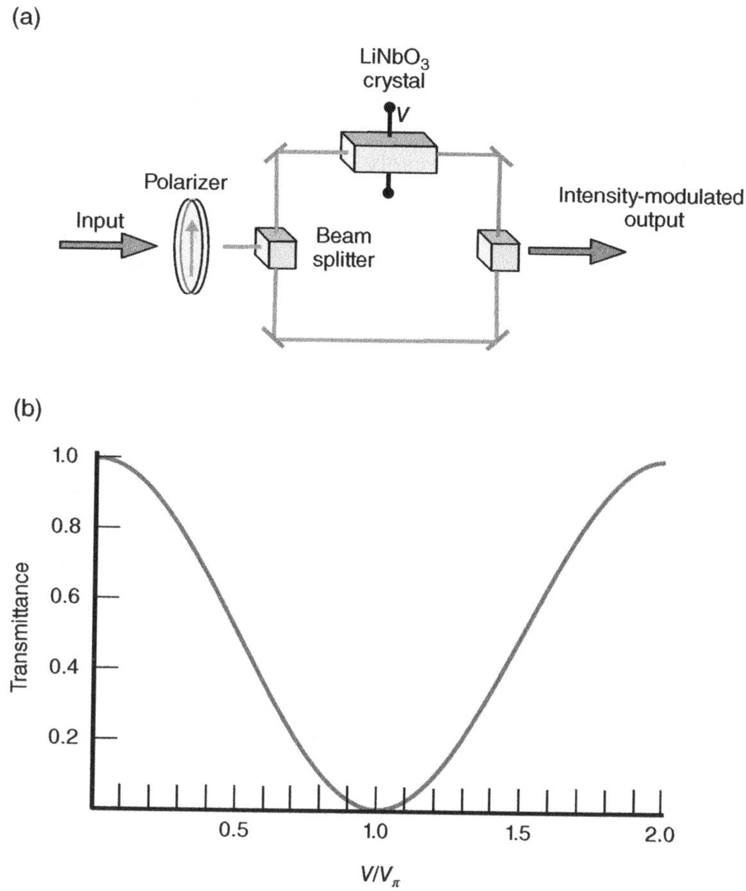

Figure 9.11 *An electro-optic intensity modulator: (a) schematic of experimental arrangement; (b) output transmittance as a function of* V/V_π

transmittance, T, will be a function of the relative phase difference between the two beams, ϕ:

$$T = \cos^2\left(\frac{\phi}{2}\right)$$

The modulating crystal will change the relative phase in the two branches by an amount $\pi V/V_\pi$ so that the transmittance becomes

$$T = \cos^2\left(\frac{\phi_0}{2} - \frac{\pi V}{2V_\pi}\right)$$

where ϕ_0 is a constant phase difference depending upon the construction of the device, which, in practice, is compensated to zero. The transmittance then has a typical cos² form (Figure 9.11b). If the external voltage is switched between 0 and V_π ($V/V_\pi = 1$), the modulator acts as an on–off switch. Operating close to ½V_π ($V/V_\pi = 0.5$) will give an approximately linear intensity modulation. If desired, operating at other voltages gives a non-linear modulation.

A more common way of building an intensity modulator uses two $LiNbO_3$ crystals in tandem. The crystals are arranged in a set-up similar to that in an amplitude modulator and are positioned so that the external field is applied parallel to the optical axis and the light beam travels perpendicular to it. However, the crystals are positioned so that the *x*- and *y*-axes are arranged at an angle of 45° to each other. The entry beam is accurately polarised vertically so that its electric vector bisects the *x*- and *y*-directions (Figure 9.12). This will split into an ordinary and extraordinary ray, each with polarisation normal to the other as it enters the first crystal. The second crystal is oriented so that the ordinary and extraordinary rays in the first crystal, defined by their relative polarisation, swap and become the extraordinary and ordinary rays in the second crystal. In this way, the ordinary and extraordinary beams travel identical optical paths. On emerging, the ordinary and extraordinary waves will have a phase difference due to the imposed values of the external electric field and will be linearly polarised perpendicular to one another.

Although two light beams polarised perpendicular to one another do not interfere or form interference patterns, two beams with parallel polarisation can. On meeting the final polariser, which again bisects the *x*- and *y*-axes but is oriented perpendicular to the first, the electric field components of the ordinary and extraordinary rays parallel to the allowed direction will pass. The phase difference will result in interference because the resultant electric field vectors are now parallel in each ray, which fulfils the interference condition. As before, the phase difference leads to a transmission function that is proportional to $\cos^2(\pi V/2V_\pi)$. The modulator can then act as a linear or non-linear intensity modulator or an on–off switch as described earlier.

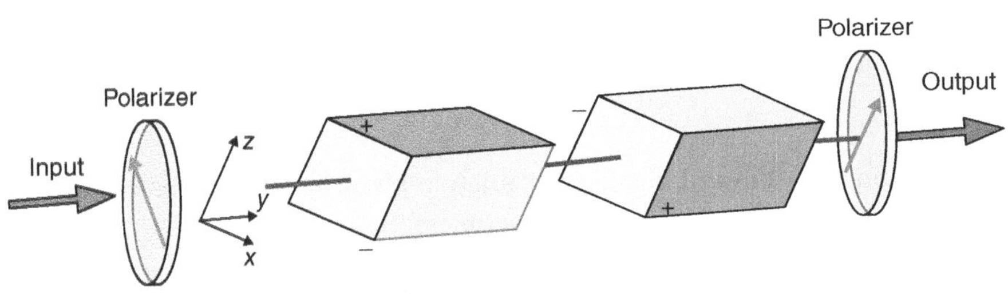

Figure 9.12 *An electro-optic intensity modulator using a pair of oriented* $LiNbO_3$ *crystals*

9.7.4 Ceramic Modulators

Ceramic materials are polycrystalline and can be fabricated in a wide variety of shapes, although discs are commonest. Ceramic perovskite units have a number of advantages over single crystals, including the fact that they are rugged and generally of lower cost than crystals and do not need to be carefully sliced in a specific orientation to give a good performance. However, they have to be carefully prepared to give transparent products (Section 9.5). A number of materials have been explored in depth for this purpose, including $Pb_{1-x}La_x(Zr_yTi_{1-y})_{1-x/4}O_3$ (PLZT), $Pb(Mg_{1/3}Nb_{2/3})O_3$–$PbTiO_3$ (PMN-PT), $Pb(Zn_{1/3}Nb_{2/3})O_3$–$PbTiO_3$ (PZN-PT) and, more recently, lead free materials derived from $K_{0.5}Na_{0.5}NbO_3$ (KNN)-based perovskites.

Ceramic PLZT has a number of structures, depending upon composition, and can show both the Pockels (linear) electro-optic effect in the ferroelectric rhombohedral and tetragonal phases and the Kerr (quadratic) effect in the cubic paraelectric state. Because of the ceramic nature of the material, the non-cubic phases show no birefringence in the as-prepared state and must be poled to become useful electro-optically (Section 6.4.1). PMN-PT and PZN-PT are relaxor ferroelectrics. These have an isotropic structure in the absence of an electric field, but this is easily altered in an applied electric field to give a birefringent electro-optic material. All of these phases, with optimised compositions, have much higher electro-optic coefficients than $LiNbO_3$ and are actively studied for device application.

9.8 Perovskite Solar Cells

Recently (2014/2015) there has been a great deal of research into photovoltaic solar cells using perovskite materials. The photovoltaic effect is the development of a voltage across a material when it is illuminated. The photovoltaic cells that were originally investigated were dye-sensitised solar cells (DSSCs). The primary step in this type of electrochemical cell is for a sensitiser (in original designs, a dye molecule) to absorb photons, thus generating electron–hole pairs. The rest of the cell apparatus is designed to separate the electrons and holes, to allow the electrons to move through an external circuit and to arrange for the dye molecules to be returned to their original state, so that the cycle can be repeated. Important components of this type of (mesoscopic) DSSC are a thin layer of mesoporous anatase, a polymorph of TiO_2, as an electron conductor, which is itself deposited upon a thin film of a transparent conducting oxide (TCO) anode on glass. In addition the cell includes a hole-transporting medium and a counter electrode to complete the cell circuit. A large number of different dyes have been tried in the role of sensitiser, but a considerable breakthrough occurred in 2009 when the dye was replaced by a layer of the organic–inorganic perovskite methylammonium lead iodide, $CH_3NH_3PbI_3$ (Figure 9.13a).

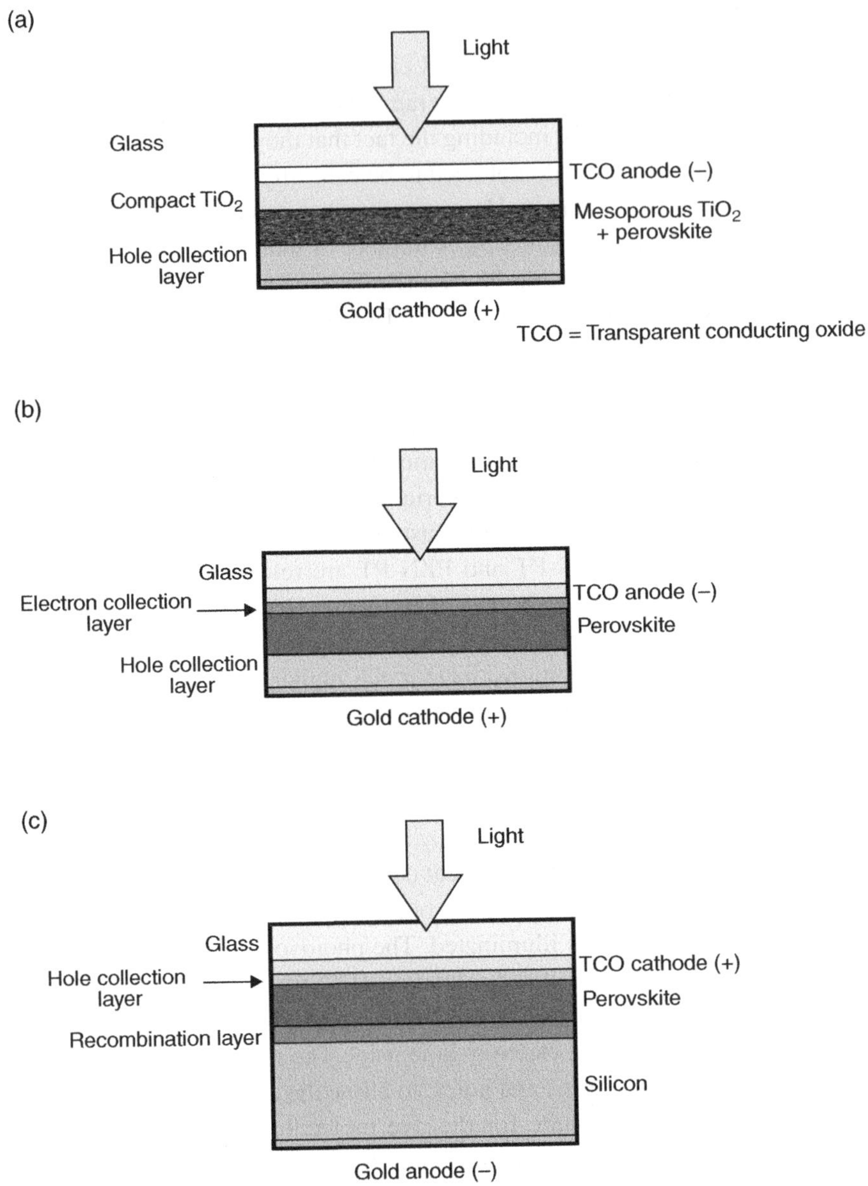

Figure 9.13 *Schematic of perovskite solar cells: (a) DSSC architecture using* $CH_3NH_3PbI_3$ *perovskite; (b) conventional solar cell architecture; (c) tandem silicon/perovskite architecture*

The conversion efficiency of these earliest cells was approximately 4%, but it was immediately clear that this preliminary figure could be substantially improved by device engineering. Since then other closely related perovskites $CH_3NH_3PbBr_3$, $CH_3NH_3PbI_2Cl$ and $CH_3NH_3Pb(I_{1-x}Br_x)_3$ have been tested.

The efficiency of these perovskite sensitised cells is superior to the dye alternatives, and the perovskite component offers a number of other advantages. The main one of these is that the electron- and hole-conducting ability of the perovskite is high enough to allow for the omission of the mesoporous TiO_2 layer, allowing a traditional photovoltaic cell design to be employed (Figure 9.13b). In addition, thin films of the active phase can be laid down from water solution, making processing relatively inexpensive. In these cells, efficiency of over 20% was achieved in 2014.

In view of the flexibility of the perovskite structure, many modifications are possible. The band gap of these phases and thus the absorption spectrum of the perovskite and its light-harvesting efficiency can be tuned by doping the organic and the halogen sites. The perovskites $CH_3NH_3Pb(I_{1-x}Br_x)_3$ have a band gap variation given by

$$E_g = 1.57 + 0.39x + 0.33x^2$$

giving a smooth band gap variation from 1.57 ($x=0$) to 2.29 ($x=1$) and hence a smooth change in the optical absorption spectrum, from the infrared into blue, over this composition range. Similarly the mixed methylammonium $(CH_3NH_3)^+$ and formamidinium $(HN{=}CHNH_3)^+$ perovskites $(CH_3NH_3)_x(HN{=}CHNH_3)_{1-x}PbI_3$ show a decrease in light harvesting at red wavelengths as the formamidinium amount increases.

It is important to note, at this stage, that the microstructures of these materials in working cells have not yet been fully understood. In particular, the perovskite film is sensitive to water vapour and humidity changes cause a significant degradation in properties and performance. Thus these cells must be sealed from atmospheric water if they are to have a commercially viable lifetime. Nevertheless further studies will undoubtedly lead to optimisation of cells for specific wavelength conversion. In addition, the construction of stacks of differing types of solar cells will give an improved output. The most investigated of these at present is a perovskite/silicon solar cell tandem (Figure 9.13c), which is anticipated to have efficiencies of the order of 33–35%.

Research on perovskite solar cells is continuing at a furious pace. Reference to the recent reviews in "Further Reading" section will provide background to this rapidly advancing area of study.

Reference

X. Zhang *et al.*, *J. Am. Ceram. Soc.*, **97**, 1389–1392 (2014).

Further Reading

The optical properties of crystals are described by

B. E. A. Saleh, M. C. Teich, *Fundamentals of Photonics*, John Wiley & sons, Inc., New York (1991), Chapter 6.

R. J. D. Tilley, *Colour and the Optical Properties of Materials*, 2nd edition, John Wiley & Sons, Ltd, Chichester (2011), Chapter 4.

The electrocaloric effect is reviewed by

S. P. Alpay *et al.*, *Mater. Res. Soc. Bull.*, **39**, 1099–1109 (2014).

J. F. Scott, *Annu. Rev. Mater. Res.*, **41**, 229–240 (2011).

For electro-optics, see

B. E. A. Saleh, M. C. Teich, *Fundamentals of Photonics*, John Wiley & sons, Inc., New York (1991), Chapter 18.

For electro-optic constants, see

G. W. C. Kaye, T. H. Laby, *Tables of Physical and Chemical Constants*, 16th edition, Longman, Harlow (1995); http://www.kayelaby.npl.co.ik/toc/

For recent reviews of perovskite solar cells, see

M. A. Green, A. Ho-Baillie, H. J. Snaith, *Nat. Photonics*, **8**, 506–514 (2014).

Materials Research Society Bulletin, **40**, August (2015), various authors.

S. Kazim *et al.*, *Angew. Chem. Int. Ed.*, **53**, 2812–2824 (2014).

S. D. Shranks, H. J. Snaith, *Nat. Nanotechnol.*, **10**, 391–412 (2015).

S. D. Stranks *et al.*, *Angew. Chem. Int Ed.*, **54**, 3240–3248 (2015).

Appendix A
The Bond Valence Model for Perovskites

Cation distribution and valence states are often difficult to ascertain in perovskite structure materials using standard diffraction methods. The bond valence model,[1] an empirical concept, is of help in these situations. The model correlates the notions of chemical valence, the relative strength of a chemical bond between two ions and the length of the bond. A short bond is stronger than a long bond and has a higher bond valence value. A long bond is weaker than a shot bond and has a lower bond valence. Because crystal structure determinations yield accurate interatomic distances, precise values of these bond valence values, or relative bond strengths, can be derived.

The formal valence of a cation, i, V_i, is equal to the formal charge on the cation. Thus, the value of V_i for an Fe^{3+} ion is +3, for an Nb^{5+} ion is +5 and so on. This value, V_i, is also taken to be equal to the sum of all of the bond valences, v_{ij}, of the ions j of opposite charge within the first coordination shell of cation i: thus

$$\sum_j v_{ij} = V_i = \text{formal charge on the cation} = \text{bond valence sum}\left(\text{BVS}\right) \quad (\text{A.1})$$

[1]The bond valence model should not be confused with the quantum mechanical model of chemical bonding called the valence bond model, which describes covalent bonds in compounds.

Perovskites: Structure–Property Relationships, First Edition. Richard J. D. Tilley.

To make use of this idea, it is necessary to relate the experimentally observable bond length, r_{ij}, between cation i and anion j with the bond valence, v_{ij}. For this, the empirical Equation (A.2) is employed:

$$v_{ij} = e^{(r_0 - r_{ij})/B} \tag{A.2}$$

where r_0 and B are empirical parameters[2] derived from crystal structures. In general the value of B is often taken as 0.037 nm (0.37 Å) for all bonds, so that this equation takes the practical form

$$v_{ij} = e^{(r_0 - r_{ij})/0.037} \tag{A.3}$$

The method can be applied to resolve two complementary questions during crystal structure determination, the site occupied by a cation of known valence or the valence of a cation in a known site. The technique is be illustrated by two examples.

A.1 Cation Location in Hexagonal 12R-$Ba_4Ti_2Mn_2O_{12}$

The structure of the hexagonal perovskite 12R-$Ba_4Ti_2Mn_2O_{12}$ is trigonal: space group R$\bar{3}$m (166), a=0.56914 nm, c=2.79186 nm. The structure consists of triplets of three face-sharing octahedra separated by a single layer of apex-sharing octahedra (Section 3.6.1, Figure 3.15a). There are three different B-sites available in the structure: M1, within the central octahedron in the triplet of face-sharing octahedra; M2, within one of the outer octahedra of the triplets; and M3, within the apex-sharing octahedra.

The six bonds from M1 to the surrounding oxygen ions, O1, are equal to 0.19051 nm. For Ti^{4+}–O^{2-}, r_0=0.185 nm; hence

$$v_{ij} = \exp\left[\frac{(0.1815 - 0.19051)}{0.037}\right] = 0.784$$

$$V_i = 6 \times 0.784 = 4.704$$

The six bonds from M2 to the surrounding oxygen ions fall into two groups: M2–O1 = 0.19888 nm and M2–O2 = 0.19206 nm. For Ti^{4+}

$$v_{ij}(\text{M2}-\text{O1}) = \exp\left[\frac{(0.1815 - 0.19888)}{0.037}\right] = 0.625$$

$$v_{ij}(\text{M2}-\text{O2}) = \exp\left[\frac{(0.1815 - 0.19206)}{0.037}\right] = 0.752$$

$$V_i = 3 \times 0.625 + 3 \times 0.752 = 4.131$$

---

[2]The units used here for r_{ij}, r_0 and B are nanometre. Crystallographers often prefer the unit angstrom, where 1 nm = 10 Å.

Table A.1 *Bond valence data for 12R-$Ba_4Ti_2Mn_2O_{12}$*

Atom pair	r_{ij} (nm)	No. of bonds	Bond valence/ bond	BVS
Ti1–O1	0.19051	6	0.784	4.704
Ti2–O1	0.19888	3	0.625	1.876
Ti2–O2	0.19206	3	0.752	2.255
Ti2–O				4.131
Ti3–O3	0.19815	6	0.638	3.828
Mn1–O1	0.19051	6	0.663	3.978
Mn2–O1	0.19888	3	0.529	1.587
Mn2–O2	0.19206	3	0.636	1.908
Mn2–O				3.495
Mn3–O3	0.19815	6	0.539	3.234

Data taken from Keith *et al.* (2004). The values of r_0 are Mn^{4+}–O: 0.1753 nm and Ti^{4+}–O: 0.1815 nm; from Breese and O'Keeffe (1991).

This is repeated for M3, for Ti^{4+} and then for all three sites for Mn^{4+}–O^{2-}, as set out in Table A.1. The results show that for Ti^{4+} the bond valence sums are M1, 4.70; M2, 4.13; and M3, 3.83 and for Mn^{4+} are M1, 3.98; M2, 3.49; and M3; 3.24. These indicate that the Ti^{4+} ions preferentially occupy the M2 and then the M3 positions as the computed Ti^{4+} valence is closest to 4 in these positions. Similarly the Mn^{4+} ions preferentially occupy the M1 position because the computed valence of the Mn^{4+} is closest to 4 for this site. The probable cation distribution has thus been resolved.

A.2 Cation Valence in Hexagonal 9R-$BaIrO_{2.96}$

The slightly oxygen-deficient phase $BaIrO_{2.96}$ has a distorted version of the 9R-$BaRuO_3$ structure (Section 3.5.1, Figure 3.9a). The stacking sequence of this structure is $(chh)_3$, and it is built of triplets of face-sharing octahedra linked by apex sharing. In $BaIrO_{2.96}$, the structure is skewed to give a monoclinic cell: space group C2/m (12), a = 1.00046 nm, b = 0.57536 nm, c = 1.51839 nm, β = 103.27°. There are four independent Ir sites in the cell: Ir1 and Ir3, which are face sharing and apex sharing, at the ends of the groups of three face-sharing octahedra, and Ir2 and Ir4, at the centres of the triplets. Bond valence calculations are carried out as described previously for Ir bonded to O^{2-}. The results are given in Table A.2 and show that the bond valence sums are Ir1, 2.91; Ir2, 2.74; Ir3, 3.91; and Ir4, 2.77. These show that only Ir3 is close to the expected Ir^{4+} state. The other cations can be regarded as Ir^{3+} and suggests that there is considerable Ir—Ir bonding taking place, especially between the central cations in the blocks of face-sharing octahedra, Ir2 and Ir4.

Table A.2 *Bond valence data for 9R-$BaIrO_{2.96}$*

Atom pair	r_{ij} (nm)	No. of bonds	Bond valence	BVS
Ir1–O1	0.211	1	0.523	
Ir1–O2	0.207	2	1.165	
Ir1–O3	0.214	1	0.482	
Ir1–O4	0.224	2	0.736	
Ir1–O				2.91
Ir2–O3	0.214	2	0.964	
Ir2–O4	0.217	4	1.778	
Ir2–O				2.74
Ir3–O1	0.191	1	0.898	
Ir3–O2	0.195	2	1.611	
Ir3–O5	0.213	2	0.991	
Ir3–O6	0.220	1	0.410	
Ir3–O				3.91
Ir4–O5	0.214	4	1.928	
Ir4–O6	0.219	2	0.842	
Ir4–O				2.77

Data taken from Cheng *et al.* (2009). The value of r_0 is Ir^{4+}–O: 0.1870 nm; from Breese and O'Keeffe (1991).

References

N. E. Breese, M. O'Keeffe, *Acta Crystallogr.*, **B47**, 192–197 (1991).
J.-G. Cheng *et al.*, *J. Am. Chem. Soc.*, **131**, 7461–7469 (2009).
G. M. Keith *et al.*, *Chem. Mater.*, **16**, 2007–2015 (2004).

Further Reading

I. D. Brown, Chapter 14, in M. O'Keeffe and A. Navrotsky (Eds.), *Structure and Bonding in Crystals*, Vol **II**, Academic Press, New York, (1981).
N. Brese, M. O'Keeffe, *Acta Crystallogr.*, **B47**, 192–197 (1991).
A. Santoro, I. N. Sora, Q. Huang, *J. Solid State Chem.*, **151**, 245–252 (2000).
M. W. Lufaso, P. M. Woodward, *Acta Crystallogr.*, **B57**, 725–738 (2001).
I. D. Brown, *The Chemical Bond in Inorganic Chemistry*, International Union of Crystallography Monographs on Crystallography No. **12**, Oxford University Press, Oxford, (2002).
H. Zhang, N. Li, K. Li, D. Xue, *Acta Crystallogr.* **B53**, 812–818 (2007).

Appendix B

Summary of the Kröger–Vink Defect Notation

The Kröger–Vink notation is principally used to describe point defects in crystals. The main points are as follows:

1. Empty positions that are normally occupied by atoms are called vacancies and are indicated by the symbol *V* (italic) to differentiate it from the chemical symbol for the element vanadium, V (roman). The atom which is absent from the normally occupied site is specified by the normal chemical symbol for the element, written as a subscript. Thus in $SrTiO_3$, the symbol V_{Sr} would represent a strontium atom vacancy.
2. A substituent atom is given its normal chemical symbol, and the site occupied is written as a subscript, using the chemical symbol for the atom normally present. Thus, a Ba substituent on an Sr site in $SrTiO_3$ is written as Ba_{Sr}.
3. Positions in a crystal not normally occupied by an atom, called interstitial positions, are denoted by the subscript i. For example, O_i would represent an interstitial oxygen atom in $La_2CuO_{4+\delta}$.
4. Associated lattice defects are indicated by enclosing the components of such a cluster in parentheses.
5. The charge on defect is taken as an effective charge, which is the charge that the defect has with respect to the charge that would be present at the same site in a perfect crystal. The superscript ′ is used for each unit of effective negative charge, and the superscript • is used for each unit of effective positive charge. Thus an La

Perovskites: Structure–Property Relationships, First Edition. Richard J. D. Tilley.

Table B.1 *The Kröger–Vink notation for defects in crystals*

Defect type	Notation	Defect type	Notation
Metal vacancy at metal (M) site	V_M	Non-metal vacancy at non-metal (Y) site	V_Y
Impurity metal (A) at metal (M) site	A_M	Impurity non-metal (Z) at non-metal (Y) site	Z_Y
Interstitial metal (M)	M_i	Interstitial non-metal (Y)	Y_i
Neutral metal (M) vacancy	V_M^x	Neutral non-metal (Y) vacancy	V_Y^x
Metal (M) vacancy with negative effective charge	V_M'	Non-metal (Y) vacancy with positive effective charge	$V_Y^\bullet$
Interstitial metal (M) with *n* positive effective charges	$M_i^{n\bullet}$	Interstitial non-metal (Y) with *n* negative effective charges	$Y_i^{n\prime}$
Free electron	e′	Free hole	$h^\bullet$
Neutral associated defects (vacancy pair)	$(V_M V_Y)$	Associated defects with positive effective charge	$(V_M V_Y)^\bullet$

substituent on an Sr site in $SrTiO_3$ would have an effective charge of +1 and be written $La_{Sr}^\bullet$. The charge balance would be maintained by the creation of a Ti^{3+} cation in place of a Ti^{4+} cation, written Ti_{Ti}'.

6. Electrons and holes that are free to move through the matrix are denoted by the symbols e′ and $h^\bullet$.
7. When an atom or an ion occupies an interstitial site, the effective charge is the same as its real charge. For example, O_i'' would represent an interstitial O^{2-} anion in $La_2CuO_{4+\delta}$.
8. A defect that has no effective charge can be given the (optional) superscript *x* to emphasis this state.

The main features of the notation are summarised in Table B.1. It is useful to note that the nomenclature is not structure sensitive, so that the same terminology can be used across the whole spectrum of perovskite and perovskite-related phases.

Further Reading

The definitions of this nomenclature and further examples are to be found in the IUPAC Red Book on the Nomenclature of Inorganic Chemistry, Recommendations 2005, International Union of Pure and Applied Chemistry, RSC Publishing, Cambridge, UK (2005).

Index

Note: Page numbers in *italics* denote tables, where located outside page ranges.

Perovskites: Structure–Property Relationships, First Edition. Richard J. D. Tilley.

Printed and bound by CPI Group (UK) Ltd, Croydon, CR0 4YY
16/09/2026
14963479-0004